| Atomic Number | Element | Atomic Mass[a] | Crystal Structure | Density Mg/m³ | Melting Point (°C)[b] | Symbol |
|---|---|---|---|---|---|---|
| 24 | Chromium | 52.00 | bcc | 7.19 | 1863 | Cr |
| 25 | Manganese | 54.94 | cubic | 7.47 | 1246 | Mn |
| 26 | Iron | 55.85 | bcc | 7.87 | 1538 | Fe |
| 27 | Cobalt | 58.93 | hcp | 8.8 | 1495 | Co |
| 28 | Nickel | 58.71 | fcc | 8.91 | 1455 | Ni |
| 29 | Copper | 63.55 | fcc | 8.93 | 1084.87 | Cu |
| 30 | Zinc | 65.38 | hcp | 7.13 | 419.58 | Zn |
| 31 | Gallium | 69.72 | orth. | 5.91 | 29.77 | Ga |
| 32 | Germanium | 72.59 | dia. | 5.32 | 938.3 | Ge |
| 33 | Arsenic | 74.92 | rhomb. | 5.78 | 603 | As |
| 34 | Selenium | 78.96 | hex. | 4.81 | 221 | Se |
| 35 | Bromine | 79.90 | | | −7.25 | Br |
| 36 | Krypton | 83.80 | | | −137.38 | Kr |

Sources: [a]R. C. Weast. *CRC Handbook for Physics and Chemistry,* 66th ed. (Boca Raton: CRC Press Inc., 1985/86).
[b]T. B. Massalski, ed. *Binary Alloy Phase Diagrams,* Vols. 1 and 2. (Metal Park, Ohio: American Society for Metals, 1986).

Table A.3    Physical Constants and Conversion Factors

### Physical Constants

| | |
|---|---|
| Avogadro's number, $N_A$ | $0.6023 \times 10^{24}$ /mole |
| Atomic mass unit, amu | $1.661 \times 10^{-24}$ g |
| Bohr radius, $a_B$ | 0.05291 nm |
| Permittivity of vacuum, $\epsilon_0$ | $8.854 \times 10^{-12}$ F/m |
| Permeability of vacuum, $\mu_0$ | $1.256 \times 10^{-6}$ H/m |
| Electron mass, $m_0$ | $0.911 \times 10^{-27}$ g |
| Electron charge, $q$ | $1.602 \times 10^{-19}$ C |
| Gas constant, $R$ | 8.314 J/mol·K |
| Boltzmann's constant, $k$ | $1.381 \times 10^{-23}$ J/K |
| Planck's constant, $h$ | $6.626 \times 10^{-34}$ J·s |
| Speed of light, $c$ | $2.998 \times 10^8$ m/s |
| Bohr magneton, $\mu_B$ | $9.274 \times 10^{-24}$ A·m² |

### SI Prefixes

tera, T $10^{12}$
giga, G $10^9$
mega, M $10^6$
kilo, k $10^3$
milli, m $10^{-3}$
micro, $\mu$ $10^{-6}$
nano, n $10^{-9}$
pico, p $10^{-12}$
femto, f $10^{-15}$

**Table B.1**    Atomic and Ionic Radii

| Element | Atomic Radius (nm) | Ionic Radius (nm) |
|---|---|---|
| Hydrogen | 0.046 | 0.154 |
| Helium | | |
| Lithium | 0.152 | 0.078 |
| Beryllium | 0.114 | 0.054 |
| Boron | 0.097 | 0.02 |
| Carbon | 0.077 | 0.02 |
| Nitrogen | 0.071 | 0.02 |
| Oxygen | 0.060 | 0.132 |
| Fluorine | | 0.133 |
| Neon | 0.160 | |
| Sodium | 0.186 | 0.098 |
| Magneisum | 0.160 | 0.078 |
| Aluminum | 0.143 | 0.057 |
| Silicon | 0.117 | 0.198 |
| Phosphorus | 0.109 | 0.039 |
| Sulphur | 0.106 | 0.174 (bivalent) |
| Chlorine | 0.107 | 0.181 |
| Argon | 0.192 | |
| Potassium | 0.231 | 0.133 |
| Calcium | 0.197 | 0.106 |
| Scandium | 0.160 | 0.083 |
| Titanium | 0.147 | 0.064 (quadravalent) |
| Vanadium | 0.132 | 0.065 (trivalent) |
| Chromium | 0.125 | 0.064 (trivalent) |
| Manganese | 0.112 | 0.091 (bivalent) |
| Iron | 0.124 | 0.067 (trivalent) |
| Cobalt | 0.125 | 0.082 (bivalent) |
| Nickel | 0.125 | 0.078 |
| Copper | 0.128 | 0.072 (bivalent) |
| Zinc | 0.133 | 0.083 |
| Gallium | 0.135 | 0.062 |
| Germanium | 0.122 | 0.044 |
| Arsenic | 0.125 | 0.04 (pentavalent) |
| Selenium | 0.116 | 0.03 (hexavalent) |
| Bromine | 0.119 | 0.196 |
| Krypton | 0.197 | |

*Source:* R. C. Weast. *CRC Handbook for Physics and Chemistry,* 66th ed. (Boca Raton: CRC Press Inc., 1985/86).

# Electronic Materials

## H. L. Kwok
### University of Victoria

## PWS Publishing Company

ITP

An International Thomson Publishing Company

Boston • Albany • Bonn • Cincinnati • Detroit • London • Madrid • Melbourne • Mexico City • New York • Pacific Grove • Paris • San Francisco • Singapore • Tokyo • Toronto • Washington

## PWS Publishing Company
## 20 Park Plaza, Boston, MA 02116-4324

I(T)P™

International Thomson Publishing
The trademark ITP is used under license.

**For more information, contact:**
**PWS Publishing Company**
**20 Park Plaza**
**Boston, MA 02116-4324**

International Thomson Publishing Europe
Berkshire House 168-173
High Holborn
London WC1V 7AA
England

Thomas Nelson Australia
102 Dodds Street
South Melbourne, 3205
Victoria, Australia

Nelson Canada
1120 Birchmont Road
Scarborough, Ontario
Canada M1K 5G4

Printed and bound in the United States of America.
97 98 99 00—10 9 8 7 6 5 4 3 2 1

Sponsoring Editor: *Jonathan Plant*
Technology Editor: *Leslie Bondaryk*
Editorial Assistant: *Monica Block*
Marketing Manager: *Nathan Wilbur*
Production Editor/Interior Designer: *Pamela Rockwell*
Cover Designer: *Julia Gecha*
Manufacturing Buyer: *Andrew Christensen*
Compositor: *Modern Graphics*
Text Printer: *Quebecor/Fairfield*
Cover Printer: *Mid-City Lithographers*

International Thomson Editores
Campos Eliseos 385, Piso 7
Col. Polanco
11560 Mexico D.F., Mexico

International Thomson Publishing GmbH
Königswinterer Strasse 418
53227 Bonn, Germany

International Thomson Publishing Asia
221 Henderson Road
#05-10 Henderson Building
Singapore, 0315

International Thomson Publishing Japan
Hirakawacho Kyowa Building, 31
2-2-1 Hirakawacho
Chiyoda-ku, Tokyo 102
Japan

**Library of Congress Cataloging-in-Publication Data**
Kwok, H. L.
    Electronic materials / H.L. Kwok.
       p.  cm.
    Includes index.
    ISBN 0-534-93948-1
    1. Electronics—Materials.  I. Title.
TK7871.K93  1997
620.1'1297—dc21
                          96-39301
                            CIP

**For my parents and my family**

# Contents

# Preface

The electronic properties of solids have become a subject of increasing importance in this age of technology and information. The study of solids and materials, although originating from disciplines in physics and chemistry, has evolved considerably in the last few decades. The classical treatment of solid-state physics, which emphasizes classifications, theories, and fundamental physical phenomena, is no longer capable of bridging the gap between materials advances and their applications. Increasingly often, developments in devices are driven not so much by the discovery of new materials, but rather by our ability to produce finer and better structures through improvements in materials growth techniques. In some instances, new applications have evolved from the use of the more established technologies in entirely different disciplines, as in the case of micromachines.

## Approach

This book intends to bridge the gap between solid-state physics, materials, and device applications and put forward an integrated text placing equal emphasis on each area. Instead of directly describing the conventional solid-state theories and trying to explain the related physical phenomena, we introduce these topics at a more elementary level, with an orientation toward design and applications. The primary purpose of such an approach is to offer to the students not only a basic understanding of physics but also an understanding of what lies around and beyond it (in the practical sense). The treatment is aimed to be introductory; consequently, we have diluted and sometimes omitted the more detailed theories as a trade-off. Also, in view of the diversity of the subject matter we have provided worked examples whenever possible after the presentation of a topic. The overall objective is to give to the students an adequate coverage of the properties of materials and solids, their usage, and the important applications. From our experience, the coupling of basic theories and applications appears to be quite essential to keep the students (especially those with an engineering inclination) interested in the topics.

## Organization

In the first two chapters of this book, we address the questions of what solids are, how they can be formed, and what their structures and physical properties are. These are the subjects that most engineers and physicists should know. In subsequent chapters, we look at the more specific types of solids and explain, in particular, those properties that are linked to applications. These solids include metals, semiconductors, insulators, magnetic solids, superconductors, and light-sensitive solids. Devices are also covered in some of the chapters. A single chapter is devoted to solid-state transducers and their applications, in view of their recent advances. This chapter also includes a brief introduction to materials processing. Another chapter describes the *p-n* junctions and transistors because of their unique importance, and a chapter on magnetism reviews the basic magnetic and superconducting properties of solids and their applications, which we feel are becoming more and more useful to engineers. Finally, we also have a chapter on the optical properties of solids, which provides a broad view of light–solid interactions and the important optoelectronic devices.

## CD-ROM Supplement

Included in this book is a CD-ROM, *Excerpts from Materials Science: A Multimedia Approach* by John C. Russ at North Carolina State University. Using the CD along with *Electronic Materials,* readers can gain a strong grounding in basic principles of materials science. These help to explain many of the physical properties observed in electronic, optical, and magnetic devices. It is assumed that readers have taken a course in materials, or that they have some exposure to bonding, atomic structure, and electronic structure before taking this course. However, since many of these concepts are difficult to master, this interactive study aid has been included to help review and reinforce these concepts. The CD contains excerpts from a materials science CD containing animations, simulations, quizzes, and interactive math problems that teach basic materials concepts. The sections on the CD included at the back of *Electronic Materials* have been chosen specifically to reinforce the materials concepts required to be successful in the electronic devices course, or to enhance the explanation of phenomena in later chapters. In addition to the animations and simulations available in the hyptertext, there are numerous live-math examples in Mathcad® (Window only) that may help the reader with the exercises in the book.

## Acknowledgments

This work was initiated a number of years ago and it evolved from my notes for a course on the electronic properties of materials at the University of Victoria. During the development of this book, there have been many iterations and numerous people have assisted. At PWS Publishing Company, these include: Jonathan Plant, Monica Block, Pamela Rockwell, and Kathleen Wilson. To all these people and the following reviewers, I would like to express my deep appreciation: Fumio S. Chuchi, *University of Washington,* and F. Xavier Spiegel, *Loyola College.*

This book was also written with inputs from many of my colleagues. I am grateful to Prof. A. Antoniou and Prof. S. S. Stuchly of the Department of Electrical and Computer Engineering at the University of Victoria for providing encouragement and to Prof. J. S. Williams of the Department of Electronic Materials Engineering at the Australian National University, Canberra, for providing me with an opportunity to visit his department. The successive generations of engineering students at the University of Victoria who took my course on the electronic properties of materials also helped significantly in changing my approach to teaching the subject. Finally, but not least, I must thank my wife, Rosy, and my sons, Stan and Shing, who have provided me with love and encouragement during the preparation of the manuscript.

*H. L. Kwok, Victoria, B. C.*

# CHAPTER 1

# The Basic Structures of Solids

## 1.1  INTRODUCTION

Solids are an important part of the materials world and, just like any other substances, solids are made up of atoms or ions (ions are atoms that have either gained or lost one or more electrons). The term **solid** usually refers to the state of a collection of atoms or ions, and it differs from terms like *liquid* and *gas,* which are substances without any rigid structure. As we know, in a solid the atoms or ions are held together by forces so that they do not fall apart easily. The entire collection of atoms or ions defines the physical boundary that gives to the solid its shape and appearance. In general, the natural (crystalline) appearance of a solid depends on the arrangement of the atoms or ions, the chemical composition of the solid, and the method of preparation. Solids can have very diverse physical properties, ranging from ductile to brittle, stable to reactive, conducting to insulating, and transparent to opaque. Most solids that we encounter frequently can be classified as either metals, semiconductors, ceramics, polymers, or composites. Structurally speaking, solids are either crystalline, polycrystalline, or noncrystalline. In a **crystalline solid,** the positions of the atoms or ions are predetermined, and similar crystals will have the same structure and properties. A noncrystalline, or **amorphous, solid,** on the other hand, has no defined crystal structure, except perhaps in the arrangement of the nearest neighboring atoms or ions. In fact, in the noncrystalline solids the atoms or ions are randomly placed, and the properties can vary from sample to sample. A **polycrystalline solid** is similar to a crystalline solid except that it is composed of small crystal grains rather than one single crystal. The physical properties of the polycrystalline solids are most frequently influenced by the presence of defects and impurities, especially those present at the grain boundaries.

In this chapter, we begin with a review of electrons, protons, and neutrons, which make up atoms or ions. We then describe the interactions of these particles and the physical origin of the forces that hold them together. In addition to these so-called short-range forces, there are also long-range interatomic forces. Interatomic forces are responsible for the formation of energy bonds, which hold the atoms or ions together. Different types of energy bonds are, therefore, responsible for the different crystal structures and their physical properties. The second part of this chapter considers the noncrystalline and polycrystalline solids. Phase diagrams are introduced to show thermodynamically how homogeneous and inhomogeneous solids can be formed. We also compare the physical properties of "real" solids versus ideal solids and examine how impurities and defects can modify the properties of the former. Such modifications as we show are very important in the study of crystal growth and thin-film deposition as well as in the process of impurity diffusion. Finally, in the last section, we provide a brief introduction to the tools commonly used in the study of crystalline and noncrystalline solids and in the determination of their chemical compositions.

## 1.2   ATOMS AND THEIR BINDING FORCES

### Atomic Structure

**Atoms** are made up of electrons, protons, and neutrons; the simplest way to visualize an atom is to think of it as a miniaturized solar system. A schematic of such an idealized (carbon) atom is shown in Fig. 1.1. In this model, the electrons are depicted as the planets orbiting around a sun. The "sun" is the nucleus of the atom and is made of protons and neutrons. Simple as it may be, we know very well that such a model of the atom is deficient; the orbiting electrons are not exactly particles but are more like electron waves. To a large extent, the model also ignores the interactions between the electrons and the nucleus. Before we consider the structure of an atom in more detail, let us look at properties of electrons, protons, and neutrons. Dimensionally, an electron is a very small unit; it has a mass $m_0$ of $9.11 \times 10^{-30}$ kg (kilogram). This mass is much smaller than the corresponding mass of a proton or a neutron, which is $1.66 \times 10^{-27}$ kg. For most purposes, therefore, the electron

**Fig. 1.1**
An idealized carbon atom

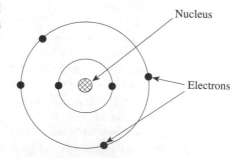

Nucleus

Electrons

mass is negligible, and the mass of an atom is essentially that of the protons and neutrons. Conventionally, the mass of a proton or a neutron is called 1 atomic mass unit (amu), which makes the mass of an atom in amu equal to the number of protons and neutrons present. The carbon atom shown in Fig. 1.1, for instance, has 6 protons and 6 neutrons and an atomic mass of 12 amu. In most practical work, however, amu is too small a unit to use, and the mass of a larger and fixed number of atoms is chosen. This number is called the **Avogadro number,** $N_{AG}$, equal to $6.02 \times 10^{23}$. The same number of atoms is also referred to as 1 mole (mol).

It is often more convenient to refer to an atom in the periodic table (given in Appendix A) by the number of protons present in the nucleus. This number is called the **atomic number.** A carbon atom having 6 protons has an atomic number of 6. In general, atoms as a whole are electrically neutral, even though both the electrons and the protons have charges (charges may be considered as the substance that gives rise to electrostatic attraction and repulsion). **Electron charge** is negative and has a value of $1.60 \times 10^{-19}$ C (coulomb). A proton has an equal but opposite (positive) charge, so all atoms have equal numbers of electrons and protons.

**Example 1.1**  Calculate the number of copper (Cu) atoms present in a cylinder that has both diameter $d$ and height $h$ equal to 1 μm. The mass density of copper, $\rho_0$, is $8.93 \times 10^3$ kg/m$^3$, and the atomic mass, $m_{atom}$, of copper is 63.55 g/mol.

*Solution:*  The volume of the Cu cylinder is $V = \pi r^2 h = 3.1416(0.5 \times 10^{-6}\ \text{m})^2 \times 10^{-6}\ \text{m} = 0.78 \times 10^{-18}\ \text{m}^3$. The mass of the cylinder is $m = V\rho_0 = 8.93 \times 10^3\ \text{kg/m}^3 \times 0.78 \times 10^{-18}\ \text{m}^3 = 7.96 \times 10^{-15}\ \text{kg} = 7.96 \times 10^{-12}\ \text{g}$. Therefore, the number of Cu atoms present in the cylinder is $mN_{AG}/m_{atom} = 7.96 \times 10^{-12}\ \text{g} \times 6.02 \times 10^{23}$ atoms/mol/63.55 g/mol $= 6.64 \times 10^{10}$ atoms.  **§**

## Electron Orbits

The planetary model of the carbon atom shown in Fig. 1.1 suggests that the electrons are revolving around the nucleus in **orbits.** This is, of course, not possible because we know from physics that a revolving electron will accelerate due to centripetal acceleration and emit electromagnetic energy. As the energy of the electron decreases, it will collapse into the nucleus. This apparent contradiction, however, can be resolved if quantum mechanical theory is applied. Based on quantum mechanics, the electrons in an atom have wavelike motion, and stable orbits are observed only when the electron waves complete an integral number of cycles per revolution around the nucleus (these cycles are sometimes referred to as *standing waves* because the same wave pattern is repeated every revolution). Under this condition, the electrons will not radiate electromagnetic energy and can stay on in the same orbit indefinitely. In principle, there can be an unlimited number of such orbits, and we normally use a set of numbers called the *quantum numbers* to label them. For instance, those orbits having the same energy will have the same principal quantum number $\tilde{n}$ and they belong to the same energy level. Fig. 1.2 shows schematically the energy levels in a hydrogen atom ($E_0$ is the ground energy level). As expected, only the lowest energy level is filled in this case.

Fig. 1.3 shows the spatial distribution of the electron orbits up to the energy

Fig. 1.2
Energy levels of a
hydrogen atom

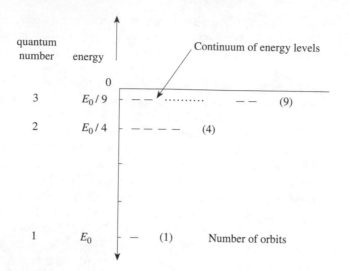

Fig. 1.3
Spatial distribution of the
orbits of a hydrogen atom

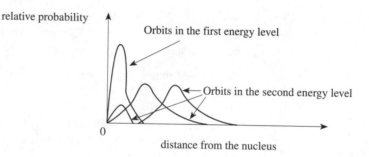

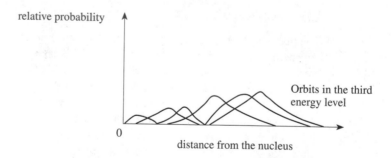

level $\tilde{n} = 3$ for the hydrogen atom. As observed in the figure, there can be more than one orbit with the same energy (in fact, the number of orbits having the same energy is equal to $\tilde{n}^2$). Thus, in the lowest energy level ($\tilde{n} = 1$), there is only 1 orbit, but in the next energy level ($\tilde{n} = 2$), there are 4. It may also be pointed out at this stage that the number of electrons present in an orbit is governed by the physical law called **Pauli exclusion principle,** which states that only 2 electrons with different spins are allowed to exist in a given orbit. Because of this restriction, we can expect a maximum of 2 electrons in the first (lowest) energy level, a maximum of 8 electrons in the second (lowest) energy level, and so on. Depending on the number of electrons present in the atom, some of the energy levels will be filled or partially filled. In the case when we have an atom with only completely filled energy levels, we refer to it as having an *inert gas configuration;* such an atom is electronically very stable. Of course, stability of an atom also depends on the spatial distribution of the electrons. Since orbit locations are determined by the angular movements of the electrons, different spatial orbits will be found at each energy level. These are labeled as the *s*-orbits, the *p*-orbits, the *d*-orbits, etc. In general, it can be shown that the higher the energy level, the larger is the number of orbits. For the same energy level, electrons in the *s*-orbits (ones that are closer to the nucleus; see Fig. 1.3) will be more stable compared with electrons in the *p*-orbits, which are further away from the nucleus.

**Example 1.2**    **(a)** Give the electronic configurations of the sodium (Na) atom and the chlorine (Cl) atom in terms of the *s*-orbits and the *p*-orbits. **(b)** Do the same for the $Na^+$ ion and the $Cl^-$ ion. Note that the *s*-orbit has 2 electrons, and the *p*-orbit has 6 electrons.

*Solution:*

   **(a)** The electronic configurations for the Na atom and the Cl atom are, for Na, $1s^2, 2s^2, 2p^6, 3s^1$, and for Cl, $1s^2, 2s^2, 2p^6, 3s^2, 3p^5$. The superscript indicates the number of electrons in the electron orbits.

   **(b)** For the $Na^+$ ion and the $Cl^-$ ion, the electronic configurations are, for $Na^+$, $1s^2, 2s^2, 2p^6$, and for $Cl^-$, $1s^2, 2s^2, 2p^6, 3s^2, 3p^6$.   **§**

Note that the transfer of the single electron in the 3*s* orbit of the Na atom to the Cl atom fills up the 3*p* orbits of the $Cl^-$ ion and forms the inert gas configuration in NACL.

**Example 1.3**    The edge of a cube containing 1 mol of MgO is $2.24 \times 10^{-2}$ m. Calculate the density of MgO.

*Solution:*    The volume of 1 mol of MgO is $V = (2.237 \times 10^{-2}$ m$)^3 = 1.12 \times 10^{-5}$ m$^3$. Its mass is $m = (24.31 + 16)$ g $= 0.0403$ kg (see Table A.2) and its density is $m/V = 0.0403$ kg$/1.12 \times 10^{-5}$ m$^3 = 3.6 \times 10^3$ kg/m$^3$.   **§**

## Energy Bonds

Since both protons and electrons carry charges, we can expect the forces acting on the atoms or ions in a solid to be electronic in nature. These forces are responsible for the binding of the atoms or ions through the formation of the energy bonds. Principally, there are four different types of energy bonds:

1. **Ionic bond,** formed when one or more electrons in the outermost energy orbit of an atom are transferred to another.
2. **Covalent bond,** formed when electrons in the outermost energy orbits of the atoms are shared between two or more neighbors.
3. **Metallic bond,** formed when there exists some form of collective interactions between the (negatively charged) electrons and the (positively charged) nuclei in the solid.
4. **Van der Waals bond,** formed when there exist distant electronic interactions between (opposite) charges present in the neighboring atoms or molecules.

In general, both ionic bonds and covalent bonds are fairly stable and are found primarily in semiconductors and insulators. Metallic bonds, on the other hand, are unique to the metals and require the presence of a large number of the loosely bound electrons. A van der Waals bond is usually much weaker and is found in polar molecules (molecules with separated charges). One example of a van der Waals bond is observed in water ($H_2O$) molecules, where negatively charged $O^{2-}$ ions are loosely bound to the positively charged $H^+$ ions.

**Example 1.4**   Give examples of different solids with: **(a)** ionic bonds, **(b)** covalent bonds, and **(c)** metallic bonds.

*Solution:*

**(a)** Sodium chloride (NaCl) is a solid formed by ionic bonds;

**(b)** silicon (Si) and germanium (Ge) possess covalent bonds; and

**(c)** copper (Cu) and aluminum (Al) are examples of solids with metallic bonds.   §

The type or types of energy bonds present in a solid often have a strong bearing on its physical properties. For instance, ionic and covalent solids, which are formed by ionic and covalent bonds, respectively, are primarily crystalline or polycrystalline, whereas metals and their alloys (alloys are mixtures of metals and other atoms, which are sometimes metals) are typically polycrystalline. In the following paragraphs, we describe some of the physical properties of the different solids.

*Ionic Solid*   An ionic solid usually involves two or more different types of atoms and is formed when some of the electrons in one type of atoms are physically transferred to another type, as illustrated in Fig. 1.4. Those atoms that have acquired additional electrons will become negatively charged and are called *anions,* as opposed to *cations,* which are atoms that have lost one or more electrons. Accordingly, cations are positively charged. In an ionic solid such as NaCl, for example, the $Na^+$ ions are the cations and the $Cl^-$ ions are the anions. An anion and a cation are

**Fig. 1.4**
Formation of an ionic bond in NaCl

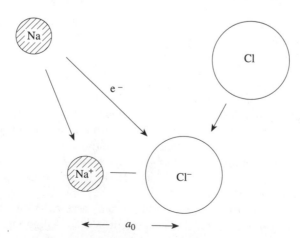

attracted to each other electrically and form an ionic bond. A three-dimensional ionic solid is, therefore, an array of alternating anions and cations. Since there are equal numbers of anions and cations, an ionic solid as a whole is electrically neutral.

In principle, it is possible to isolate an anion and cation pair in a solid and quantitatively examine the bond energy of the pair by summing up the **electrostatic** force and nuclear forces acting on the two ions. There will be an attractive force $F_a(x)$ newtons (N), which according to Coulomb's law is given by

$$F_a(x) = \frac{K_0 z_1 z_2 q^2}{x^2} \qquad (1)$$

where   $K_0$ is a physical constant and is equal to $9 \times 10^9$ V·m/C in vacuum,

$q \, (= 1.6 \times 10^{-19}$ C) is the electron charge,

$z_1$ and $z_2$ are the respective charge states of the two ions, and

$x$ is the ion separation in meters (we have assumed $x = 0$ to be the position of one of the ions).

Using this notation, $F_a(x)$ will be negative for oppositely charged ions. In addition, the anion and the cation are also held apart by repulsive forces $F_r(x)$, which include both the nuclear forces and the forces exerted by the overlapping electrons in similar energy orbits. In general, $F_r(x)$ in N may be expressed as

$$F_r(x) = F_0 \exp(-\alpha_e x) \qquad (2)$$

where   $F_0$(N) and $\alpha_e$ (/m) are also physical constants.

Fig. 1.5 shows how the attractive and repulsive forces will balance out for different values of $x$, the ion separation. At **equilibrium,** we know that the anions and the cations will be stationary, so there should be no net forces acting on these ions. If we assume the equilibrium ion separation to be $x_0$, then the sum of the forces, denoted by $F_s(x)$, will be zero when $x = x_0$. This leads to

$$F_s(x_0) = F_a(x_0) + F_r(x_0) = 0 \qquad (3)$$

**Fig. 1.5**
Attractive and repulsive forces versus ion separation

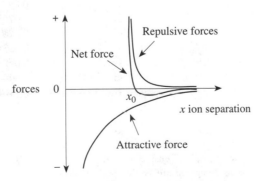

Equations (1)–(3) can now be used to solve for $x_0$, the equilibrium ion separation. A second approach to determining $x_0$ is to consider the **potential energy** $E_{PE}(x)$ of the ions. Since, by definition, $E_{PE} = \int_x^\infty [F_s(x)]\,dx$, the force $F_s(x)$ acting on a specific ion is given by

$$F_s(x) = d(E_{PE}(x))/dx \qquad (4)$$

At equilibrium, $F_s(x_0) = 0$ and Eq. (4) may be used to determine $x_0$ provided $E_{PE}(x)$ is known (see Example 1.6). Furthermore, by substituting $x_0$ into the equation for $E_{PE}(x)$, the equilibrium **binding energy** $E_{PE}(x_0)$ may also be found. In general, $x_0$ is expressed as the sum of the ionic radii $r_0$ of the anions and cations. The values of the more common ions are listed in Table 1.1.

**Example 1.5**  Compute the coulombic attractive force between a $Na^+$ ion and a $Cl^-$ ion at $T = 300$ K.

**Solution:**  From Table 1.1, $r_0(Na^+) = 0.098$ nm and $r_0(Cl^-) = 0.181$ nm. Therefore, the equilibrium ion separation is $x_0 = 0.098$ nm $+ 0.181$ nm $= 0.278$ nm. Based on Eq. (1),

$$F_a = -9 \times 10^9 \text{ V·m/C} \times (1 \times 1.6 \times 10^{-19} \text{ C})^2/(0.278 \times 10^{-9} \text{ m})^2$$
$$= -2.98 \times 10^{-9} \text{ N}. \quad \S$$

**Example 1.6**  If the potential energy $E_{PE}(x)$ of a pair of ions varies as $E_{PE}(x) = Ax^2 - Bx + 1$, where $A$ and $B$ are constants, determine **(a)** the equilibrium ion separation $x_0$; and **(b)** the binding energy $E_{PE}(x_0)$ in terms of $A$ and $B$.

**Table 1.1**  Values of $r_0$ for the more common ions at $T = 300$ K

| Ion | $r_0$ (nm) | Ion | $r_0$ (nm) |
|-----|-----------|-----|-----------|
| $Li^+$ | 0.078 | $Sc^{2+}$ | 0.083 |
| $Be^{2+}$ | 0.054 | $Ti^{4+}$ | 0.064 |
| $B^{3+}$ | 0.020 | $V^{4+}$ | 0.061 |
| $C^{4+}$ | 0.020 | $Cr^{3+}$ | 0.064 |
| $N^{5+}$ | 0.010 | $Mn^{2+}$ | 0.091 |
| $O^{2-}$ | 0.132 | $Fe^{2+}$ | 0.087 |
| $F^-$ | 0.133 | $Co^{2+}$ | 0.082 |
| $Na^+$ | 0.098 | $Ni^{2+}$ | 0.078 |
| $Mg^{2+}$ | 0.078 | $Cu^+$ | 0.096 |
| $Al^{3+}$ | 0.057 | $Zn^{2+}$ | 0.083 |
| $Si^{4+}$ | 0.039 | $Ga^{3+}$ | 0.062 |
| $P^{5+}$ | 0.030 | $Ge^{4+}$ | 0.044 |
| $S^{2-}$ | 0.174 | $Br^-$ | 0.196 |
| $Cl^-$ | 0.181 | $Y^{3+}$ | 0.106 |
| $K^+$ | 0.133 | $Zr^{4+}$ | 0.087 |
| $Ca^{2+}$ | 0.106 | $Nb^{4+}$ | 0.074 |

(*Source*: R. A. Flinn and P. K. Trojan, *Engineering Materials and Their Applications* (Boston: Houghton Mifflin Company, 1975).)

*Solution:*

(a) Based on Eq. (4), $d(E_{PE}(x))/dx = 2Ax - B$. Since $d(E_{PE}(x))/dx|_{x=x_0} = 0$, $x_0 = B/2A$.

(b) By setting $x = x_0 = B/2A$, we find $E_{PE}(x_0) = -B^2/4A + 1$. §

**Example 1.7** For a one-dimensional ionic solid, if the electrostatic attraction of the ions results in a potential energy given by $E_{PE}(x) = -1.75 \, q^2/(4\pi\epsilon_0 x)$, where $q$ is the electron charge, $\epsilon_0$ ($8.85 \times 10^{-12}$ F/m) is the permittivity of vacuum, and $x$ is the ion separation, (a) determine $x_0$ if $E_{PE}(x_0) = -0.662$ eV (1 eV $= 1.6 \times 10^{-19}$ J); and (b) if $F_r(x) = A/x^4$, evaluate $A$.

*Solution:*

(a) Based on the given equation we can write $x_0 = -1.75 q^2/[4\pi\epsilon_0 E_{PE}(x_0)] = 1.75 \times (1.60 \times 10^{-19} \text{ C})^2/(4 \times 3.14 \times 8.85 \times 10^{-12} \text{ F/m} \times 0.662 \times 1.60 \times 10^{-9} \text{ J}) = 3.8 \times 10^{-9}$ m $= 3.8$ nm.

(b) $E_{PE}(x) = -1.75 q^2/(4\pi\epsilon_0 x) + A/x^4$. At equilibrium, $d(E_{PE}(x))/dx|_{x=x_0} = 1.75 q^2/(4\pi\epsilon_0 x_0^2) - 4A/x_0^5 = 0$. This result gives $A = 5.52 \times 10^{-57}$ J·m$^4$. §

In addition to the nearest-neighbor forces we just discussed, there are also forces involving the more distant anions and cations. To account for these long-range forces, a correction factor known as the **Madelung constant**, $\alpha_M$, is sometimes added to the coulombic force term of Eq. (1). For a three-dimensional cubic crystal, $\alpha_M \approx 1.75$. The Madelung constant therefore increases the overall binding force between the ions and gives rise to a more compact crystal structure.

**Example 1.8** Estimate the Madelung constant $\alpha_M$ in a one-dimensional ionic chain. Note that in this case, $\alpha_M$ does not increase the compactness of the chain, even though it will do so in a three-dimensional structure.

*Solution:* Based on Eq. (1), the potential energy $E_{PE}(x)$ of a given ion has the form $-A/x$, where $A$ is a constant. If we add the contributions of all the other ions in the one-dimensional chain, we get

$$E_{PE}(x) = -2A(1/x - 1/2x + 1/3x - 1/4x + \cdots) = -2 \log(2) A/x.$$

This suggests $\alpha_M = 2 \log(2) = 0.602$. §

In an ionic solid, the maximum number of the nearest neighbors an ion can have is given by the **coordination number** ($\eta_c$), which is defined only for the **close-packed (crystal) structure**—i.e., when all the ions are touching one another. In determining $\eta_c$, the smaller of the two ions is always placed at the center of the structure. The size disparity of the ions will impose an upper limit on $\eta_c$. Table 1.2 lists the values of $\eta_c$ for different ratios of the ionic radii ($r/R$).

**Example 1.9** Calculate the minimum radius ratio for two different types of ions in a solid if the coordination number is 8.

*Solution:* Let $r$ be the radius of the smaller ion placed at the center of a cube of side $a_0$. The eight nearest neighbors, each of radius $R'$, are located at the

**Table 1.2**  Coordination numbers and the range of the ratios of the ionic radii

| $\eta_c$ | Ratio of the Atomic Radii | Example |
|---|---|---|
| 2 | $0 < r/R' < 0.155$ | $B_2O_3$ |
| 3 | $0.155 < r/R' < 0.225$ | |
| 4 | $0.225 < r/R' < 0.414$ | $SiO_2$ |
| 6 | $0.411 < r/R' < 0.732$ | MgO |
| 8 | $0.732 < r/R' < 1$ | CaO |
| 12 | $r/R' = 1$ | |

corners of the cube. For the minimum radius ratio, the body diagonal of the cube is $2(r + R') = \sqrt{3}a_0$. Since $a_0 = 2R'$, $2(r + R') = \sqrt{3}2R'$, or $r/R' = 0.732$.  §

Note that if $r/R' < 0.732$, the corner atoms in the cube will still be in contact with each other, but they will no longer be touching the center atom.

*Covalent Solid*   A covalent bond is formed by the sharing of the valence electrons of the atoms in the solid. This situation usually results in a higher concentration of electrons between the adjacent atoms. Fig. 1.6 shows a typical covalent solid, in which the electron orbits of the neighboring atoms overlap with one another and the electrons in these orbits are being shared equally. In some solids, this can result in completely filled energy levels and the stable inert gas configuration. Covalent bonds are found primarily in semiconductors, insulators, and organic molecules. In the hydrocarbon molecules, in particular, covalent bonds are responsible for the well-known polymerization effect, which binds together the carbon skeletons of the molecules and gives rise to long molecular chains known as *polymers.*

Table 1.3 summarizes the values of the **bond energies** and the **bond lengths** for some typical covalent bonds.

**Example 1.10**   Calculate the maximum length of a polyethylene molecule, $-(C_2H_4)-n$, when $n = 1000$.

*Solution:*   According to Table 1.3, the carbon–carbon (C—C) bond length is 0.154 nm, and the bond angle in a tetrahedral structure is 109.5°. Assuming that the molecule is fully stretched, the effective distance between any two C

**Fig. 1.6**
A covalent bond between two Cl atoms

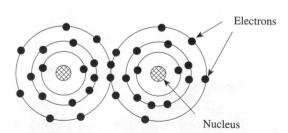

Electrons

Nucleus

Table 1.3    Values of the bond energies and the bond lengths for some typical covalent bonds

| Covalent Bond | Energy (kJ/mol) | Bond Length (nm) |
|---|---|---|
| C—C | 370 | 0.154 |
| C=C | 680 | 0.13 |
| C≡C | 890 | 0.12 |
| C—H | 435 | 0.11 |
| O—H | 500 | 0.10 |
| O—Si | 375 | 0.16 |
| O—N | 250 | 0.12 |
| H—H | 435 | 0.074 |

(*Source*: L. Van Vlack, *Elements of Materials Science and Engineering*, © 1989 by Addison-Wesley Publishing Company, Inc. Reprinted by permission of Addison-Wesley Publishing Company, Inc.)

atoms along the length of the molecule $d_{C-C}$ (shown in Fig. E1.1) is 0.154 nm $\times$ sin(109.5°/2) = 0.129 nm. For $n = 1000$, the full length of the molecule is $(n - 1) L_{C-C} = (1000 - 1) \times 0.129$ nm = 129 nm.    §

Fig. E1.1

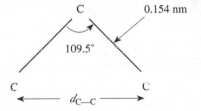

*Metallic Solid*    Metallic bonds are found solely in metals that have a large number of electrons. Because of this, quite a few of the electrons in the outer energy orbits are far away from the nuclei and may spatially overlap with the neighboring nuclei, as illustrated in Fig. 1.7. These electrons have negligible excitation energy because

Fig. 1.7

Spatial distribution of an electron in a metallic solid

A single electron distributed over several ion cores

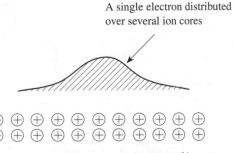

An array of ion cores

they are essentially detached from their own nuclei. They are commonly called **free electrons.** Free electrons are distinguished from **valence electrons,** which are located in the inner orbits of the atoms. A metallic bond, therefore, is formed by the coulombic interactions of the free electrons and the positively charged nuclei, as illustrated in Fig. 1.8. Due to the charge neutralization effect, the overall potential energy of a metal is minimized, and the nuclei are held together tightly in a close-packed structure. Metallic bonds are therefore known to be quite strong, which is reflected in the values of the heat of vaporization (i.e., the amount of heat energy required to vaporize one mole of atoms at a fixed temperature). For instance, the heat of vaporization of Al (with metallic bonds) is 293 kJ/mol, whereas the value for C (with covalent bonds) is 356 kJ/mol.

**Example 1.11**   If each atom in copper contributes one free electron to the solid, determine the total number of free electrons present in a volume of $1 \times 10^{-6}$ m$^3$.

*Solution:*   From Example 1.1, we know that there are $6.64 \times 10^{10}$ Cu atoms in a cylinder with volume $V = 0.785 \times 10^{-18}$ m$^3$. Thus, the density of copper $\rho_0$ is Cu atoms present/$V = 6.64 \times 10^{10}$ atoms/$0.785 \times 10^{-18}$ m$^3$ = $8.45 \times 10^{28}$ atoms/m$^3$. The number of free electrons is $8.45 \times 10^{28}$ atoms/m$^3$ $\times$ 1 electron/atom $\times 1 \times 10^{-6}$ m$^3$ = $8.45 \times 10^{22}$. **§**

*van der Waals Solid*   A van der Waals bond is formed by the electronic attraction between permanent or induced positive and negative charges present in the atoms or the molecules making up the solid. Essentially, these are **dipoles** (positive and negative charges separated by a fixed distance) in the solid, and the dipoles are attracted to one another. A van der Waals bond is usually quite weak and is frequently found in the pairing of the inert gas ions, such as in a He—He molecule. The heat of vaporization in this case, for instance, is only 84.5 J/mol. As an example, Fig. 1.9 shows schematically the van der Waals bond in a pair of polar molecules. As expected, the distance of separation between the molecules is a good measure of the bond strength.

Table 1.4 compares the values of the bond energies and the bond lengths for ionic, covalent, metallic, and van der Waals solids.

**Fig. 1.8**
Electrons and nuclei in a metal

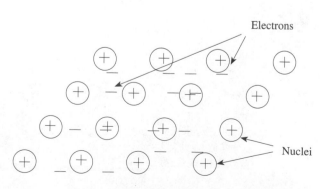

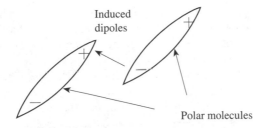

Fig. 1.9
A van der Waals bond in
polar molecules

Induced
dipoles

Polar molecules

Table 1.4   Values of the bond energies and the bond lengths for ionic, covalent, metallic, and van
der Waals solids

| Solid | Bond Type | Bond Energy (kJ/mol) | Bond Length (nm) |
|---|---|---|---|
| NaCl | Ionic | 748 | 0.282 |
| Al | Metallic | 326 | 0.152 |
| C—C | Covalent | 370 | 0.154 |
| FeO | Covalent | 509 | 0.216 |
| Ar | van der Waals | 1 | 0.382 |

**Example 1.12**   If the energy-position relationship of a pair of Argon (Ar) atoms is given by $E_{PE}(x) = -10.37 \times 10^{-78}/x^6$ J·m$^6$ + $16.16 \times 10^{-135}/x^{12}$ J·m$^{12}$, estimate the equilibrium bond length.

***Solution:***   Based on Eq. (4), $d(E_{PE}(x))/dx|_{x=x_0} = 0$. This leads to $6 \times 10.37 \times 10^{-78}/x_0^7$ J·m$^6$ − $12 \times 16.16 \times 10^{-135}/x_0^{13}$ J·m$^{12}$ = 0, or $x_0 = [2 \times 16.16 \times 10^{-135}/(10.37 \times 10^{-38})]^{1/6}$ m = $0.382 \times 10^{-9}$ m = 0.382 nm.   **§**

To sum up what we have discussed, ionic solids are good insulators primarily because their electrons are tightly bound to the nuclei and the energy levels are either completely filled or completely empty. These solids are very stable and are often transparent to visible light because of the large energy difference between the filled and unfilled energy orbits. Quite a few insulators, however, are known to have native defects and impurities that can produce structural and electrical instabilities. Some of these native defects and impurities, for instance, are also responsible for the colors found in many otherwise transparent insulators. One example is the ruby crystal (Al$_2$O$_3$:Cr$^{3+}$), which would be colorless in the absence of the Cr$^{3+}$ impurities.

Unlike the ionic solids, covalent solids are either insulators or semiconductors. Since covalent bonds are tetrahedral, each atom in the solid will have four nearest neighbors. Through sharing of the outermost electrons, these atoms can achieve inert gas configuration, which makes them very stable. The sharing of the electrons also lowers the excitation energy of the electrons and can result in light absorption down to the infrared (IR) range. This is the reason why many semiconductors are opaque to visible light and have a dull metallic appearance. For the same reason, semiconductors are also sensitive to temperature changes. Experimentally, it is possible to distinguish between an ionic solid and a covalent solid simply through a measurement of the **dielectric constant** $\epsilon_r$ (a parameter that quantifies the charge separation or polarization in the atoms or the molecules; see Section 2.10). At a low frequency, an ionic solid will usually exhibit a large $\epsilon_r$ ($>> 1$) because of the polarization of the anions and the cations. Such a polarization effect is absent in a purely covalent solid.

Metallic bonds are found only in metals that have a significant number of free electrons. Generally speaking, the free electrons in a metal are not directly attached to the individual nuclei but may be considered to be electrically bound to the array of positively charged nuclei. Because of this, free electrons can move more easily inside the metal. They are responsible for the excellent thermal and electrical conduction properties.

Normally, free electrons do not readily interact with light, which is partially responsible for the high reflectivity of a polished metal surface. The physical properties of a metal—in particular, those related to hardness and ductility—can also be traced to the close-packed crystal structure. Some metals, known as transition metals and rare-earth metals, possess magnetic properties that give rise to effects known as ferromagnetism (strong magnetization) and antiferromagnetism (weak magnetization). The transition and rare-earth metals fall between the Group II and Group III **elements** in the periodic table; they have partially filled energy levels, which also tend to favor spin alignment in the orbiting electrons. These topics are discussed further in Chapter 5.

## 1.3    CRYSTAL STRUCTURES

A **crystal** is a solid whose atoms or ions are arranged in a periodic array. As a result, the positions of the atoms or ions are fixed; i.e., there is a constant spacing between the atoms or ions, and the same atoms or ions will reappear after a given translation along a crystal direction. Periodicity is one of the most important properties of a crystal; it allows us to focus on a smaller unit of the three-dimensional array. Periodicity also gives rise to properties known as **symmetries,** which can effectively reduce the complexity of a crystal structure through mathematical transformations. In fact, it is often more convenient to examine a crystal through its crystal symmetries.

For instance, the crystal shown in Fig. 1.10 has translational symmetry along

**Fig. 1.10**
A three-dimensional
lattice showing
translational symmetries

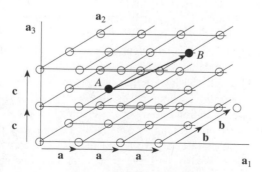

the $\mathbf{a}_1$ direction, which means that the same crystal structure will reappear after all the atoms in the crystal make a shift of one atomic spacing in the $\mathbf{a}_1$ direction. The same result is observed for several such shifts. A similar argument applies to a translation in the $\mathbf{a}_2$ direction as well as to one in the $\mathbf{a}_3$ direction. In this case, $\mathbf{a}_1$, $\mathbf{a}_2$, and $\mathbf{a}_3$ are called the unit (translational) vectors. Thus, in this particular crystal, translational symmetries exist that allow us to move from one atomic or ionic site, say, site A, to another, such as site B. Mathematically, the translation vector $\mathbf{a}_{AB}$ linking sites A and B is given by

$$\mathbf{a}_{AB} = n_1\mathbf{a}_1 + n_2\mathbf{a}_2 + n_3\mathbf{a}_3 \tag{5}$$

where   $\mathbf{a}_1$, $\mathbf{a}_2$, and $\mathbf{a}_3$ are the unit (translational) vectors along the principal **crystal directions** (sometimes called bases), and

$n_1$, $n_2$, and $n_3$ are the integers, or units, of translation.

In addition to translational symmetries, other types of symmetries are also present in crystalline solids:

1. *Inversion symmetry* is best explained by the invariance of a crystal after an inversion of all the atoms or ions at a given atomic or ionic site. Mathematically, the inversion operation is represented by $\mathbf{a}_{AB} = -\mathbf{a}_{AB}$; i.e., for every site B (see Eq. (5) with site A being the origin), there is a corresponding site B′ at the position $n_1'$ $(= -n_1)$, $n_2'$ $(= -n_2)$, and $n_3'$ $(= -n_3)$. Thus, with inversion symmetries, a crystal will reappear identically the same before and after the inversion process. Pictorially, this is illustrated in Fig. 1.11 for a two-dimensional crystal structure. As expected, the majority of the simple crystal structures have one or more inversion sites.

2. *Reflection symmetry* refers to the invariance of the crystal structure after reflection from an atomic plane within the crystal. It is analogous to light reflection from a mirror. Fig. 1.12 shows a crystal structure with reflection symmetry at plane

**Fig. 1.11**
Crystal symmetries for a
two-dimensional crystal

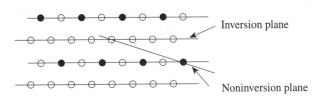

**Fig. 1.12**
Crystal structure with
reflection symmetry

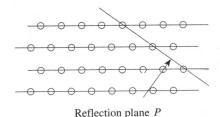

Reflection plane *P*

*P*. A simple cubic crystal structure, for instance, has nine reflection symmetry planes: three are parallel to the faces of the cube and six others each pass through the opposite edges.

3. *Rotational symmetry* is best illustrated by viewing the invariance of the crystal structure after a rotation of all the atoms or ions around a given axis in the crystal by a fixed angle. The angle of rotation is always a fraction of $2\pi$ rad (i.e., a full circle). Fig. 1.13 shows the top view of a crystal structure with rotational symmetry around the *c* axis. The angle of rotation is $2\pi/3$ rad (or 120°) and in this case, the *c* axis is called a threefold axis since a full revolution will require three such rotations. As an example, a cubic crystal structure has three fourfold rotational axes normal to the faces of the cube and four fourfold rotational axes, each passing through the two opposite corners.

Symmetries are most frequently used to classify the different crystal structures found in solids. In general, one can generate 14 basic crystal structures through symmetries. These are called **Bravais lattices** and are shown in Fig. 1.14. Any crystal structures can be reduced to one of these 14 Bravais lattices; as an example, Fig. 1.15 shows the NaCl crystal structure and it is composed of two face-centered cubic lattices superimposed on each other.

**Example 1.13**    Show that the triclinic crystal structure shown in Fig. 1.14 has no reflection planes, whereas the monoclinic crystal structure (also in the same figure) has one reflection plane.

*Solution:*    By inspection, it can be shown that the triclinic crystal structure has no reflection planes. The reflection plane in the monoclinic crystal structure cuts across the crystal midway between and parallel to the bases (see Eq. (5)).    §

**Example 1.14**    Show that the triclinic crystal structure has no rotational symmetry and that the monoclinic crystal structure has one twofold rotational axis (see Fig. 1.14).

*Solution:*    By inspection, it can be shown that the triclinic crystal structure has no rotational symmetries. The two-fold rotational axis of the monoclinic crystal structure is normal to the base.    §

**Fig. 1.13**
Top view of a crystal structure with rotational symmetry

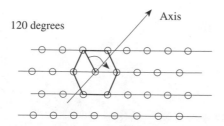

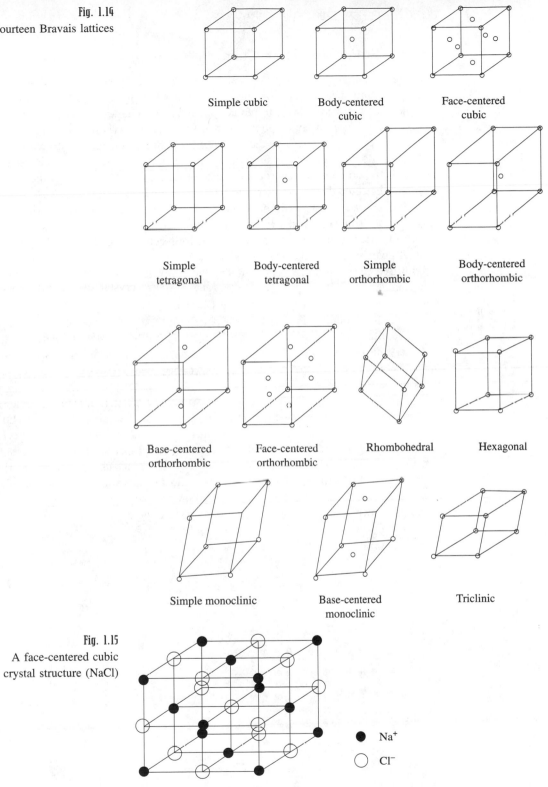

Fig. 1.14
Fourteen Bravais lattices

Simple cubic

Body-centered cubic

Face-centered cubic

Simple tetragonal

Body-centered tetragonal

Simple orthorhombic

Body-centered orthorhombic

Base-centered orthorhombic

Face-centered orthorhombic

Rhombohedral

Hexagonal

Simple monoclinic

Base-centered monoclinic

Triclinic

Fig. 1.15
A face-centered cubic crystal structure (NaCl)

● Na⁺
○ Cl⁻

17

## Lattice Points and Unit Cells

The representation of a complex crystal structure frequently involves the display of a significant number of atoms or ions. Instead of showing all the atoms or ions, it is sometimes more convenient to replace them by discrete points, or lattice points. An entire collection of lattice points is then known as a **crystal lattice,** or simply a lattice. The lattice contains all the spatial information of the crystal structure, and each lattice point is depicted by its **coordinates** (a convenient lattice point is usually treated as the origin of the coordinates). For instance, Fig. 1.16 displays the lattice points of the three-dimensional crystal structure shown in Fig. 1.10.

With the crystal symmetries mentioned earlier, we can also divide a crystal into smaller, yet identical, units. Each of these units serves the same role as a single brick in a brick wall. These units are called **unit cells;** depending on the choice, unit cells can have different sizes and shapes. A minimum-volume unit cell is called a **primitive cell.** A primitive cell contains the equivalence of a single lattice point (i.e., one lattice point per primitive cell). When a lattice point falls at the boundary between two or more unit cells, only that fraction of the lattice point falling within the unit cell is counted. For instance, a lattice point at the corner of a cubic unit cell is counted as 1/8. Fig. 1.17 shows the relationship between a unit cell and the entire crystal lattice. To fully specify a unit cell, both the cell dimensions and the angles between the principal crystal axes have to be given. These values are referred to as the **lattice constants.**

The packing density of atoms or ions in a unit cell is often of interest because it is a good indicator of the mechanical properties of the solid. A more tightly packed crystal structure will tend to be mechanically harder and less fragile. In addition, the packing density also affects such crystal properties as light refraction and impurity diffusion. The compactness of a crystal is usually measured by the **atomic packing factor (APF),** which is the ratio of the volume occupied by the atoms or ions in the unit cell divided by the volume of the unit cell. Mathematically, it can be shown that most crystals have APFs falling between 0.52 and 0.74.

**Fig. 1.16**
A three-dimensional array
of lattice points

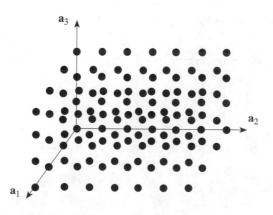

Fig. 1.17
Unit cells and primitive
cells

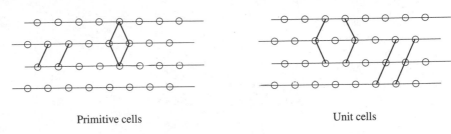

Primitive cells                                    Unit cells

**Example 1.15**   Show that in two dimensions, the area-centered square lattice shown in Fig. E1.2(a) can be reduced to a simple square lattice.

*Solution:*   As shown in Fig. E1.2(b), the diagonals of the area-centered square lattice will cut out a square lattice bounded by $a'b'c'd'$.   §

Fig. E1.2

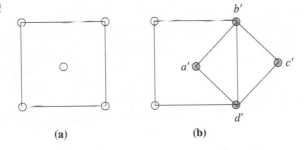

(a)                                    (b)

## Lattice Directions and Lattice Planes

We recall that the lattice vector $\mathbf{a}_{AB}$ (see Eq. (5)) joins the two lattice sites A and B, and the principal **lattice directions** are $\mathbf{a}_1$, $\mathbf{a}_2$, and $\mathbf{a}_3$. To simplify the notation, we pick site A to be the origin of the coordinates and give $\mathbf{a}_{AB}$ by the coordinates of site B (placed in square brackets). For example, the lattice direction of the body diagonal of a cube is given by [1 1 1], and the unit vector $\mathbf{a}_1$ is given by [1 0 0]. Because of crystal symmetries, lattice directions are not unique. For a cubic crystal, the lattice directions [1 0 0], [0 1 0], [0 0 1], [$\bar{1}$ 0 0], [0 $\bar{1}$ 0], and [0 0 $\bar{1}$] are all equivalent (this result is evident if we rotate the cubic crystal by $\pi/4$ rad around each of the edges). The entire family of these equivalent lattice directions can be represented by one of its members put in angular brackets, such as <1 0 0>. To illustrate this point, Fig. 1.18 shows the <1 1 1> family of lattice directions.

Lattice planes are needed to define crystal surfaces and interfaces. A **lattice**

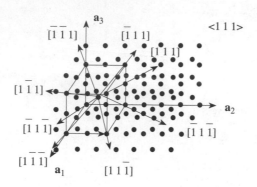

**Fig. 1.18**
Equivalent lattice directions in the <1 1 1> family

**plane** is described by a set of numbers *h, k, l,* known as **Miller indices,** where *h, k,* and *l,* are integers. Miller indices can be obtained in the following manner:

1. Determine the intercepts of the lattice plane on the principal lattice directions.
2. Take the reciprocals of the values of these intercepts and reduce them to the three smallest integers having the same ratio.
3. Enclose the integers in parentheses: (*h k l*).

Fig. 1.19 shows pictorially how Miller indices can be determined. Since the plane *P* shown in the figure intercepts the principal lattice directions at the points (1 0 0), (0 3 0), and (0 0 3), the reciprocals of these intercepts will be 1, 1/3, 1/3. Taking the smallest integers having the same ratio will give the Miller indices:

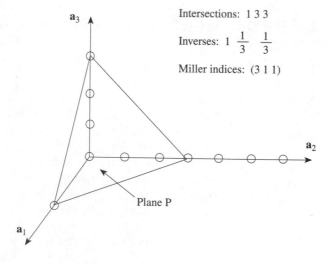

**Fig. 1.19**
Miller indices for plane P

Intersections: 1 3 3

Inverses: $1 \quad \dfrac{1}{3} \quad \dfrac{1}{3}$

Miller indices: (3 1 1)

Plane P

(3 1 1). The same Miller indices also represent the set of planes that are parallel to $P$, the entire family of such planes is represented by $\{3\ 1\ 1\}$. (Planes such as (1 3 1) and (1 1 3) also belong to the same family).

In the study of crystal structures it is often necessary to compute the shortest distance $d_{hlk}$ between two members of the same lattice plane family $\{h\ k\ l\}$. For a cubic crystal structure, $d_{hlk}$ is given by

$$d_{hlk} = \frac{a_0}{\sqrt{h^2 + l^2 + k^2}} \tag{6}$$

where    $a_0$ is the length of one side of the cube in meters.

As an example, the shortest distance between the $\{1\ 1\ 1\}$ family of planes is given by $d_{111} = a_0/\sqrt{1^2 + 1^2 + 1^2} = a_0/\sqrt{(3)}$.

**Example 1.16**   Label the nine reflection planes in a cubic unit cell and show the three fourfold rotational axes.

*Solution:*   As shown in Fig. E1.3, the reflection planes are (1 0 0), (0 1 0), (0 0 1), (1 1 0), ($\bar{1}$ 1 0), (1 0 1), ($\bar{1}$ 0 1), (0 1 1), and (0 $\bar{1}$ 1). The fourfold rotational symmetries can be observed if the cubic structure is rotated by $\pi/2$ (90°) along each principal axis.

**Fig. E1.3**

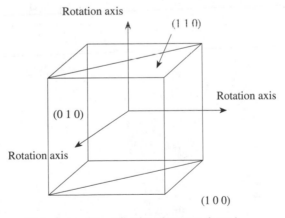

(Only some of the reflection planes are shown.)    §

## Common Crystal Structures

Most elements and compounds in their crystalline state have well-defined structures. In the following we outline some of the more common crystal structures and their structural properties:

1. The **face-centered cubic (fcc) crystal structure** exists primarily in metals and in ceramics. The unit cell is shown in Fig. 1.20. In the Bravais lattice, there are 4 lattice points per unit cell, and in the fcc metals, the APF is 0.74 (the highest possible value in the packing of atoms or ions of an identical size). An example of an fcc metal is Al.

2. The **body-centered cubic (bcc) crystal structure** shown in Fig. 1.21 is less compact when compared with the fcc crystal structure. It is also commonly found in metals. There are 2 lattice points per unit cell, and the APF of the bcc metal is only 0.68. Examples of the bcc crystal structures include α-Fe, V, Cr, Mo, and W.

Table 1.5 summarizes the structural properties of cubic lattices, including the

**Fig. 1.20**
The face-centered cubic
crystal structure (Al)

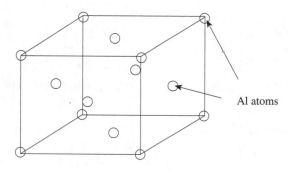

Al atoms

**Fig. 1.21**
The body-centered cubic
crystal structure (Mo)

Mo atoms

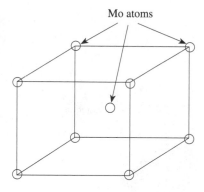

Table 1.5    Structural properties of cubic lattices ($a_0$ is the length of one side of the unit cell)

|  | sc | bcc | fcc |
| --- | --- | --- | --- |
| Volume of unit cell | $a_0^3$ | $a_0^3$ | $a_0^3$ |
| Lattice points per cell | 1 | 2 | 4 |
| Volume of primitive cell | $a_0^3$ | $a_0^3/2$ | $a_0^3/4$ |
| Number of nearest neighbors | 6 | 8 | 12 |
| Distance of nearest neighbors | $a_0$ | $0.865a_0$ | $0.707a_0$ |
| APF | 0.523 | 0.68 | 0.74 |

simple cubic (sc) lattice, the face-centered cubic (fcc) lattice, and the body-centered cubic (bcc) lattice.

3. The **diamond crystal structure** shown in Fig. 1.22 actually has the fcc crystal structure. Its unit cell contains 8 atoms. The **non-Bravais lattice** of the diamond crystal structure can be reduced to the fcc lattice by associating 2 atoms with each lattice point. In the diamond crystal structure, the nearest neighbor atoms exist in a tetrahedron (shown in the figure), and there are four nearest neighbors per atom. The majority of the elemental semiconductors, such as Si and Ge, have this type of crystal structure.

4. The **zinc-blende crystal structure** is similar to the diamond crystal structure, except that there are two different types of atoms or ions present in every unit cell. ZnS, for instance, has the zinc-blende crystal structure, and the $Zn^{2+}$ ions are located at the corners and the faces of the unit cell. However, the $S^{2-}$ ions are found along the body diagonals (the exact coordinates of the $S^{2-}$ ions are $(\frac{1}{4}\,\frac{1}{4}\,\frac{1}{4})$; $(\frac{1}{4}\,\frac{3}{4}\,\frac{3}{4})$; $(\frac{3}{4}\,\frac{1}{4}\,\frac{3}{4})$; and $(\frac{3}{4}\,\frac{3}{4}\,\frac{1}{4})$. The zinc-blende crystal structure is primarily found in compound semiconductors, like GaAs and InSb.

5. The **hexagonal close-packed (hcp) crystal structure** is found in metals and in semiconductors. In a way, it resembles the fcc crystal structure. The main difference lies in the fact that the highest density planes in the hcp crystal structure are repeated every other atomic or ionic layer (rather than every third layer, as in the fcc crystal structure). This difference is illustrated in Fig. 1.23. In the hcp crystal structure, the APF is 0.74 and the ratio $c/a_0$—i.e., the height of the unit cell divided by the length of one side of the hexagon in the base plane—has a value of 1.63. Outstanding examples of solids having a hcp crystal structure are CdS and ZnO.

**Example 1.17**   For **(a)** the fcc crystal structure, and **(b)** the bcc crystal structure, compute the APFs if the atoms or ions have the same radius $r$.

**Fig. 1.22**
The diamond crystal
structure (Si)

Two superimposed fcc structures

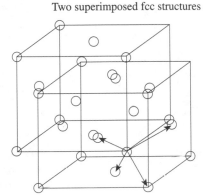

Tetrahedral nearest neighbors

**Fig. 1.23**
Hexagonal close-packed
crystal structure

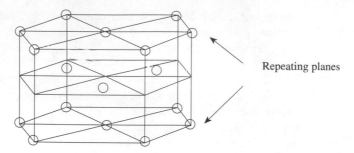

Repeating planes

***Solution:***    See Figs. E1.4 and E1.5.

**(a)** For the fcc crystal structure, the face diagonal is $4r = \pi(a_0^2 + a_0^2)$, where $a_0$ is one side of the unit cell, so $a_0 = \sqrt{8}r$. Since the unit cell contains 4 atoms and its volume is $a_0^3$, APF $= (4/3)4\pi r^3/a_0^3 = 16\pi/(3 \times 8 \times \sqrt{8}) = 0.74$.

**(b)** For the bcc crystal structure, the body diagonal is $4r = \sqrt{3}a_0^2$, so $a_0 = \sqrt{\frac{16}{3}}r$. Since there are 2 atoms per (unit) cell, APF $= \frac{4}{3} \times 2\pi \times r^3/a_0^3 = 8\pi/[3 \times (\sqrt{\frac{16}{3}})^3] = 0.68$.   **§**

**Fig. E1.4**

Face diagonal

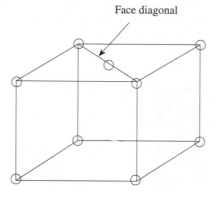

**Fig. E1.5**

Body diagonal

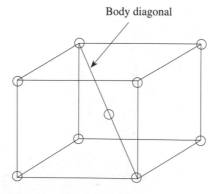

**Example 1.18**   **(a)** Determine the number of lattice points present in an fcc unit cell, and **(b)** draw a primitive cell within the unit cell.

*Solution:*
    **(a)** There are $6 \times \frac{1}{2} \, (= 3)$ lattice points on the faces of the fcc unit cell and $8 \times \frac{1}{8} \, (= 1)$ lattice point at the corners, for a total of 4 lattice points in the unit cell.

Fig. E1.6

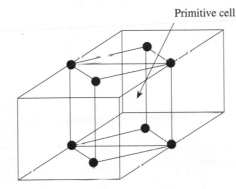

Primitive cell

    **(b)** The primitive cell is shown in Fig. E1.6.   §

**Example 1.19**   For a cubic cell with unity dimensions (i.e., $a_0 = 1$), compute the atomic radius $r$ if the solid has **(a)** a bcc crystal structure, and **(b)** a diamond crystal structure.

*Solution:*   See Figs. E1.7 and E1.8.
    **(a)** For the bcc crystal structure, the body diagonal is $4r =$ $\sqrt{a_0^2 + a_0^2 + a_0^2} = \sqrt{1 + 1 + 1} = \sqrt{3}$. Therefore, $r = \sqrt{\frac{3}{4}} = 0.433$.

Fig. E1.7

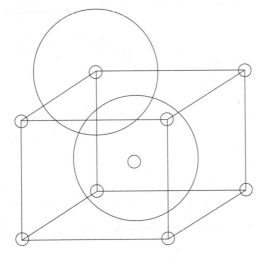

    **(b)** For the diamond crystal structure, the body diagonal is $4(2r) = \sqrt{3}$. Therefore, $r = \sqrt{\frac{3}{8}} = 0.266$.   §

Fig. E1.8

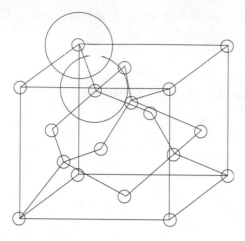

**Example 1.20**   (a) Give the <1 0 0> family of lattice directions in the cubic system, and (b) sketch these lattice directions.

*Solution:*
   (a) The family is [1 0 0], [0 0 1], [0 1 0], [0 $\bar{1}$ 0], [0 0 $\bar{1}$], and [0 0 $\bar{1}$].
   (b) The lattice directions are shown in Fig. E1.9.   §

Fig. E1.9

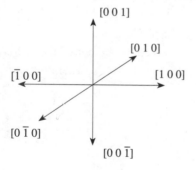

**Example 1.21**   For a cubic crystal structure, draw (a) the [1 1 1], [1 3 3], and [2 3 6] lattice directions and (b) the (1 1 1), (2 3 4), and (3 2 1) lattice planes.

*Solution:*   See Fig. E1.10.   §

**Example 1.22**   (a) Determine which planes in the fcc crystal structure have the highest density of atoms, and (b) evaluate its value for copper. The lattice constant $a_0$ of Cu is 0.361 nm.

*Solution:*
   (a) The highest density exists in the {1 1 1} planes since they have the shortest nearest-neighbor distance.

Fig. E1.10

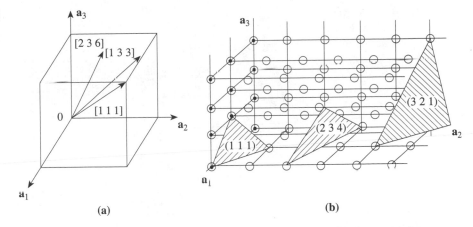

(a)                                             (b)

(b) There is a total of 2 atoms on the face planes of the unit cell. The density of the Cu atoms on these planes is $2/(a_0^2)$ = 2 atoms/$(0.361$ nm$)^2$ = 15 atoms/nm$^2$.  §

# 1.4   CRYSTALLINE SOLIDS

## Metals and Their Oxides (Ceramics)

Although metals may have more than one type of crystal structure (for example, $\alpha$-Fe has a bcc crystal structure, whereas $\gamma$-Fe has a fcc crystal structure (see Section 1.6)), in general, they have either the bcc, the fcc, or the hcp crystal structure. Examples of bcc metals include $\alpha$-Fe, V, Cr, Mo, and W; fcc metals include Al, Ni, $\gamma$-Fe, Cu, Ag, Pt, and Au; and hcp metals include $\alpha$-Ti, Zn, and Zr. Figs. 1.24, 1.25, and 1.26 show the crystal structures of $\alpha$-Fe (bcc), Cu (fcc), and Zn (hcp), respectively. Metal **compounds** and oxides of metals, on the other hand, are primarily ionic solids, and the oxides are often called ceramics.

Most ceramics are insulators and are best known for their hardness and heat resistance. Unlike metals, ceramics are not ductile and cannot be easily molded into different shapes. Many ceramics are transparent to light and a few, in particular, have superior light-transmission properties. An outstanding example of a transparent

Fig. 1.24
Body-centered cubic
structure of $\alpha$-Fe

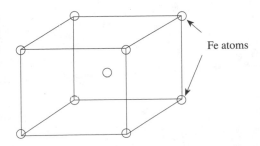

Fe atoms

Fig. 1.25
Face-centered cubic
structure of Cu

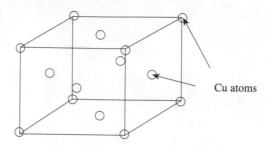

Fig. 1.26
Hexagonal close-packed
structure of Zn

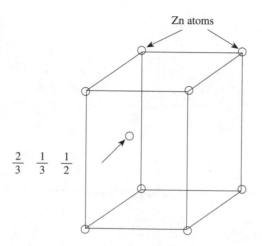

ceramic is quartz, which is widely used in optical transmission systems. For the better-known ceramics, there are several different crystal structures. Examples of fcc ceramics include MgO, CaO, FeO, and NiO; fcc ceramics with non-Bravais lattices include $UO_2$, $TeO_2$, and $SiO_2$. Figs. 1.27 and 1.28 show the crystal structures of FeO (NaCl structure) and $SiO_2$ (it has an fcc Bravais lattice), respectively.

Fig. 1.27
Crystal structure of FeO

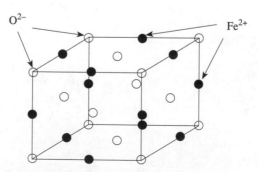

Fig. 1.28
Crystal structure of $SiO_2$

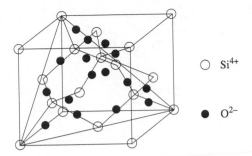

○  $Si^{4+}$

●  $O^{2-}$

Other ceramic crystal structures include the rhombohedral crystal structure of $Al_2O_3$ and $Cr_2O_3$ and the spinel crystal structure (modified fcc structure) of the ferrites and the ferromagnetic solids. These are illustrated in Figs. 1.29 and 1.30.

Fig. 1.29
Rhombohedral structure of $Al_2O_3$

● $Al^{3+}$

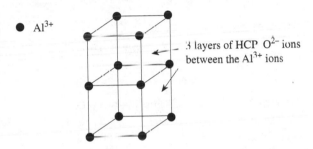

3 layers of HCP $O^{2-}$ ions
between the $Al^{3+}$ ions

Fig. 1.30
Spinel structure of ferrites

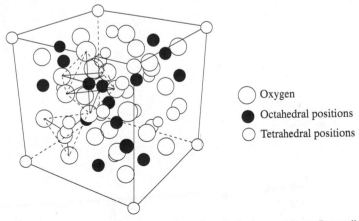

○  Oxygen

●  Octahedral positions

○  Tetrahedral positions

(Source: F. G. Brockman, "Magnetic Ceramics—A Review and Status Report," *Bulletin of the American Ceramic Society,* 1968. Reprinted by permission of the American Ceramic Society.)

**Example 1.23** Al has an fcc crystal structure. Compute its density if the atomic weight of Al is 26.98 gm/mol and the lattice constant $a_0$ is 0.404 nm. (See Fig. E1.11.)

Fig. E1.11

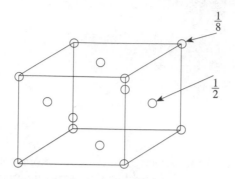

**Solution:** The mass of 1 mol ($6.02 \times 10^{23}$ atoms) of Al is 26.98 g, so the mass of each Al atom is $m_{Al} = 4.49 \times 10^{-23}$ g. In the Al unit cell there are 4 atoms, and the volume of the unit cell is $V = (4.04 \times 10^{-10})^3$ m$^3 = 6.59 \times 10^{-29}$ m$^3$. The atomic density is $\rho_{atom} = 4$ atoms/$V =$ $4/(6.59 \times 10^{-29})$ atoms/m$^3 = 6.07 \times 10^{28}$ atoms/m$^3$. The density of Al is $\rho_0 = m_{Al} \times \rho_{atom} = 4.49 \times 10^{-23}$ g $\times 6.07 \times 10^{28}$/m$^3 = 2.72 \times 10^6$ g/m$^3$. **§**

Note that the specific gravity of Al is 2.70.

**Example 1.24** Calculate the atomic density of $\alpha$-Fe, which has a bcc structure.

**Solution:** See Fig. E1.12. For a bcc crystal structure, the ratio of the lattice constant $a_0$ to the atomic radius $r_{Fe}$ is $a_0/r_{Fe} = 4/\sqrt{3}$. Since $r_{Fe} = 0.124$ nm, $a_0 = 4/\sqrt{3} \times 0.124$ nm $= 0.286$ nm (see Appendix B). There are 2 atoms per unit cell, and the atomic density of $\alpha$-Fe is $\rho_{atom} = 2$ atoms/$a_0^3 =$ 2 atoms/$(0.286 \times 10^{-9})^3$/m$^3 = 8.55 \times 10^{28}$ atoms/m$^3$.

Fig. E1.12

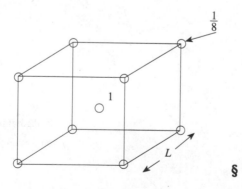

**§**

**Example 1.25** Fig. E1.13 shows the unit cell of an hcp crystal structure. Given that $c/a_0$ is 1.63, calculate the APF, assuming that all the atoms have the same radius $r$.

**Solution:** The volume of the unit cell is $V =$ base area $\times$ height $=$ $(a_0 \times a_0/\sin(60°)) \times 1.63 \times a_0 = 1.41 \times a_0^3$. Since $a_0 = 2r$ in a close-packed

Fig. E1.13

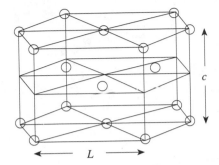

structure, $V = 1.41 \times (2r)^3 = 11.3 \times r^3$. There are 2 atoms per unit cell, and APF $= 2 \times 4\pi/3 \times r^3/V = 2 \times 4\pi/3 \times r^3/(11.3 \times r^3) = 0.74$. §

**Example 1.26**   Calculate the planar density of the ions in the (1 1 1) plane of CaO. CaO has an fcc crystal structure, and there are 2 atoms in the (1 1 1) plane within the unit cell.

*Solution:* The lattice constant of CaO is $u_0 = 2\sqrt{(2)}(r_{Ca}^{2+} + r_O^{2-}) = 2\sqrt{(2)}(0.132 \text{ nm} + 0.106 \text{ nm}) = 0.673$ nm. Thus, the area of the triangle in the unit cell intersected by the (1 1 1) plane is $A_{111} = (a_0/2) \times \sqrt{3} \times (a_0/2) = (0.673 \text{ nm}/2) \times \sqrt{3} \times (0.673 \text{ nm}/2) = 0.196 \text{ nm}^2$. The ion density in the (1 1 1) plane is 2 ions/$A_{111}$ = 2 ions/0.196 nm$^2$ = 10.2 ions/nm$^2$. §

## Semiconductors

Most common **semiconductors** have either a modified fcc crystal structure or an hcp crystal structure, as shown in Fig. 1.31. Some outstanding examples are Si, Ge, and GaAs. Si and Ge have a diamond crystal structure (a non-Bravais fcc crystal structure), whereas GaAs has a zinc-blende crystal structure (similar to the diamond crystal structure except that more than one type of atoms are present in the unit cell). In both cases, the nearest neighbors are bonded tetrahedrally. An alternative form of the zinc-blende crystal structure is the *wurtzite crystal structure,* which is built on a non-Bravais hexagonal lattice with 4 atoms at each lattice point. The same tetrahedral arrangement of the atoms is also present. Table 1.6 lists the crystal structures and the lattice spacings for the more important semiconductors.

Table 1.6.   Crystal structures and the lattice spacings for the important semiconductors

| Semiconductor | Crystal Structure | Lattice Spacing (nm) |
|---|---|---|
| Si | Diamond | 0.543 |
| Ge | Diamond | 0.563 |
| GaAs | Zinc-blende | 0.565 |
| InP | Zinc-blende | 0.586 |
| CdS | Wurtzite | 0.583 |
| ZnS | Wurtzite | 0.542 |

Fig. 1.31
Different semiconductor
structures

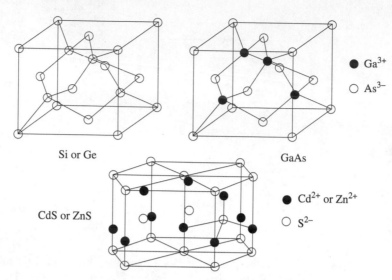

Si or Ge

GaAs

● Ga³⁺

○ As³⁻

CdS or ZnS

● Cd²⁺ or Zn²⁺

○ S²⁻

**Example 1.27**    Calculate the planar density of atoms in the (1 1 1) plane of Ge. (See Fig. E1.14.)

*Solution:*    Since the lattice constant of the unit cell of Ge is $a_0 = 0.563$ nm, the face diagonal $d_{fd} = \sqrt{2}a_0 = \sqrt{2} \times 0.563$ nm $= 0.796$ nm. The area of the triangle in the unit cell intersected by the (1 1 1) plane is $A_{111} = (d_{fd}/2) \times \sqrt{3} \times (d_{fd}/2) = (0.796$ nm$/2) \times \sqrt{3} \times (0.796$ nm$/2) = 0.274$ nm². Since there are 2 atoms within this area, the planar density of atoms in the (1 1 1) plane is 2 atoms$/A_{111} = 2$ atoms$/(0.274 \times 10^{18}$ m²$) = 7.29 \times 10^{18}$ atoms/m².

Fig. E1.14

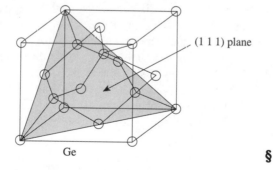

(1 1 1) plane

Ge

§

## 1.5    POLYCRYSTALLINE AND NONCRYSTALLINE SOLIDS

Both **polycrystalline solids** and noncrystalline solids can be deposited on a suitable substrate either in bulk form or in the form of thick films or thin films (<50 μm). The majority of the devices, especially electronic devices, are prepared on thin films to reduce cost; sometimes, however, this preparation is done to allow for a larger surface area, as in the case of a solar cell. For proper adhesion, a rigid substrate, such as a glass plate or a metal sheet, is most frequently used, although thin insulating films are also deposited on semiconductors to provide insulation. Generally speaking, the properties of bulk solids and thick films are quite similar, while the thin films are more susceptible to surface conditions and contamination by the environment. In the following subsections we examine in greater detail the physical properties of the different types of polycrystalline and noncrystalline solids and thin films.

## Polycrystalline Solids and Thin Films

Polycrystalline solids and thin films are made up of small crystal grains (single crystals with a grain size ranging from tens of nanometers to several microns), and their structures are critically dependent on the method of preparation, including the deposition temperature and the texture of the **substrate.** Most polycrystalline films are deposited by **vapor-phase deposition (VPD),** and the crystal size can vary considerably, depending on the deposition rate, the substrate temperature, and the film thickness. Finer structures are formed whenever there is sufficient time for nucleation (the gradual growth of smaller crystallites consisting of only a few hundred atoms) to occur. Nucleation is also facilitated at a higher substrate temperature. All polycrystalline thin films are sensitive to atmospheric and surface contamination, including the absorption of gases, such as $O_2$, $N_2$, and moisture. Gas absorption occurs reversibly, depending on the atmospheric conditions. Usually, more reliable thin films are deposited in an inert gas atmosphere or in a vacuum. Fig. 1.32 shows the schematic of a hypothetical polycrystalline thin film with different grain sizes due to progressive nucleation. As we shall see in later chapters, grain size can play a very important role in affecting the electronic properties of polycrystalline solids; this situation is especially important when the thin films are used for metallization (interconnection lines).

## Amorphous Solids and Thin Films

Not all solids have well-defined crystal structures. In amorphous solids, the atoms or ions are arranged randomly. Except for the nearest neighbors, the placement of

**Fig. 1.32**
A hypothetical thin film showing variations in grain size and morphology

Increasing
nucleation

$\longrightarrow$

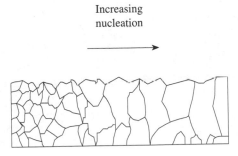

atoms or their ions is totally irregular. As a result, there are **dangling bonds** (incomplete energy bonds due to missing neighboring atoms or ions) and **voids** (empty spaces in a solid where groups of atoms are missing). A simple microscopic model of an amorphous solid is shown in Fig. 1.33. As shown in the figure, there is little long-range order, even though, on the average, the atoms or ions still maintain roughly the same number of nearest neighbors. Amorphous solids and thin films are, in general, stable only up to a certain temperature; beyond that, they will recrystallize—i.e., they will become polycrystalline. Sometimes, under more favorable conditions, as in the case of intense localized heating caused by a laser beam, an amorphous thin film deposited on a suitable substrate may recrystallize into a single crystal. Generally speaking, amorphous solids have the same physical properties, such as solubility and melting point, as the single crystals, even though properties linked to their structures, such as thermal conductivity and thermal expansivity, are quite different.

As mentioned earlier, amorphous solids are relatively easy and inexpensive to prepare; their major weakness has been that their properties can vary considerably from sample to sample. Amorphous thin films, such as $SiO_2$ and $Si_3N_4$, are well-known insulators and have been widely used for electrical insulation and surface passivation (as protective coatings) in devices and circuits. More recently, fairly reliable amorphous semiconductor thin films have been prepared using the glow discharge technique; they are competing with polycrystalline semiconductors in the manufacture of low-cost, large-area devices such as solar cells and thin-film transistors (TFTs). Similar to polycrystalline thin films, amorphous thin films are also sensitive to surface and atmospheric contamination.

## Polymers

Another form of noncrystalline solids receiving increasing attention are polymers. **Polymers** are solids consisting of long chains of repeating units. These are molecules of a complex structure, as shown in Fig. 1.34. Most polymers have a carbon skeleton that can be partially crystalline and partially amorphous. The chains are irregular in shape and can be up to several hundred nanometers long, usually folded back and forth within the solid. An example of a simple polymer is polyethylene

**Fig. 1.33**
Atomic structures for crystalline and polycrystalline solids

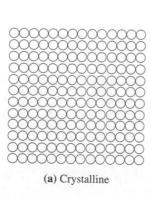

**(a)** Crystalline

Voids

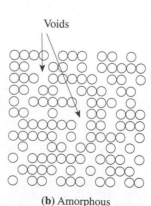

**(b)** Amorphous

Fig. 1.34

A long chain of solid
polyethylene

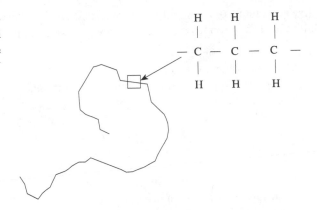

$-(C_2H_4)-_n$, which is made up of repeating units of $-(C_2H_4)-$, where $n$ stands for the total number of units in the chain. *Polymerization* is a term used to describe the breaking up of complex energy bonds of the C atoms to form simple energy bonds, which is followed by the linking up of the C atoms in shorter chains, as shown in Fig. 1.35. In some polymers, the chains can be bonded into fibers of significant strength and are used in composite materials. Nevertheless, due to the random orientation of the chains, most polymers have a low packing density (about 50%), and they are insulators. More recently, conducting polymers have been formed; they offer potential applications in the replacement of metal interconnections in microelectronic circuits. Some polymers have rectification properties, which have potential applications in light detection and luminescence display.

In general, polymer thin films are relatively simple to prepare. The preparation usually involves a **spin-on technique** (polymers can also be evaporated on a suitable substrate). During preparation, a purified polymer precursor is first deposited onto a suitable substrate and then spun at a high speed to ensure uniformity in the thickness of the precursor. A low-temperature bake is then applied to remove the solvent, and the residue forms the polymer.

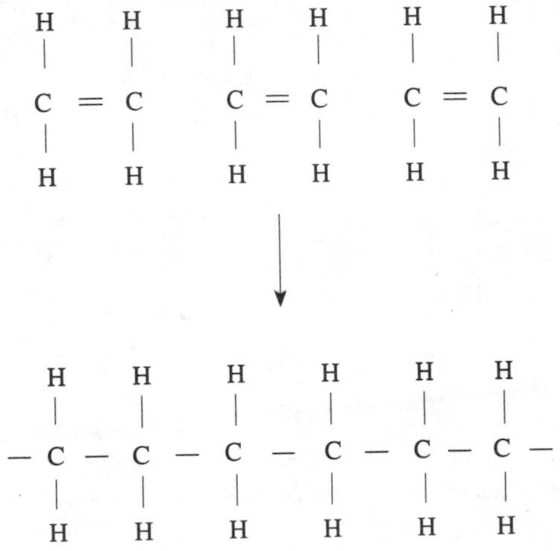

Fig. 1.35

Polymerization of vinyl
chloride

**Example I.28**     Give an example of a naturally occurring solid that is **(a)** crystalline; **(b)** polycrystal-line; and **(c)** amorphous. Structurally, how can we distinguish between these solids?

*Solution:*     An example of
(a) a crystalline solid is a diamond;
(b) a polycrystalline solid is common iron ore; and
(c) an amorphous solid is limestone (a form of calcium carbonate).
A crystal has a well-defined shape regardless of its size. A polycrystalline solid does not have any defined shape. An amorphous solid does not have any long-range structure.   §

**Example I.29**     Compute the volume of the polyethylene unit $-(C_2H_4)-$, which has the structure shown in Fig. E1.15. The dimensions of the unit cell are 0.74 nm $\times$ 0.494 nm $\times$ 0.255 nm.

**Fig. E1.15**
Unit cell of polyethylene

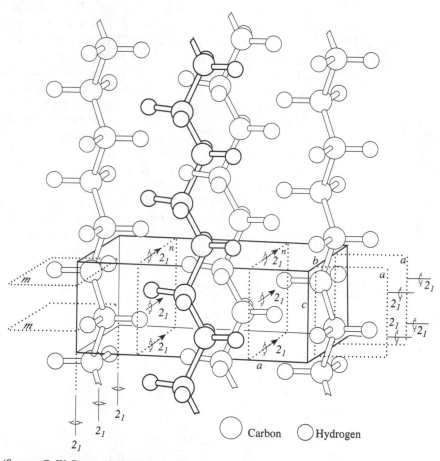

○ Carbon   ○ Hydrogen

(Source: C. W. Bunn, *Chemical Crystallography—An Introduction to Optical and X-ray Methods*, Oxford, Clarendon Press, 1961. Reprinted by permission of Oxford University Press). The cell dimensions are: length = 0.741 nm; width = 0.494 nm; and height = 0.255 nm.

*Solution:*   According to Fig. E1.15, the volume of the polyethylene unit is
$V = 0.741$ nm $\times$ 0.494 nm $\times$ 0.255 nm $= 0.0933$ nm$^3$.   §

**Example 1.30**   How many polyethylene units —$(C_2H_4)$— are present in 1 kg of polyethylene that is 60% crystalline by volume? The density of polyethylene is $0.94 \times 10^3$ kg/m$^3$.

*Solution:*   The volume of 1 kg of polycrystalline polyethylene is $0.6 \times 1/(0.94 \times 10^3)$ m$^3 = 0.532 \times 10^{-3}$ m$^3$. The volume of a single unit of polyethylene is $V = 0.0933$ nm$^3$ (see Example 1.28). The number of such units present in 1 kg of polyethylene is $(0.532 \times 10^{-3}$ m$^3)/(0.0933 \times 10^{-27}$ nm$^3) = 5.7 \times 10^{3}$.   §

## 1.6   PHASE DIAGRAMS

In simple terms, a **phase diagram** is a graph used to provide information on the phases of a material under different physical conditions, such as changes in temperature, pressure, and chemical composition. A phase diagram is prepared for the conditions at equilibrium; the word *phase* itself refers to a chemically homogeneous portion of material, which may be a liquid, a solid, a gas, or a mixture of the three (often only two of the three phases are considered). In setting up a phase diagram, we have to identify the components present, which can be either elements or compounds. Most systems of interest to us have either two components or three components. Such systems are called binary systems and ternary systems, respectively.

Fig. 1.36 shows the phase diagram of a binary system with two miscible compo-

**Fig. 1.36**
Phase diagram of a binary system with two miscible components

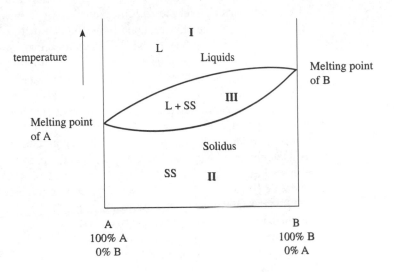

nents (i.e., components that are completely soluble in each other in both the liquid phase and the solid phase). Such isomorphic components are Cu and Ni, for example. In this figure, temperature is plotted along the $y$ axis and the chemical composition is shown along the $x$ axis. From the phase diagram, we can identify three different regions. Region I (immediately above the line called the *liquidus,* where both components A and B are in the liquid phase) occurs at a high temperature. Region II (below the line called the *solidus*) is similar to region I, except that the components are in the solid phase. It occurs at a low temperature. Region III contains mixed phases, i.e., the simultaneous existence of the liquid phase and the solid phase of both components. This region is called a *two-phase* region. Finally, the endpoints where the liquidus meets the solidus correspond to the melting points of the two components. The phase diagram shown in Fig. 1.36, therefore, provides the information on the demarcation between the different phases of the material and the physical conditions: namely, the temperature and the chemical composition.

More often we use a phase diagram to determine the physical conditions that allow for the formation of a particular microstructure in a given material. Theoretically, our ability to control the physical conditions without changing the phases is governed by a parameter called the degrees of freedom, $F$. Mathematically, $F$ is given by Gibbs' phase law, i.e.,

$$F = n_c - p + s \tag{7}$$

where     $n_c$ is the number of components in the system,

           $p$ is the number of phases in a region, and

           $s$ is the number of noncompositional variables.

We can now evaluate $F$ in the three regions shown in Fig. 1.36. In regions I and II (where there are only single phases), $F = 2 - 1 + 1 = 2$, and we expect that both the chemical composition and the temperature can be varied independently. At the endpoints (i.e., at the melting points), $F = 1 - 1 - 0 = 0$, which implies that any change in the physical conditions will automatically induce a phase change. In region III, $F = 2 - 2 + 1$, and we can expect change in only one of the physical conditions. For instance, a change in temperature will simultaneously change the phase composition (but not necessarily the microstructure).

Since region III contains both the solid phase and the liquid phase of the two components, it is often useful to know precisely the relative chemical compositions at a given temperature, say, $T_1$. To do this, we must first draw a horizontal line (called a *tie line*) across $T = T_1$ on the phase diagram, as shown in Fig. 1.37. From its interception with the liquidus, we can find the percentages or fractions of A and B in the liquid phase (i.e., $c_1$ wt% and $1 - c_1$ wt%). Similarly, from the interception of the tie line with the solidus, we can determine the percentages or fractions of A and B in the solid phase (i.e., $c_2$ wt% and $1 - c_2$ wt%). These chemical compositions can then be translated to the respective masses through what is called the *lever rule.* For instance, the masses of A and B combined in the two phases—i.e., $m_l$ and $m_s$ in kilograms—are given, respectively, by

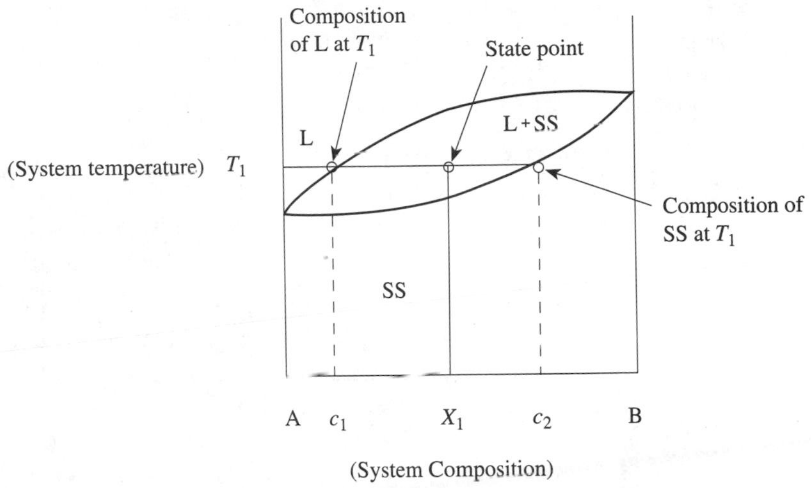

Fig. 1.37

The composition of the different phases in a binary phase diagram

(System temperature)  $T_1$

(System Composition)

$$m_l = \frac{m_{\text{sys}}(c_2 - c_0)}{c_2 - c_1} \tag{8}$$

$$m_s = \frac{m_{\text{sys}}(c_0 - c_1)}{c_2 - c_1}$$

where $m_{\text{sys}}$ is the mass of the entire system in kilograms and $c_0$ is the overall composition of A and B in wt%.

**Example 1.31**  The temperature of a 50%-50% binary alloy is lowered until it is in the two-phase region. If the composition of one of the components in the liquid phase is 20 wt% and in the solid phase is 70 wt%, determine the fractional amount of the alloy in each phase.

*Solution:*  Based on Eq. (8), the fraction in the liquid phase is $m_l/m_{\text{sys}} = (c_2 - c_0)/(c_2 - c_1) = (70 - 50)/(70 - 20) = 0.4$. Similarly, the fraction in the solid phase is $m_s/m_{\text{sys}} = (c_0 - c_1)/(c_2 - c_1) = (50 - 20)/(70 - 20) = 0.6$.  §

Some systems, such as Al and Si, are not isomorphic. The phase diagram for such materials is shown in Fig. 1.38. In this case, we can again identify the regions in between the liquidus and the solidus, where both the liquid phase and the solid phase coexist. Since the two components are immiscible, the (bottom) region containing the solid phase actually contains two different phases (corresponding to different crystal structures of the two components). One of the more important features in this phase diagram is the temperature where the liquidus meets the

**Fig. 1.38**
Phase diagram for
nonisomorphic
components

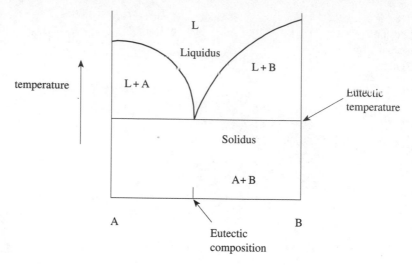

**Fig. 1.38**
Phase diagram for
nonisomorphic
components

solidus. This temperature is called the *eutectic* temperature; at this temperature, the material is fully melted. The eutectic temperature can be much lower than the melting point of either of the components in the system and is often chosen to be the temperature needed to alloy two immiscible solids. In the Au–Si system, the melting points of Au and Si are 1063 °C and 1414 °C, respectively, but the eutectic temperature of the Au/Si alloy is only 363 °C.

Many components have limited solubility, and they can have phase diagrams such as the one shown in Fig. 1.39. In this phase diagram, two additional solid phases (α-phase and β-phase) are identified. Both α-phase and β-phase have distinct crystal structures; an example of these distinctions is the Pb–Sn system. More complicated systems will have even more complex phase diagrams.

**Fig. 1.39**
Phase diagram for two
components with limited
solubility

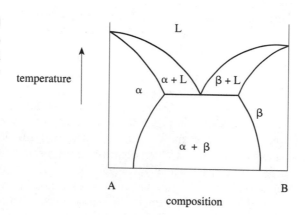

# 1.7  TECHNIQUES FOR CRYSTAL GROWTH AND THIN-FILM DEPOSITION

The growth of high-purity crystals is an extremely important issue in device applications. In particular, semiconductors often have to be prepared with a minimum amount of defects and impurities; insulators are required to have a high breakdown voltage and very low leakage; and metals must be stable and have very low resistivity. Single-crystal semiconductors are by far the most difficult solids to prepare; significant efforts have been spent to improve their quality. In the following paragraphs we highlight the different methods of crystal growth and thin-film deposition techniques commonly adopted by industry.

## Czochralski Growth of Semiconductors

**Czochralski crystal growth** is used in almost all the commercial growth of single crystals of Si, GaAs, and InP. In the Czochralski method, the molten semiconductor is contained in a large crucible inside a furnace, as shown in Fig. 1.40. A seed crystal is first brought into contact with the melt and then is slowly withdrawn. This action results in a single crystal boule drawn on the retreating seed. Fig. 1.41 shows the schematic of a Czochralski crystal-growth facility. The main concern in the growth process is nonuniformity and the presence of built-in **defects.** Nonuniformity also applies to the distribution of the **dopants** (**impurities** added intentionally to the melt to change the conductivity of the semiconductor). Control of the heat flow and the rate at which the seed is withdrawn from the melt are the two most important parameters that determine the quality of the crystal. Boule diameters up to 20 cm (8 in.) in Si and 7.6 cm (3 in.) in GaAs can be easily achieved, even though larger diameters have also been grown. The most common impurities found in Czochralski-grown crystals are O in the case of Si and C in the case of GaAs.

**Fig. 1.40**
Basic features in a Czochralski growth

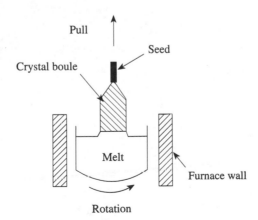

**Fig. 1.41**
A schematic of a crystal-growth facility

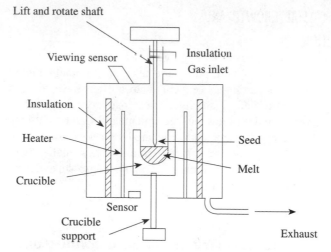

## Thermal Oxidation Growth Techniques

Some semiconductors can be directly oxidized to form surface passivation layers as in the case of Si. The process is called **thermal oxidation,** and Si is oxidized to amorphous $SiO_2$ either in an oxygen-enriched atmosphere or in the presence of water vapor. In both cases, a high temperature (between 850 and 1100 °C) is required. Amorphous $SiO_2$ has a rather open structure because 1 μm of Si produces roughly 2.27 μm of $SiO_2$. The kinetics of the growth process are quite complicated. The growth rate is divided into a linear region and a parabolic region, as shown in Fig. 1.42. High-quality amorphous $SiO_2$ may also be produced by dry oxidation in the absence of water vapor. The $SiO_2$ formed by thermal oxidation usually has internal stress developed at the $SiO_2$–Si interface due to the large volume change from Si to $SiO_2$, and this can lead to missing atoms (point defects) and sometimes missing rows of atoms (dislocations) at the interface. The internal stress is minimized if the oxidation process takes place at a relatively high temperature, such as 1000 °C. Any

**Fig. 1.42**
Growth rate of an $SiO_2$ thin film

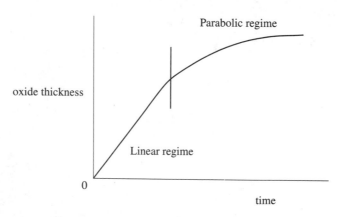

dopants present in the Si prior to oxidation will redistribute during the oxidation process. The redistribution rate depends on the types of dopants present. For instance, B will tend to move into the oxide layer, whereas P and Ar will tend to move away from the oxide-semiconductor interface.

## Chemical Vapor-Phase Deposition

The deposition of oxide films in the vapor phase is particularly useful in the formation of insulating thin films. These oxide films are frequently used in passivation (surface protection of a solid from the atmosphere or from other forms of contaminants). Passivation normally requires a thicker layer of oxide that is impermeable to mobile ions. The oxide could also be formed at a low temperature, such as a few hundred degrees Celsius, to avoid heating of the substrate. One of the more popular techniques used in the deposition of thin oxide films is known as **chemical vapor-phase deposition (CVD).** In the case of $SiO_2$, a carrier gas like $SiH_4$ is used. When $O_2$ is added, the chemical reaction results in the formation of $SiO_2$ and water vapor. To ensure a smooth oxide surface and a uniform coverage of the substrate, a reduced pressure (below atmospheric pressure) is often used. CVD also allows dopants to be added during the oxide growth, which will sometimes change the physical properties of the oxide layer. For instance, the addition of $PH_3$ during the deposition of the $SiO_2$ will form **phosphosilicate glass (PSG).** PSG acts as a smooth and conformal passivation layer (a surface layer with a smooth contour). $Si_3N_4$ layers are also formed using CVD. In this case, $NH_3$ is used in place of $O_2$. Often, to reduce the deposition temperature, glow discharge (a form of gas ionization caused by current discharge and accompanied by light emission of the excited ions) during deposition has been found to activate the reactants and enhance the reaction rates. This state is called *plasma-enhanced CVD*. Two typical CVD chambers are shown in Fig. 1.43.

**Fig. 1.43**
Chemical vapor-phase deposition chambers

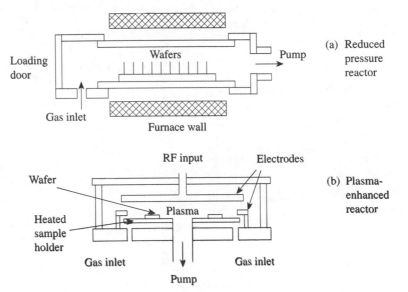

(a) Reduced pressure reactor

(b) Plasma-enhanced reactor

Other than Si, very few semiconductors possess native oxides. GaAs, for instance, will form $Ga_2O_3$ by direct oxidation in an $O_2$-enriched atmosphere. The oxide, however, is known to give a significant leakage current when used for electrical insulation. It also has a low breakdown voltage and plenty of surface defects. Improved oxide quality is obtained if $As_2O_3$ is formed. Other types of insulating layers that can be deposited on a GaAs substrate using CVD include $SiO_2$, $Si_3N_4$, AlN, and $Al_2O_3$. The quality of these insulating films are usually much better than those of the native oxides.

## Liquid-Phase Epitaxy

Single crystals grown in boules often do not have the quality needed for high-performance devices, particularly in the case of compound semiconductors. Since most electronic devices are fabricated on the top of several microns of the substrate, it is sometimes desirable simply to grow a high-quality surface layer on top of the substrate. The process is known as *epitaxial growth*. Epitaxial layers can be formed from **liquid-phase epitaxy (LPE).** The process involves the formation of a supersaturated solution (melt), such as a supersaturated solution of As in Ga in the case of GaAs LPE. When a suitable substrate is placed near the melt, a thin layer of high-quality GaAs will be precipitated onto it. Fig. 1.44 shows a schematic of the LPE growth process. In the process, a temperature well below the melting point of the compound is often used. This makes it possible to control the properties of the compound through changes in the **stoichiometry** (chemical composition). Both **ternary semiconductors** (compound semiconductors consisting of three different elements) and **quarternary semiconductors** (compound semiconductors consisting of four different elements) can be grown using LPE. In the case when the epitaxial layer is identical to the substrate, the interface is normally defect-free. For an epitaxial layer different from the substrate, the lattice mismatch will generate a misfit (distortion of the lattice) over a number of lattice spacings near the interface.

**Fig. 1.44**
Liquid-phase growth process

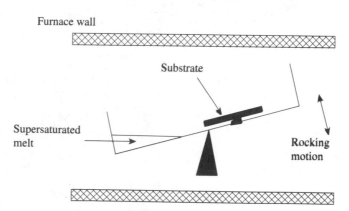

Table 1.7 shows the maximum theoretical thicknesses of epitaxial layers that may be formed for different percent mismatches with the substrate. From Table 1.7, it is obvious that extremely good lattice matching is required in order to form an appreciable thickness of the epitaxial layer. For example, in a 1-$\mu$m-thick GaAs epitaxial layer, the percent **lattice mismatch** between GaAs and the substrate must be less than 0.1. In addition, dislocations at the interface will also be found because of stress resulting from the difference in the thermal expansion coefficients of the epitaxial layer and the substrate. These dislocations happen frequently when the sample is cooled down after the epitaxial growth.

## Vapor-Phase Deposition

**Vapor-phase deposition (VPD)** is used to deposit thin films and is best represented by the processes of evaporation and sputtering. The quality of the deposited films depends on both the chemical composition of the vapor and the substrate temperature. The latter also controls the chemistry and the growth of the microstructure (which determines the grain size). As an example, consider the deposition of Al thin films on a semiconductor, a process frequently used for electrical interconnections. In this case, the Al films can be deposited either by evaporation or sputtering. During evaporation, Al is first melted in a tungsten holder heated to above 1500 °C. As the partial vapor pressure of Al exceeds the ambient pressure, Al evaporates and condenses on a substrate placed directly above the holder. The thickness of the Al film (and also to some extent the grain size) can be controlled by the time of evaporation and the distance of the substrate away from the holder. **Sputtering** is similar to evaporation. It is facilitated by the bombardment of the Al source using an ionized plasma gas. The Al atoms then vaporize and are deposited on the substrate. Al thin films formed by the vapor-phase deposition techniques have electrical properties quite similar to crystalline Al. The defect density, however, is much higher, and the grain size is also quite small.

The CVD technique mentioned earlier is also effective in the growth of epitaxial layers. CVD is also much more suitable for mass production, and the technique is well established for the growth of Si epitaxial layers. In the case when metallo-organic chemicals are used as the source materials, the technique is known as **metallo-organic chemical vapor deposition (MOCVD)**. MOCVD can produce

**Table 1.7**   Maximum thicknesses of the epitaxial layers for different percent mismatches

| % Mismatch | Epitaxial GaAs (nm) | Epitaxial InP (nm) |
| --- | --- | --- |
| 1 | 0.005 | 0.005 |
| 01 | 0.92 | 0.9 |
| 0.01 | 1.35 | 1.24 |
| 0.001 | 17.4 | 15.3 |

(*Source*: C. R. M. Grovenor, *Microelectronic Materials,* Adam Hilger, Boston and Philadelphia: 1989. Reprinted by permission of IOP Publishing Ltd.)

very high quality epitaxial layers, and the composition can be changed from layer to layer by switching between different metallo-organic sources.

## Molecular Beam Epitaxy

**Molecular beam epitaxy (MBE)** is one of the more advanced techniques used for epitaxial growth of semiconductors. The process requires a very high vacuum and can produce high-quality epitaxial layers of different types of compounds. Fig. 1.45 shows a schematic of the deposition chamber. The chamber consists of a rotating substrate, which is heated and placed across from a set of crucibles, each containing a source or element to be deposited. Shutters in the chamber are used to control the flux of the molecular beams vaporized from the crucibles and direct them to the substrate. Very low temperature deposition is possible (down to about 300 °C).

MBE is quite similar to the evaporation process except for the high vacuum and the low deposition temperature. When the elements are deposited on the substrate, chemical reactions take place and the required semiconductor is formed. Because of the low deposition rate (0.1 to 0.5 μm/h), extremely thin epitaxial layers (down to a few nanometers) can be deposited. The MBE technique is particularly suitable for the deposition of alternating thin semiconductor layers used in the construction of quantum-well structures (alternating submicron layers of related semiconductors).

**Fig. 1.45**
An MBE deposition chamber

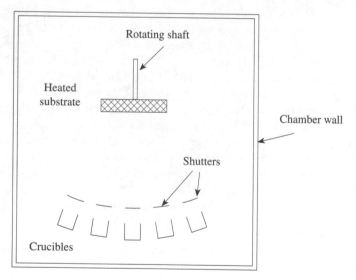

## 1.8 CRYSTAL IMPERFECTIONS

Crystal imperfection is due to the presence of defects and impurities. Defects are the by-products of imperfect crystal growth, and impurities are foreign atoms or ions present in a solid other than the host atoms or ions. During crystal growth, impurities may be deliberately added to the melt, may be embedded in the source chemicals, may be from the environment, or may be introduced unintentionally in the process. Defects and impurities are in many ways interrelated, and they often act together to modify the properties of a solid. Defects, in general, are divided into either structural defects, which can be detected and measured by sensitive instruments, or complexes (unstable defects), which are less easy to detect due to their instability and association with other impurities. In the following subsections, we discuss some of the common structural defects and impurities.

## Structural Defects and Their Properties

Structural defects exist in many different forms; their presence is related primarily to missing or misplaced atoms or ions. Once formed, defects will disrupt the lattice potential and increase the scattering probability of any moving electrons (see Section 2.5). As a result, we can expect a change in the electrical properties of the solid. Defect-filled samples are commonly found in heavily doped semiconductors and in radiation-damaged solids. Microscopically, defects sometimes act as capture and recombination centers. A free electron, when captured by a defect (acting as a **trap** or capture center), is temporarily inactivated and will not take part in the conduction process (even though the captured electron will eventually be released). A **recombination center,** on the other hand, captures a free electron and allows it to recombine with a hole (a lattice site where there is a missing electron). This effectively removes the free electron and reduces the free electron density.

Structural defects are also responsible for the redistribution of impurities. Inside a solid, the migration of impurities along a density gradient depends on the density of imperfect lattice sites, such as vacancies (lattice sites where there are missing atoms). The **diffusion** process (discussed in the next section) is usually enhanced when there is a high density of vacancies. In association with diffusion, defects and impurities are also known to interact with one another in some very intricate ways. For instance, impurities such as the heavy metal ions are naturally attracted to structural defects in semiconductors, and the process has been used as a means of removing heavy metal ions during device fabrication. This process is commonly called **gettering.** During gettering, a heavily damaged layer is intentionally created in the semiconductor, and harmful impurities are allowed to migrate to the damaged layer by raising the temperature. After the harmful impurities have been collected, the damaged layer is either removed by chemical etching or is set aside away from the active areas where devices are to be formed.

More commonly, structural defects are classified according to their dimensionality. **Point defects** are local defects that are either vacancies or interstitials. A **vacancy** is a lattice site where an atom or ion is missing, and an **interstitial** is an extra atom or ion squeezed in between filled lattice sites. In a solid, typical point defects include the **Schottky defect,** a vacancy created because a pair of oppositely charged ions is missing, and the **Frenkel defect,** a coupled pair of one vacancy and one interstitial. These defects are illustrated schematically in Fig. 1.46.

The density of the point defects is affected by temperature since they are usually produced as a result of thermal vibrations of the lattice atoms or ions. If we assume that the energy needed to generate a point defect in J is $\Delta E_d$, then the ratio of the

**Fig. 1.46**
Point defects in a solid

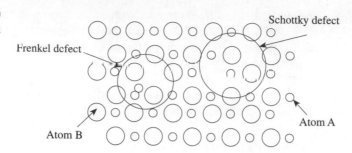

density of the point defects $N_d$ to the density of the ideal lattice sites $N_{\text{site}}$ is given by

$$N_d/N_{\text{site}} = K_d \exp(-\Delta E_d/kT) \tag{9}$$

where   $K_d$ is a dimensionless physical constant (for most semiconductors, it lies between 1 and 10),

$k\,(= 1.38 \times 10^{-23}$ J/K) is the Boltzmann constant, and

$T$ is the absolute temperature in K.

According to Eq. (9), the density of the thermally activated point defects increases exponentially with increasing temperature. In addition, since Eq. (9) is derived from thermodynamic principles, the following empirical rule has been observed to apply in the case of a metal:

$$\Delta E_d \approx 10kT_m \tag{10}$$

where   $T_m$ (in K) is the melting temperature of the metal.

Equations (9) and (10) suggest that the density of the point defects in a metal is exponentially dependent on the melting point. Table 1.8 lists the melting points of some of the common metals.

**Table 1.8**   Melting points of the common metals

| Metal | $T_m$ (K) |
|-------|-----------|
| Al | 933 |
| Cu | 1356 |
| Zn | 419 |
| Sn | 505 |
| Na | 371 |
| Cr | 2130 |
| Au | 1337 |
| Ag | 1235 |

(*Source*: T. B. Massalski, ed., *Binary Alloy Phase Diagrams,* Vols. 1 and 2. (Metals Park, Ohio: American Society for Metals, 1986).)

**Example 1.32**  Assuming that $K_d = 3$ and $\Delta E_d = 2.4$ eV, compute the ratio $N_d/N_{\text{site}}$ at **(a)** $T = 300$ K, and **(b)** $T = 1000$ K.

*Solution:*
**(a)** Based on Eq. (9), $N_d/N_{\text{site}} = K_d \exp(-\Delta E_d/kT)$. At 300 K, $N_d/N_{\text{site}} = 3 \times \exp[-2.4 \times 1.6 \times 10^{-19} \text{ J}/(1.38 \times 10^{-23} \text{ J/K} \times 300 \text{ K})] = 1.7 \times 10^{-40}$.
**(b)** At 1000 K,

$$N_d/N_{\text{site}} = 3 \times \exp[-2.4 \times 1.6 \times 10^{-19} \text{ J}/(1.38 \times 10^{-23} \text{ J/K} \times 1000 \text{ K})]$$
$$= 2.3 \times 10^{-12}. \quad §$$

Note that for a semiconductor such as Si with $N_{\text{site}} = 5 \times 10^{28}/\text{m}^3$, there is only a negligible density of vacancies at 300 K.

**Example 1.33**  Compute the ratio of $N_d$ for Al at 400 K versus its value at 300 K. The melting temperature of Al is 933 K.

*Solution:*  Based on Eq. (9),

$$N_d/N_{\text{site}} = K_d \exp(-\Delta E_d/kT) \text{ and } N_d(400 \text{ K})/N_d(300 \text{ K}) = \exp(-\Delta E_d/$$
$$[1.38 \times 10^{-23} \times 400)]/\exp[-\Delta E_d/(1.38 \times 10^{-23} \times 300)].$$

Since $\Delta E_d \approx 10kT_m = 10 \times 86.2 \times 10^{-6} \times (273 + 660) \text{ eV} = 0.80$ eV, this gives $N_d(400 \text{ K})/N_d(300 \text{ K}) = \exp[-0.8/(1.38 \times 10^{-23} \times 0.625 \times 10^{19} \times 400)]/\exp[-0.8/(1.38 \times 10^{-23} \times 300)] = 2285. \quad §$

Note that there are almost two thousand times as many defects at 400 K as the number at 300 K.

Defects that exist in one dimension are known as linear defects, or **dislocations.** These are normally generated as a result of mechanical deformation during crystal growth. A dislocation is defined as a missing or extra row of atoms or ions over a certain distance in the crystal lattice, as illustrated in Fig. 1.47. Dislocations near

**Fig. 1.47**
A linear defect in a solid

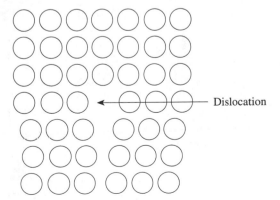

Dislocation

the surface of a solid can often be detected visually by etching the solid with chemicals; these will appear as tiny "pits" when the etched surface is viewed under a microscope. Carefully grown solids, such as single-crystal Si used in integrated circuits, can have a dislocation density as low as $10^8/m^2$, although most solids will have a much higher dislocation density.

Dislocations themselves involve the straining of the lattice atoms or ions; in some cases, the strain energy can be very high ($\sim 1 \times 10^{-8}$ J/m). This fact makes it almost impossible for dislocations to be caused by thermal vibrations alone. Along a dislocation, there are also dangling bonds (broken energy bonds at the surface of a solid) that will attract free electrons. In some cases, the dangling bonds form negative **space-charge regions,** which impede the movement of the free electrons, as illustrated in Fig. 1.48. Dislocations sometimes form recombination centers, and they are also used as gettering sites for the elimination of harmful ions.

*Planar defects* are two-dimensional. They include the grain boundaries in polycrystalline solids, stacking faults, and twins. **Grain boundaries** in polycrystalline solids are space-charge regions filled with disoriented atoms, as shown in Fig. 1.49. These regions are also places where contaminants and impurities accumulate. Just like the dislocations, planar defects are effective recombination and gettering centers. Due to their space charge, grain boundaries can create potential barriers, which are

**Fig. 1.48**
A dangling bond in a diamond lattice

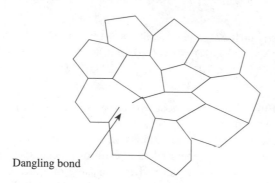

Dangling bond

**Fig. 1.49**
Grain boundary of a polycrystalline solid

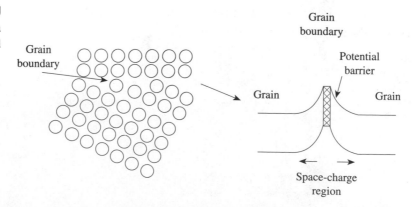

known to be responsible for the high resistance found in polycrystalline solids. Table 1.9 compares the resistivities between the common metal thin films and their bulk crystals. Stacking faults and twins are dislocations in two dimensions, as shown in Fig. 1.50. They affect the solid much as linear dislocations do.

*Amorphous solids* or glasses are examples of solids with three-dimensional defects. In an ideal amorphous solid, there is a complete lack of crystal structure, as shown in Fig. 1.51. In addition, we can expect the presence of a large number of voids (empty spaces due to groups of missing atoms or ions), contaminants, and precipitates (clusters of a single species of atoms or ions). These structural imperfections are known to make the properties of amorphous solids highly variable. Practical amorphous solids are quite insensitive to the addition of impurities; intrinsically, amorphous solids are poor current conductors because many of the free electrons are bound to the defect sites and can move from site to site only via a mechanism called hopping. **Hopping** produces very slow free electrons. The electrical properties

**Table 1.9**   Differences in resistivity between common metallic thin films and bulk crystals

| Metal | Thin-Film Resistivity ($\mu\Omega\cdot$m) | Bulk Resistivity ($\mu\Omega\cdot$m) |
|---|---|---|
| Al | 0.0267 | 0.028–0.033 |
| Cu | 0.017 | 0.018 |
| Au | 0.022 | 0.024 |
| Ni | 0.069 | 0.12 |
| W | 0.054 | 0.054–0.2 |
| Mo | 0.057 | 0.10 |
| Ta | 0.135 | 0.15 |

(*Source*: C. R. M. Grovenor, *Microelectronic Materials*, Bristol and Philadelphia: Adam Hilger, 1989. Reprinted by permission of IOP Publishing Ltd.)

**Fig. 1.50**
A perfect diamond cubic lattice and one with the c—C plane missing to form a stacking fault

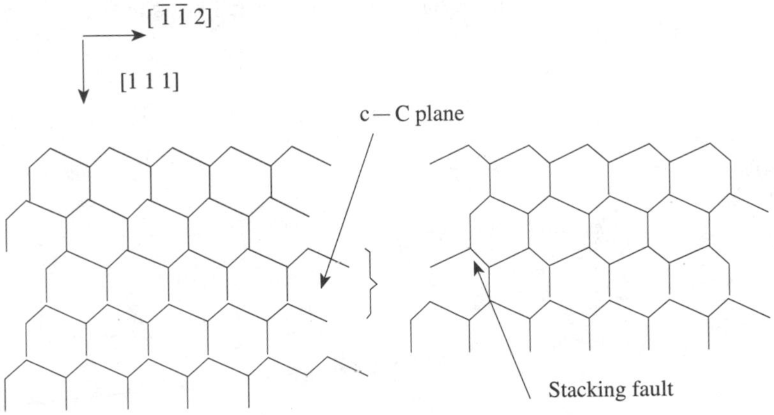

(Source: C. R. M. Grovenor, *Microelectronic Materials*, Adam Hilger, Bristol and Philadelphia: 1989. Reprinted by permission of IOP Publishing Ltd.)

**Fig. 1.51**
Structure of an amorphous
solid

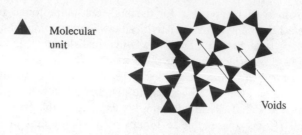

of an amorphous solid are also sensitive to temperature changes, and amorphous solids will readily recrystallize (into the polycrystalline form) at a sufficiently high temperature.

## Impurities

The properties of most crystalline solids are also affected by the presence of impurities. Let us begin by looking at a metal with a "significant" amount of impurities. Such a solid, when formed by a mixture of two similar types of atoms, is called a *solid solution*. One example is the Au–Cu solid solution, which is a mixture of Au and Cu atoms. When there are more Cu atoms than Au atoms, the Au atoms are called the *solute* and the Cu atoms, the *solvent*. An **ordered solid solution** is a solid solution that contains a fixed ratio of the two types of atoms and in which the atoms are located at specific lattice sites. In the case of the Au–Cu$_3$ system, the Au atoms replace the Cu atoms at the corners of the fcc lattice. This situation is illustrated in Fig. 1.52.

For any two types of atoms to form an ordered solid solution, they must first be miscible—i.e., they do not separate out into different phases. This requirement is qualitatively summarized in the **Hume-Rothery rules:**

**1.** The constituents must have less than 15% difference in the atomic radii.
**2.** They must have the same crystal structure.

**Fig. 1.52**
Disordered and ordered
solid solutions

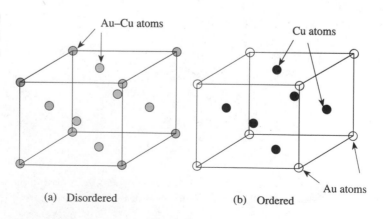

(a)  Disordered          (b)  Ordered

**3.** They must have similar electronegativities or electropositivities (i.e., tendency to form ions).

**4.** They must have the same valence (i.e., the same ionization state).

As expected, an ordered solid solution can exist only at a low temperature since randomization of the atoms will take place when there is enough thermal energy. In a random solid solution, there are no specific lattice sites for the atoms, as illustrated in Fig. 1.53. In the case when the radii of the atoms are comparable, the replacement of the host atoms usually takes place substitutionally, i.e., by direct replacement. When the radii are very different, the solute usually enters the lattice randomly and is dispersed among the host atoms. This case is known as an interstitial solid solution. One example of such a solid solution is steel, where the C atoms are randomly dispersed in the Fe lattice. Charge neutrality is an important factor governing the solubility of impurities. Substitution is possible only if the impurities and the host atoms have the same ionization state (i.e., the same valence). As a whole, a solid solution must be neutral.

Most of the time, the presence of impurities will change the electrical properties of a solid. Since the changes are different depending on the crystal structures and the physical properties of the host atoms, we shall consider the different types of solids separately.

A metal is generally less sensitive to the addition of small amounts of impurities because of its close-packed crystal structure and the presence of a large number of free electrons. It is only in the case of a metallic alloy, where there are major differences in the lattice parameters, that the electrical properties are altered. For instance, a metal alloy can have a resistance considerably higher than those of its constituents. Normally, the presence of impurities will affect only the mechanical properties, such as those related to fracture and slip.

A semiconductor is more sensitive to the presence of impurities. We can broadly classify impurities in a semiconductor as either native impurities or dopants. Native impurities are incorporated unintentionally during crystal growth, whereas dopants are added to change the electrical properties of the solid. Native impurities include atmospheric contaminants such as C, N, and O atoms and any other impurities that may exist in the growth apparatus. Most of the native impurities will not directly

**Fig. 1.53**
Random solid solution of
NiO and MgO

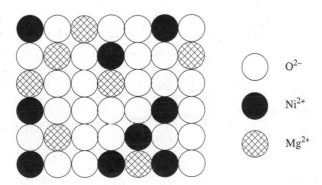

$O^{2-}$

$Ni^{2+}$

$Mg^{2+}$

affect the conduction properties of the semiconductor, but rather they act as capture and recombination centers. Undesirable impurities can sometimes be removed by the gettering process, as mentioned earlier. Dopants will increase the number of free electrons or holes in a semiconductor (holes are created when electrons are missing at the lattice sites). They are effective only at a reasonably high temperature (say, room temperature or higher), depending on the **ionization energies** of the dopants. Dopants that produce free electrons are called *n*-type dopants, and those that produce holes are called *p*-type dopants. When they exist in sufficient quantities, dopants can dramatically change the conductivity of a semiconductor (by many orders of magnitude). Table 1.10 lists the common dopants in semiconductors and their ionization energies.

Not all semiconductors are affected by the presence of impurities. For instance, the conduction properties of some Group II through Group VI compound semiconductors are not affected by the addition of dopants. Normally, when dopants are added to a semiconductor, free electrons, or holes, are created and the conductivity will increase proportionally. In the Group II through Group VI compound semiconductors, a comparable number of recombination centers will also be formed, and they act to remove the excess free electrons or holes. The overall effect is that the densities of the free electrons and holes remain essentially unchanged. This situation is called *self-compensation* and is the reason why some Group II through Group VI compound semiconductors are not suitable for making devices. An excessively large amount of dopants added to a semiconductor will also result in an effect known as *segregation*. During segregation, the dopants move out from their substitution sites (they actually precipitate out in clusters), which increases the resistance of the semiconductor. Segregation is commonly observed in heavily doped semiconductors, especially when the solubility of the dopant is lowered due to a rise in temperature.

The electrical properties of an insulator are particularly sensitive to the presence of impurities and defects. We take $SiO_2$ as an example because it is most widely used in devices. A $SiO_2$–Si interface, for instance, is affected by the presence of defects called surface state traps, or **surface states.** Surface state traps have a wide range of energies and can remove free electrons and holes near the interface by capturing them. The capture process involves the free electrons or holes falling into the traps, and the excess energy is lost to the lattice (in the form of heat). A neutral

**Table 1.10**    Common dopants in semiconductors and their ionization energies

| Semiconductor | Dopant | Type | Ionization Energy (meV) |
| --- | --- | --- | --- |
| Si | P | *n* | 45 |
| | As | *n* | 54 |
| | B | *p* | 45 |
| GaAs | Si | *n* | 6 |
| | Cr | *p* | 63 |
| | Zn | *p* | 31 |
| GaP | Se | *n* | 102 |
| | S | *n* | 104 |
| | Mg | *p* | 53 |

trap will become charged when a free electron or a hole is captured. Surface state traps have been found to compensate partially for the surface charge present in a $SiO_2$–Si capacitor so that the Si substrate is no longer responsive to an applied voltage (this makes the device less functional). Because of this, a lot of effort has been spent to reduce the surface state traps in $SiO_2$. Normally, the density of the surface state traps found in $SiO_2$ per unit energy is of the order of $10^{14-15}/m^2 \cdot eV$, even though carefully prepared samples can have values as low as $10^{13}/m^2 \cdot eV$ (1 eV $= 1.6 \times 10^{-19}$ J).

In addition to surface state traps, positively charged defects (for instance: $O^{2-}$ vacancies) are also found in $SiO_2$. These charges are called *fixed charges,* and their densities can be as high as $10^{14-16}/m^2$. Fixed charges have been found to be responsible for changing the threshold properties of metal-oxide-semiconductor (MOS) devices (for details, see Chapter 3). Mobile charges (as opposed to fixed charges) are also present in $SiO_2$, and they usually belong to ions of the lighter elements, such as $Na^+$, $K^+$, and $Li^{2+}$. These ions are frequently introduced (unintentionally) during the cleaning or etching process. The alkali ions will move rapidly through the oxide and are responsible for instabilities found in electronic devices (i.e., devices with different electrical characteristics at different times). To minimize ionic contamination, $Si_3N_4$ is sometimes coated on top of a $SiO_2$ layer to minimize mobile ion penetration. In addition, a phosphosilicate glass (phosphorous-doped $SiO_2$) layer can also be used as the segregation sites for the removal of alkali ions as in the process of gettering.

## 1.9   DIFFUSION IN SOLIDS

Diffusion occurs naturally when there are variations in the impurity density within a solid. The actual movement of the impurities depends on the size difference between the impurities and the host atoms (or ions). If their radii are comparable, diffusion will be via vacancies in the host lattice. If, however, the radius of the impurities is much smaller, they will move in between the host atoms (or ions). This is called *interstitial diffusion*. In both cases, diffusion is temperature dependent and can be enhanced by raising the temperature. Fig. 1.54 illustrates the process of diffusion involving impurities of different radii.

Other than the temperature dependence, diffusion of impurities in solids also depends on the impurity density gradient. This is governed by **Fick's first law.** To

**Fig. 1.54**
Diffusion in a solid

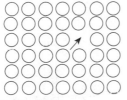

Via vacancy

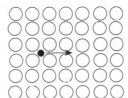

Interstitial diffusion

describe the diffusion process mathematically, we define $c_i(x)/m^3$ as the impurity density at position $x$ and $\phi_i(x)/m^2 \cdot s$ as the flow rate of impurities per unit area at position $x$ ($\phi_i(x)$ is also called the **flux**). In one dimension, Fick's first law is given by

$$\phi_i(x) = -D_i d(c_i(x))/dx \qquad (11)$$

where    $D_i$ is the **diffusivity** in $m^2/s$, and

$d(c_i(x))/dx$ is the gradient of the impurity densities in the $x$ direction in $/m^4$ (the subscript $i$ identifies the type of impurities).

According to Eq. (11), diffusion of impurities, therefore, will continue as long as there is a density gradient. The diffusivity $D_i$ is a measure of how effective the impurities can move through the solid. Impurities will move faster at a higher temperature because of their higher kinetic energy, and $D_i$ is temperature dependent. The time evolution of the impurity density profile $d(c_i(x, t))/dt$, however, obeys **Fick's second law.** In one dimension and along the $x$ direction, it is given by

$$d(c_i(x, t))/dt = -d(\phi_i(x))/dx \qquad (12)$$

Fick's second law states that the impurity density will change when there is a spatial difference in the flow rate (the flux). Since both Eqs. (11) and (12) involve the same set of variables, i.e., $\phi_i(x)$ and $c_i(x)$, they can be combined to give a single equation describing the variations of $c_i(x, t)$ as a function of time and position if the boundary conditions are known. For instance, when there is a constant source ($= c_{is}$), $c_i(x, t)$ is given by

$$c_i(x, t) = c_{io} + (c_{is} - c_{io})[1 - \mathrm{erf}(x/2\sqrt{D_i t})] \qquad (13)$$

where    $c_{is}$ is the impurity density at $x = 0$ in./$m^3$,

$c_{io}$ is the impurity density at $x = \infty$ in./$m^3$, and

$\mathrm{erf}(x/2\sqrt{(D_i t)}$ is called the error function (see Table 1.11).

Equation (13) describes the impurity density profile of a simple diffusion system, and Fig. 1.55 shows how the impurity profile changes with time for given values of $c_{is}$, $c_{io}$, and $D_i$. Fick's laws are also applicable to other types of diffusion species, including the diffusion of free electrons and holes.

**Example 1.34**    In an IC fabrication process, a 100-nm-thick boron (B)-doped surface layer has to be formed on Si. The required B density within the surface layer must be no less than $1 \times 10^{24}/m^3$. If the density of B at the upper surface of Si is kept constant at $1 \times 10^{25}/m^3$, how long will it take for the surface layer to be formed if **(a)** $T =$

Table 1.11   Values of erf(z), where $z = x/2\sqrt{D_i t}$ (See Eq. (10).)

| z | erf (z) | z | erf (z) |
|---|---|---|---|
| 0 | 0 | 0.7 | 0.6778 |
| 0.01 | 0.0113 | 0.75 | 0.7112 |
| 0.02 | 0.0226 | 0.8 | 0.7421 |
| 0.03 | 0.0338 | 0.85 | 0.7707 |
| 0.04 | 0.0451 | 0.9 | 0.7969 |
| 0.05 | 0.0564 | 0.95 | 0.8209 |
| 0.1 | 0.1125 | 1.0 | 0.8427 |
| 0.15 | 0.1680 | 1.1 | 0.8802 |
| 0.2 | 0.2227 | 1.2 | 0.9103 |
| 0.25 | 0.2763 | 1.3 | 0.9340 |
| 0.3 | 0.3286 | 1.4 | 0.9523 |
| 0.35 | 0.3794 | 1.5 | 0.9661 |
| 0.4 | 0.4284 | 1.6 | 0.9763 |
| 0.45 | 0.4755 | 1.7 | 0.9838 |
| 0.5 | 0.5205 | 1.8 | 0.9891 |
| 0.55 | 0.5633 | 1.9 | 0.9928 |
| 0.6 | 0.6039 | 2.0 | 0.9953 |
| 0.65 | 0.6420 | | |

(*Source*: M. Abramowitz and I. A. Stegun, eds., *Handbook of Mathematical Functions,* National Bureau of Standards, Applied Mathematics Series 55, Washington, D.C., 1972.)

Fig. 1.55  Normalized impurity density profile

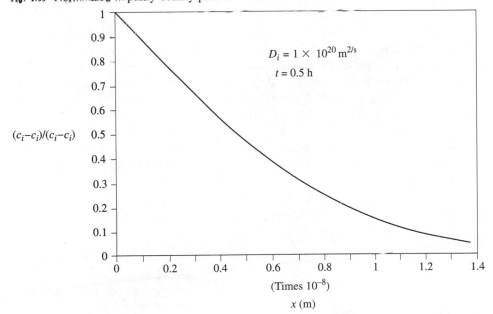

$(c_i - c_i)/(c_i - c_i)$

$D_i = 1 \times 10^{20} \, \text{m}^{2/s}$

$t = 0.5 \, \text{h}$

(Times $10^{-8}$)

$x$ (m)

750 °C and **(b)** 1100 °C? Fig. E1.16 shows a plot of the diffusivity of B as a function of temperature. Assume the original B density in Si is negligible.

*Solution:*   From Fig. E1.16, $D_i(T = 750\,°C) = 1 \times 10^{-21}$ m²/s and $D_i(T = 1100\,°C) = 1 \times 10^{-17}$ m²/s. Since there is no impurity prior to the fabrication process, $c_0 = 0$, and $c_i(x)/c_{is} = 1 - \mathrm{ert}(x/2\sqrt{D_i t})$.

   **(a)** $c_i(x)/c_{is} = 0.1$ at $x = 100$ nm and $T = 750\,°C$. This gives $0.1 = 1 - \mathrm{erf}[1 \times 10^{-7}/(2\sqrt{1 \times 10^{-21} \times t})]$, or $t = 1.86 \times 10^{6}$ s (516 h).

   **(b)** If diffusion is at 1100 °C, $0.1 = 1 - \mathrm{erf}[1 \times 10^{-7}/(2\sqrt{1 \times 10^{-17} \times t})]$, or $t = 186$ s (0.052 h).   **§**

**Fig. E1.16**

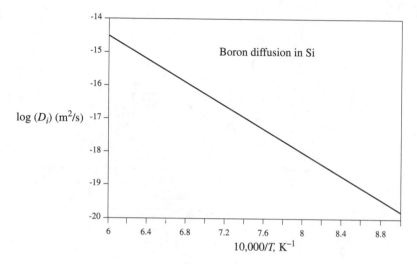

Note that the diffusion process can be significantly enhanced at the higher temperature.

Diffusion of impurities also occurs at the interface between two different solids. Physically, interdiffusion of atoms (or ions) smears out the boundary layer and forms an extended region of solid solution, as discussed earlier. The process is enhanced at a higher temperature and for a thin surface layer, which may eventually lead to perforation in the surface layer. One example illustrating the adverse effects of interdiffusion is in the formation of an Al–Si contact, which takes place at around 300 °C. The out-diffusion of the Si atoms along the surface defects will result in cracks (voids) in the Si substrate. When these cracks are filled by the Al atoms, they become Al spikes embedded in the Si surface, as illustrated in Fig. 1.56. Such

**Fig. 1.56**
Al spikes in a *p-n* junction

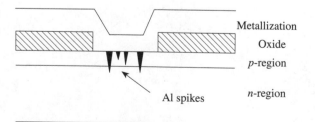

Al spikes are responsible for the anomalously large currents observed in Al–Si contacts, and the process is known eventually to lead to the thermal destruction of the contacts. One way to reduce or remove this problem is to first saturate the Al source at the preparation stage with Si in order to minimize the out-diffusion of Si in the substrate. This is frequently achieved by adding 1 atomic percent of Si to Al before the Al deposition.

Diffusion of impurities is also affected by the presence of an electric field, current flow, etc. A form of self-diffusion known as *electromigration* of atoms is found to be responsible for the breakup of Al interconnections in microcircuits. In this case, a unidirectional current flow along the interconnections can upset the migration pattern of the Al atoms (especially) along the grain boundaries, as shown in Fig. 1.57. Over a long period of time, this can result in the accumulation and depletion of Al atoms along curvatures and discontinuities. The hillocks and voids so created will produce regions of concentrated current flow and overheating. Such an electro-migration effect is minimized by either increasing the grain size of Al in order to reduce the grain boundary effect or by reducing any unnecessary geometrical constrictions in the interconnections (such as bends and sharp curvatures) along the major current paths.

# 1.10  PHYSICAL DIAGNOSTIC TOOLS

One of the most important tasks related to the study of solids is physical diagnostics. Because of the small dimensions of the samples involved, a large number of sophisticated tools are needed in the examination of the crystal structures, their defects, and the impurities. We shall look at a few of the more common diagnostic tools.

## X-Ray Diffraction

Diffraction is the result of light (or radiation) being scattered by a regular array of centers spaced about the same distance as the wavelength of the incident light. A common example of light diffraction is from a diffraction grating, when white light is decomposed into its color components (due to the different diffraction angles experienced by light of different wavelengths). Atoms in a crystal behave just like

**Fig. 1.57**
Nonuniform Al flow along grain boundaries

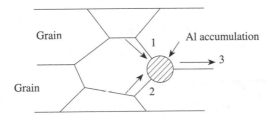

Current flow in regions 1 and 2 is greater than in region 3.

an array of scattering centers, and it is possible to obtain a diffraction pattern from them. The basic mechanism of light diffraction is illustrated in Fig. 1.58.

In this case, each atom in the array acts as an individual center, scattering the incoming light preferentially. In solids, an X-ray light source has to be used since the atomic spacing is only a few tenths of a nanometer. The incident X ray penetrates the successive crystal planes and the diffracted (or scattered) beam is reinforced if the wavefronts are in phase. For a given wavelength, **X-ray diffraction** occurs only at specific angles. The condition is given by **Bragg's law:**

$$2d' \sin(\theta) = n\lambda_L \tag{14}$$

where   $d'$ is the interplanar spacing in nm,

$\theta$ is the angle of incidence with respect to the crystal surface (also called the *Bragg angle*),

$\lambda_L$ is the X-ray (light) wavelength in nm, and

$n$ is an integer called the *order of diffraction.*

Equation (14) can be satisfied by different combinations of $d,'$ $\theta$, $\lambda_L$, and $n$. For a cubic crystal, it is sometimes convenient to relate $d'$ to the Miller indices $(h\ k\ l)$ (see Section 1.3), as given by the relationship

$$d_{hkl} = a_0/\sqrt{h^2 + k^2 + l^2} \tag{15}$$

where   $d_{hkl}$ is the interplanar spacing between the $\{h\ k\ l\}$ planes in nm, and

$a_0$ is the lattice constant of the cubic crystal in nm.

For the different crystal structures, there are also rules that govern the permissible diffractions: (1) For a bcc structure, $h + k + l$ must be an even number; (2) for an fcc structure, $h,\ k,$ and $l$ must be unmixed (i.e., they must be even numbers or odd numbers), and (3) for an hcp structure, $h + 2k$ cannot be a multiple of 3 when $l$ is odd.

As mentioned earlier, diffraction in a solid involves more than just one set of crystal planes, and the diffraction pattern is dependent on the crystal symmetries. Normally, the diffraction pattern of a crystal appears as a series of lines radiating

**Fig. 1.58**
X-ray diffraction in a
crystalline solid

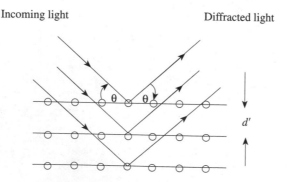

Incoming light                          Diffracted light

from a center point with periodic bright spots, as shown in Fig. 1.59. These bright spots are due to the constructive wavefronts at specific diffraction angles. The number of radiating lines is directly linked to the crystal symmetries. Special machines built to measure the diffraction patterns are called diffractometers. In these machines, the output is a plot of the refracted light intensity as a function of the diffraction angle. By examining the diffraction patterns of a given crystal, experienced crystallographers are able to reconstruct the three-dimensional crystal lattice with a high degree of accuracy.

**Example 1.35**   Using Bragg's law, calculate the diffraction angles $2\theta$ for the peaks observed in the Al powder diffraction pattern. The lattice constant of Al is $a_0 = 0.404$ nm. Assume $n = 1$.

*Solution:*   Since Al has an fcc structure, $h$, $k$, and $l$ must be all odd or even (0 is an even number). Also, $a_0 = 0.404$ nm and $\lambda_L = 0.1542$ nm. Using Eqs. (13) and (15), we can compute $d_{hkl}$ and $2\theta$. The results are as follows.

$$d_{111} = 0.234 \text{ nm} \qquad 2\theta = 38.6°$$
$$d_{200} = 0.202 \text{ nm} \qquad 2\theta = 44.8°$$
$$d_{220} = 0.142 \text{ nm} \qquad 2\theta = 65.3°$$
$$d_{311} = 0.121 \text{ nm} \qquad 2\theta = 79.2°$$
$$d_{222} = 0.117 \text{ nm} \qquad 2\theta = 82.7°$$
$$d_{400} = 0.101 \text{ nm} \qquad 2\theta = 99.5°$$
$$d_{331} = 0.093 \text{ nm} \qquad 2\theta = 112°$$
$$d_{420} = 0.090 \text{ nm} \qquad 2\theta = 118° \quad \S$$

A polycrystalline sample will give a different diffraction pattern, which is made up of bright circular rings, as shown in Fig. 1.60. This pattern occurs because the

**Fig. 1.59**
Diffraction pattern from a single crystal

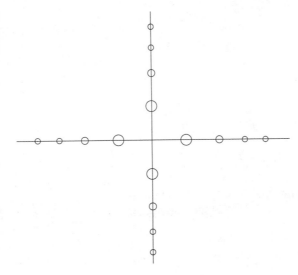

**Fig. 1.60**
Diffraction pattern from a
polycrystalline solid

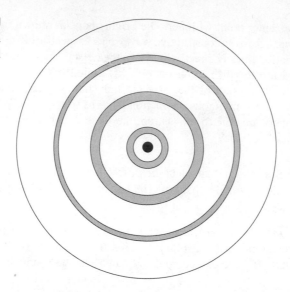

crystals in a polycrystalline sample are randomly oriented, and the incident X ray simply picks up those crystals with the right orientation to satisfy the diffraction law at each diffraction angle. For this reason, the diffraction pattern of a polycrystalline sample has no radial dependence.

**Example 1.36**    In a diffraction experiment involving a polycrystalline sample (see the setup shown in Fig. E1.17), the (1 1 1) diffraction ring measured on the photographic plate is 0.01 m from its center point. The sample is 0.03 m away from the photographic plate on which the diffraction pattern is projected, and the lattice constant is 0.42 nm. Calculate the diffraction angle $2\theta$ and the Bragg angle $\theta$ assuming that the crystals have a cubic lattice structure and $n = 1$.

**Fig. E1.17**

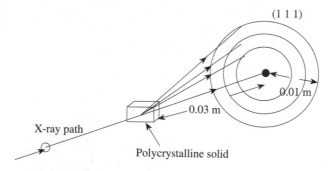

**Solution:**    The diffraction angle is $2\theta = 180° - \tan^{-1}(0.01 \text{ m}/0.03 \text{ m}) = 161.6°$. Thus, the Bragg angle is $\theta = 161.6°/2 = 80.8°$. **§**

**Example 1.37**    An experiment was used to determine the lattice spacing of a polycrystalline solid known to have a simple cubic lattice. The diffraction pattern is a series of consecutive rings measured at a distance 0.01 m from the sample. If the X-ray wavelength is 0.1 nm and the $n$th-order radii and the $(n + 1)$st-order rings are 5 mm and 7 mm, respectively, from the center of the diffraction pattern, compute the lattice spacing.

*Solution:* We are given that the distance from the source is 0.01 m = 10 mm and the ring radii are $r_1$ = 5 mm and $r_2$ = 7 mm, respectively. Therefore, $\theta_1$ = $\tan^{-1}$(5 mm/10 mm) = 26.56°, and $\theta_2$ = $\tan^{-1}$(7 mm/10 mm) = 34.99°. The two rings are of the order $n$ and $n + 1$, respectively. Thus, $n\lambda L = 2d'\sin(26.56°)$ and $(n + 1)\lambda L = 2d'\sin(34.99°)$. Since $\lambda L$ = 0.1 nm, eliminating $n$ gives the lattice spacing $d' = 0.1/\{2[\sin(34.99°) - \sin(26.56°)]\}$ nm = 0.397 nm. **§**

## Electron Microscopy

The most common tools used to study solids are the **scanning electron microscope (SEM)** and the **transmission electron microscope (TEM)**. Schematics of their constructions are shown in Fig. 1.61. These microscopes make use of an electron beam as the probing source (versus a light beam in the optical microscope) to study the fine features near the surface of a sample or in a cross-sectional slice. The magnification in the electron microscopes can be up to 100,000 times; this magnification is far superior to the 2000 times found in a conventional optical microscope. SEM and TEM also allow for a spatial resolution of 1 nm (versus a resolution of 250 nm in the optical microscope).

In the operation of an SEM, a focused electron beam of a high intensity is allowed to scan across the surface of the sample. These energetic electrons are absorbed by the surface atoms and are emitted at a lower energy. If the solid surface has a given contour, then the emitted electrons will follow the contour and appear in a distribution of varying intensities. The electron distribution provides a measure of the surface

**Fig. 1.61**
Schematics of (a) an SEM and (b) a TEM

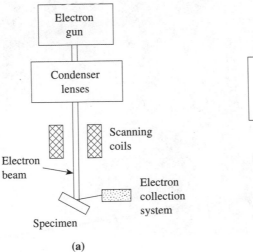

(a)

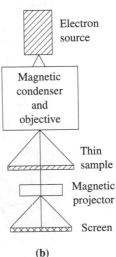

(b)

topology. Fig. 1.62 shows a typical SEM micrograph of pinholes in a $Cd_yZn_{1-y}S$ thin film. The energy of the emitted electrons can also be measured by a detector and is used to identify the types of atoms present in the solid. The characteristics of the atoms are reflected in the energy difference between the incident beam and the emitted beam. Because of the limited penetration power of the electrons, SEM can reach a depth of only several micrometers. Oftentimes, when the sample is an electronic device, a biasing voltage is applied to provide voltage contrast on the micrograph (a negative showing the intensities of the emitted electrons). The part of the device biased at a more positive voltage will give a higher emission intensity (due to the larger electron beam current). This situation is known as **voltage contrast microscopy (VCM).**

In SEM measurement, X rays are emitted when the excited electrons relax to their initial energy states. Fig. 1.63 shows a typical atom with an excited electron relaxing from a higher-energy orbit. As in the case of atomic spectroscopy, the emitted X ray has an energy that is characteristic of the atoms present at the surface of the sample. The X ray intensities will reveal the atomic composition; the technique is called **energy-dispersive X-ray analysis (EDX).** SEM can also be used to study surface defects by measuring the surface current induced by the electron beam (**electron beam–induced current (EBIC)**), as shown in Fig. 1.64. In the neighborhood where defects are present, there will be a high concentration of recombination centers, which shows up as a dip in the electron beam current as it scans through the surface.

**Fig. 1.62**
SEM of pinholes in a $Cd_y Zn_{1-y}S$ thin film

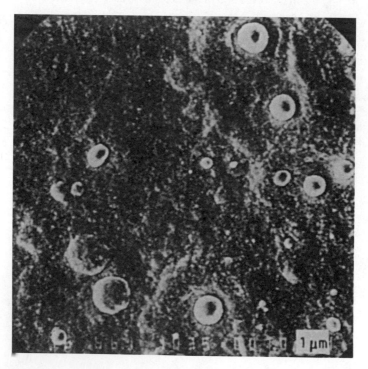

**Fig. 1.63**

The absorption and emission of electrons by an atom

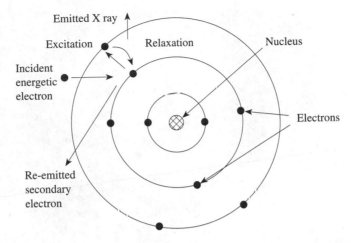

**Fig. 1.64**

EBIC study as the electron beam scans across a semiconductor sample

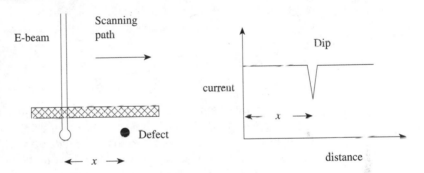

In contrast to SEM, TEM measurement allows for a direct viewing of the defects present in a thinned sample. During the measurement, the electron beam forming the source is allowed to penetrate through the solid, and the resulting diffraction pattern is recorded on a photographic plate. Any lattice defects or dislocations will appear as shaded areas in the image since the lattice defects will diffract electrons differently from the rest of the solid. The diffraction pattern is particularly useful in identifying areas of structural irregularities. To demonstrate this, Fig. 1.65 is a typical TEM micrograph of a plant cell showing its internal structures.

One major difficulty in the TEM measurement lies in sample preparation; most of the time, the sample has to be extremely thin (less than 500 nm) for enough electrons to pass through it. Poorly prepared samples often are incapable of providing the right gray-scale resolution (contrast). Another popular use of TEM is in cross-sectional microscopy. By cutting a thin cross section from a multilayer structure, it is possible to study the defects present in the different layers, a feature that is not possible with the SEM.

Fig. 1.65
A TEM micrograph of a plant cell ($\times$ 12,000). N-nucleus, M-mitochronian; S-starch granules, and L-lipid body.

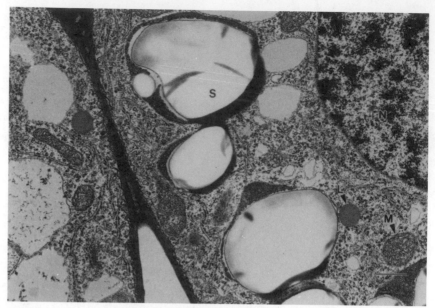

Courtesy of Dr. C. L. Singla, Department of Biology, University of Victoria.

## Secondary Ion Mass Spectroscopy

When a solid is bombarded with an ion beam, considerable damage and surface sputtering will occur. Inelastic effects will produce light emission, secondary ions, and secondary electrons. Secondary ions are produced with a yield of approximately 100 ppm; there will be about 10% of these ions in the 0.5–1.5-keV energy range. At a low beam energy, the main interaction of the ion beam will be with the outermost atomic layers in the sample. As the beam energy increases, sputtering (emission of ions) will occur. The emitted ion species may be neutral or charged; **secondary ion mass spectroscopy (SIMS)** is based on the detection and the analyses of these (secondary) ions. Since in a SIMS measurement the incident ion beam progressively removes the solid surface, the time scale of the emitted ion species can be directly translated into a depth profile. The relative yields of the secondary ions differ considerably from element to element, which produces a very sensitive measure of the chemical species present in the sample. The ion yield is sometimes affected by the presence of $O^{2-}$ ions. Since the $O^{2-}$ ions are electronegative, the ion yield for electropositive ions can be increased significantly in an $O^{2-}$ atmosphere. $Cs^{2+}$ ions, being electropositive, have the opposite effect. Table 1.12 shows the detection limits for the common ion species using different primary ions in Si.

SIMS is frequently used in the determination of the dopant distribution in a semiconductor, which is possible as long as the dopant density is not too high. The total dose can be computed by integrating the area under the SIMS profile. The conversion is essentially linear and can have an accuracy down to $\pm 5\%$. When very deep samples are studied, the effect of the interactions between the primary ions

Table 1.12   Detection limits for the common ion species using different primary ions

| Element | Primary Species | Detection Limit ($\times 10^{21}/m^3$) | ppm |
|---------|-----------------|----------------------------------------|-----|
| B | $O_2^+$ | 1 | 0.02 |
| P | $Cs^+$ | 10 | 0.2 |
| As | $Cs^+$ | 10 | 0.2 |
| Sb | $Cs^+$ | 10 | 0.2 |
| Li | $O_2^+$ | 0.1 | 0.002 |
| Na | $O_2^+$ | 0.1 | 0.002 |
| K | $O_2^+$ | 0.1 | 0.002 |
| Fe | $O_2^+$ | 100 | 2 |
| Cu | $O_2^+$ | 10 | 0.2 |
| Al | $O_2^+$ | 1 | 0.02 |
| F | $O_2^+$ | 100 | 2 |
| Cl | $O_2^+$ | 10 | 0.2 |

(*Source*: SM. Sze, ed., *VLSI Technology*, New York: McGraw-Hill Publishing Company, 1988.
Reprinted with permission of AT & T Bell Laboratories Inc.)

and the dopants frequently result in distortions at the trailing edge of the measured
profile. SIMS is used primarily for studying semiconductors, although high-current
SIMS has been recently used to study insulating materials such as phosphosilicate
glass (PSG).

Fig. 1.66 shows a SIMS profile of Si (as a dopant) in GalnAsP/InP. A depth
profile up to several micrometers can be measured. Surface analysis using SIMS
has also been used; Fig. 1.67 shows a SIMS mass scan on a cleaned Si surface.
Impurities such as Al and F are clearly visible.

Finally, there are also other techniques used to study the chemical composition

Fig. 1.66
SIMS profile of a
multilayered structure

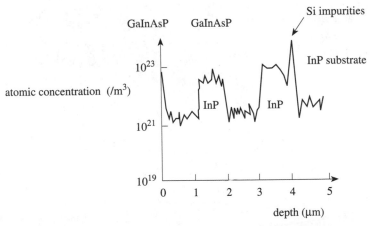

(*Source:* C. R. M. Grovenor, *Microelectronic Materials,* Bristol and Philadelphia:
Adam Hilger, 1989. Reprinted by permission of IOP Publishing Ltd. Courtesy of Dr.
G. Spiller and British Telecom Research Laboratories.)

Fig. 1.67
SIMS scan over a Si
surface

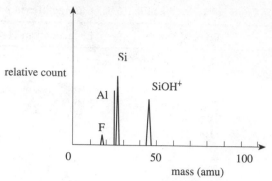

(*Source:* S. M. Sze, ed., *VLSI Technology,* New York: McGraw-Hill
Publishing Company, 1988. Reprinted with permission of AT&T Bell
Laboratories Inc.)

of a solid. These include Auger electron spectroscopy (AES) and **Rutherford back
scattering (RBS).** The first technique, like SIMS, requires a combination of etching
and electron beam spectroscopy to provide measurements of the atomic profiles.
RBS, on the other hand, does not require etching since it measures the backscattered
ions from a sample using a MeV ion source, such as a He⁺ source. The energy of
the backscattered ions are used to identify the atomic or ionic species as well as
their respective depths within the sample. These are measured in increments of 5
keV, for example, and are labeled by a parameter called the channel number. For
instance, a channel number of 300 implies an incident ion energy of $(5 \times 300)$ keV,
or 1.5 MeV. The actual counts of the backscattered ions in a given channel are then
used to identify the atomic or ionic species present in the sample. Fig. 1.68 shows
a typical RBS profile of a TaSi–Si sample showing the energies of the different ion
species.

Fig. 1.68
RBS profile of a TaSi–Si
sample

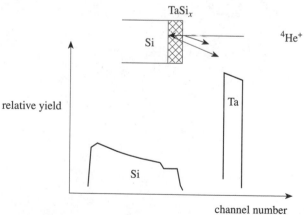

(*Source:* S. M. Sze, ed., *VLSI Technology,* New York: McGraw-Hill
Publishing Company, 1988. Reprinted with permission of AT&T Bell
Laboratories Inc.)

# GLOSSARY

| | |
|---|---|
| **solid** | the state of a collection of atoms or ions. |
| **crystalline solid** | a solid in which the atoms or ions have known positions and the structure is repetitive. |
| **amorphous solid** | a solid in which most (if not all) of the atoms or ions are randomly placed. |
| **polycrystalline solid** | a solid made up of crystal grains of different sizes. |
| **atom** | a collection of electrons, protons, and neutrons. |
| **Avogadro number** | the number of atoms or molecules present in 1 mol of a solid. |
| **atomic number** | the number of protons present in an atom. |
| **electron charge** | the negative charge present in an electron. |
| **orbit** | the electron track around the nucleus of an atom. |
| **Pauli exclusion principle** | states that no two electrons with the same energy and spin can coexist in the same location. |
| **ionic bond** | formed when one or more electrons in the outermost energy orbit of an atom are physically transferred to another atom. |
| **covalent bond** | formed when two or more electrons in the outermost energy orbits of two atoms are shared equally. |
| **metallic bond** | formed by the interactions between the negatively charged free electrons and the positively charged nuclei in a metal. |
| **van der Waals bond** | formed when there are distant interactions between (opposite) charges in the neighboring atoms or molecules. |
| **electrostatic force** | the attractive force or repulsive force due to stationary charges, or ions. |
| **equilibrium** | a state that occurs when there is no net energy exchange with the environment. |
| **potential energy** | the energy related to the physical state (other than the motion) of a particle. |
| **binding energy** | the energy holding two or more atoms or ions together. |
| **Madelung constant** | a factor used to correct for the electrostatic forces of the more distant ions in an ionic solid. |
| **coordination number** | the maximum number of nearest neighbors an ion can have in a solid when the ions are in a given crystal hexagonal structure. |
| **close-packed (crystal) structure** | the state of a solid when all the neighboring atoms or ions are physically touching one another and arranged in hexagonal layers. |
| **bond energy** | the energy involved in the formation of the bond. |
| **bond length** | the equilibrium distance between two atoms or ions after an energy bond has been formed. |
| **free electron** | one that behaves as if it is in vacuum. |
| **valence electron** | an electron that is closely attached to the nucleus. |
| **dipole** | a pair of positive and negative charges loosely bound together by electrostatic forces. |
| **dielectric constant** | a parameter that measures the polarization (induced charge separation) in a solid. |
| **element** | a solid made of one type of atom. |
| **crystal** | a solid with the atoms or ions arranged in a periodic array. |
| **symmetries** | special properties of a solid that allow the solid to retain the same structure after specific geometrical transformations. |
| **crystal direction** | the direction inside a crystal along which atoms or ions are located. |
| **Bravais lattice** | a lattice that shows only the positions of the lattice points in the crystal. |
| **crystal lattice** | an array of points giving the positions of the atoms or ions in the crystal. |
| **coordinates** | the major crystal directions in a solid. |
| **unit cell** | a unit structure in a crystal confined by the lattice points that can be used to reconstruct the entire crystal. |
| **primitive cell** | a minimum volume unit cell. |
| **lattice constants** | the physical dimensions of a unit cell (including the angles between the principal axes). |

| | |
|---|---|
| **atomic packing factor (APF)** | the ratio of the volume occupied by the atoms or ions in a unit cell divided by the volume of the unit cell. |
| **lattice direction** | a vector joining two lattice points. |
| **lattice plane** | a plane or a cross section of a crystal. |
| **Miller indices** | a set of three of more numbers used to denote a lattice plane. |
| **face-centered cubic (fcc) crystal structure** | a cubic structure with additional lattice points located at the centers of each of the faces of the cube. |
| **body-centered cubic (bcc) crystal structure** | a cubic structure with an additional lattice point located at the center of the cube. |
| **diamond crystal structure** and **zinc-blende crystal structure** | different forms of the face-centered cubic structure. |
| **non-Bravais lattice** | a lattice that displays all the atoms or ions in the crystal. |
| **hexagonal close-packed (hcp) crystal structure** | a close-packed structure with the atoms or ions arranged in layers, within which every atom or ion has six nearest neighbors. |
| **compound** | a solid made up of more than one type of element. |
| **semiconductor** | a solid that has conduction properties halfway between a conductor and an insulator. |
| **polycrystalline solid** | a solid made up of small crystal grains. |
| **substrate** | a passive support structure on which a thin film is deposited. |
| **vapor-phase deposition (VDP)** | a technique used to prepare solid, thin films by first vaporizing the solid and then redepositing it onto a suitable substrate. |
| **dangling bond** | a broken energy bond present in those parts of a solid where an atom or an ion is missing. |
| **void** | a space where many atoms are missing. |
| **polymer** | a solid made up of long chains of repeating C units. |
| **spin-on technique** | a technique for forming a thin uniform layer of a suspension on a substrate through the action of a spinning wheel. |
| **phase diagram** | a graph used to provide information on the phase relationship of a material under different physical conditions. |
| **Czochralski crystal growth** | a popular technique used for commercial crystal growth from a melt. |
| **defects** | regions of imperfection within a crystal. |
| **dopant** | an atom or ion deliberately added to a semiconductor to change its conduction properties. |
| **impurities** | foreign atoms or ions present in a solid. |
| **thermal oxidation** | the formation of an oxide by heating a solid to a high temperature in an $O_2$ atmosphere. |
| **chemical vapor-phase deposition (CVD)** | the technique of thin-film deposition through chemical reactions in the vapor phase near a substrate surface at a high temperature. |
| **phosphosilicate glass (PSG)** | a $SiO_2$ layer heavily doped with P to allow for the formation of a smooth contour. |
| **liquid-phase epitaxy** | the growth of crystalline thin films from a solution source at an elevated temperature. |
| **stoichiometry** | the ratio of the different atoms or ions in a solid. |
| **ternary semiconductor** | a compound semiconductor of three different elements. |
| **quarternary semiconductor** | a compound semiconductor of four different elements. |
| **lattice mismatch** | the incompatibility of the lattice structure at the interface of two solids with different lattice constants. |

| | |
|---|---|
| **Vapor-phase deposition (VPD)** | a technique used to deposit thin films by evaporation or sputtering. |
| **sputtering** | the vaporization of a solid by bombarding it with an ionized plasma gas. |
| **metallo-organic chemical vapor deposition (MOCVD)** | a chemical-vapor deposition technique using metallo-organic chemicals as the sources. |
| **Molecular beam epitaxy (MBE)** | thin-film deposition using molecular beams as the sources. |
| **trap** | an impurity or a defect that is capable of removing temporarily free electrons in a solid. |
| **recombination center** | an impurity or a defect that can sequentially capture a free electron and a hole and remove them from the solid. |
| **diffusion** | the process of migration of impurities or particles inside a solid due to the presence of a density gradient. |
| **gettering** | the process of creating a damaged layer inside a solid with the intention of using it to remove harmful impurities or ions. |
| **point defect** | a local defect; either a vacancy or an interstitial. |
| **vacancy** | a lattice site that has a missing atom or ion. |
| **interstitial** | an atom or ion in a solid squeezed in between occupied lattice sites. |
| **Schottky defect** | a vacancy created by missing a pair of oppositely charged ions. |
| **Frenkel defect** | a coupled pair of one vacancy and one interstitial. |
| **dislocation** | a defect that exists in one dimension due to a missing or extra row of atoms or ions. |
| **space-charge region** | a region where there is some net charge present. |
| **grain boundaries** | space-charge regions filled with disoriented atoms or ions. |
| **hopping** | the mechanism whereby electrons jump from one lattice site to another, giving rise to a reduced current flow. |
| **ordered solid solution** | a solution that contains a fixed ratio of atoms and in which the atoms are located at specific lattice sites. |
| **Hume-Rothery rules** | a set of rules that govern conditions for solid solubility. |
| **ionization energy** | the energy needed to transform an atom into an ion. |
| **surface states** | a particular type of surface defect found in solids (and also at their interface). |
| **Fick's laws** | laws governing the diffusion process in a solid. |
| **flux** | the number of particles crossing a unit area cross section in a unit time. |
| **diffusivity** | (or diffusion constant) the particle flux per unit density gradient. |
| **X-ray diffraction** | the scattering of an X-ray beam by the atoms or ions inside a solid. |
| **Bragg's law** | a law stating the relationship between the wavelength of the incoming light (radiation), the angle of incidence, and the lattice spacing during diffraction. |
| **scanning electron microscope (SEM)** | a microscope for surface microscopy that uses a scanning electron beam as the source. |
| **transmission electron microscope (TEM)** | a microscope using an electron beam as the source and a very thin sample. |
| **voltage contrast microscopy (VCM)** | electron microscopy in the presence of a voltage bias applied to the sample. |
| **energy-dispersive X-ray analysis (EDX)** | the technique used to identify atoms or ions in a solid by their X-ray emission signatures. |
| **electron beam–induced current (EBIC)** | the electron microscopy technique used to study surface defects in a solid through the observed changes in the beam current. |
| **secondary ion mass spectroscopy (SIMS)** | the study of the chemical composition of a solid using an incident ion beam as the source for sputtering. |
| **Rutherford back scattering (RBS)** | a technique for studying ion species through backscattering of a MeV ion beam. |

**REFERENCES**   Callister, W. D. *Materials Science and Engineering, An Introduction,* 2d ed. New York: John Wiley & Sons, Inc., 1990.

Grovenor, C. R. M. *Microelectronic Materials.* Bristol and Philadelphia: Adam Hilger, 1989.

Omar, M. A. *Elementary Solid State Physics.* Reading, Mass.: Addison-Wesley Publishing Company, 1975.

Shackelford, J. F. *Introduction to Materials Science for Engineers,* 3d ed. New York: MacMillan Publishing Co., 1992.

Smith, W. F. *Principles of Materials Science and Engineering,* 2d ed. New York: McGraw-Hill Publishing Co., 1990.

**EXERCISES**   ## 1.2   Atoms and Their Binding Forces

1. Compute the atomic weight of carbon based on the fact that carbon has 6 protons, 6 neutrons, and 6 electrons.

2. Give the electronic configurations of Si (atomic number 14) and As (atomic number 33). Based on their atomic structures and energy orbits, determine the energies needed to excite electrons from their outermost partially filled orbits to the next empty orbits. Assume that the energy of the orbits varies as $E_{orbit} = 2.41 \times 10^{-28} \, Z^2/\tilde{n}^2$ J, where $Z$ is the atomic number and $\tilde{n}$ is the quantum number associated with the orbit.

3. Compute the attractive force between a $Zn^{2+}$ ion and a $S^{2-}$ ion. Estimate the binding energy between the two ions. The interatomic distance between $Zn^{2+}$ and $S^{2-}$ is 0.257 nm.

4. If the inverse-square force law on electrostatic attraction is replaced by an inverse-cube force law, assuming all the other parameters are the same, suggest qualitatively what will happen to the equilibrium distance and the binding energy of an ion pair.

5. Compute the energy consumed or released if a polyethylene unit (i.e., $-C_2H_4-$) is reduced to ethane, $C_2H_6$. The binding energy for the $C{=}C$ bond is 680 kJ/mol; for the $C{-}C$ bond it is 370 kJ/mol; and for the $C{-}H$ bond it is 435 KJ/mol.

6. Verify that the range of the ratio of the atomic radii $r/R'$ for a coordination number of 6 in a solid lies between 0.414 to 0.732, where $r$ is the radius of the smaller atom and $R'$ is the radius of the larger atom.

7. In Ag (atomic number 47), what is the ratio of the number of free electrons to the total number of electrons present in the atom? The free electron density in Ag is $7.4 \times 10^{28}/m^3$. The mass density of Ag, $\rho_0$, is 10.5 Mg/m$^3$, and the atomic weight of Ag is 107.87 amu. Given: 1 amu = $1.66 \times 10^{-27}$ kg.

8. A pair of biomolecules 1 μm apart is held together by van der Waals bonds. What will be the total amount of charge present on each of the molecules if the bond energy is 5 kJ/mole?

## 1.3   Crystal Structures

9. Show whether the following crystal structures have (i) reflection symmetry, (ii) inversion symmetry, and (iii) rotational symmetry: **(a)** body-centered tetragonal structure, **(b)** rhombohedral structure, and **(c)** base-centered monoclinic structure.

10. Assuming the lattice constant is $a_0$, compute the volume of the unit cells for (a) a body-centered cubic lattice, (b) a face-centered cubic lattice, and (c) a hexagonal close-packed lattice (with a cell height $= c$).

11. If the lattice constant is $a_0$, compute the nearest-neighbor distance in a solid with the diamond crystal structure.

12. What are the crystal directions of the face diagonal and the body diagonal in the cubic crystal structure? From the spinel structure (see Fig. 1.30), give the crystal directions for each of the following: (a) an oxygen atom, (b) an atom in the octahedral position, and (c) an atom in the tetrahedral position. Assume the origin of the coordinates is at one corner of the cubic structure.

13. Draw the (1 3 1) plane. What is the atomic density in this plane for the simple cubic structure if the side of the cube is equal to $1 \times 10^{-9}$ m? Name the plane in the cubic structure that has the highest atomic density.

14. Draw the primitive cell in the diamond crystal structure. If the lattice constant is $a_0$, compute the volume of the tetrahedron formed by the nearest neighbor. How many tetrahedrons can be found within the fcc unit cell shown in Fig. 1.22?

15. Determine the atomic packing factor (APF) in the body-centered cubic structure.

## 1.4  Crystalline Solids

16. Give examples (two for each) of metals that have (a) body-centered cubic structures, (b) face-centered cubic structures, and (c) hexagonal close-packed structures.

17. Give examples of ceramics that have the following properties: (a) brittleness, (b) good optical transparency, (c) a large dielectric constant, and (d) a good insulator.

18. Draw the unit cell of the $MgAl_2O_4$ spinel structure. What is its APF?

19. Determine the mass densities of (a) Si and (b) InP from their unit cell structures. The lattice constant for Si is 0.542 nm and for InP it is 0.587 nm. The atomic weights of Si, In, and P are 28.09 amu, 114.82 amu, and 30.97 amu, respectively (1 amu $= 1.66 \times 10^{-27}$ kg).

## 1.5  Polycrystalline and Noncrystalline Solids

20. Assuming that polycrystalline solids are made up of spherical grains, estimate the approximate grain size of Ge if each grain has 4200 Ge atoms, which are arranged in the diamond structure. The nearest-neighbor distance of Ge is 0.244 nm. Assuming that the intergranular regions are filled with amorphous Ge that has a weight density one-third that of single-crystal Ge, compute the mass density of polycrystalline Ge. The mass density of single-crystal Ge is 5.32 $Mg/m^3$.

21. The surface area of a polycrystalline solid is an important parameter in determining the surface adsorption of gases. Compute the ratios of surface area to volume if the grain size in the polycrystalline solid is (a) 1 µm, (b) 5 µm, and (c) 20 µm. If the dimensions of the entire polycrystalline thin film is 100 µm $\times$ 1 cm $\times$ 1 cm, compute the total exposed surface area in each of the preceding three cases. Consider only the exposed surface area of the outermost atomic layer.

22. If each crystal grain in a polycrystalline Si sample is covered with a monolayer of $O^{2-}$ ions, estimate the potential barrier at the grain surface if the average grain size is 3 µm. The radius of an $O^{-2}$ ion is 0.132 nm and the relative permittivity of Si $\epsilon_r$ is 11.7.

Assume that the surface electric field is equal to the surface charge density divided by $\epsilon_r\epsilon_0$, where $\epsilon_0$ ($= 8.854 \times 10^{-12}$ F/m) is the permittivity of vacuum.

**23.** If amorphous Si is only half the weight of single-crystal Si, suggest what the density of the dangling bonds in amorphous Si will be.

**24.** A hypothetical polyethylene chain has a spiral C skeleton with a pitch distance of 1 mm and a radius of 100 nm. Compute the total number of C atoms present if the spiral has a height of 1 cm. The bond length of the —C—C— is 0.154 nm.

## 1.6   Phase Diagrams

**25.** Fig. P1.1 shows the phase diagram for $H_2O$. Suggest how $H_2O$ changes as temperature or pressure increases (start at low values). According to Gibbs' phase rule, what will be the degrees of freedom when $H_2O$ is at its melting point and at its freezing point?

Fig. P1.1
Phase diagram for $H_2O$

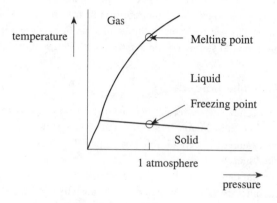

**26.** Fig. P1.2 shows the phase diagram for Ge–Si. If the initial atomic percentages of Ge and Si are 40 and 60, respectively, compute the compositions in the liquid phase and the solid phase at 1200 °C.

Fig. P1.2
Ge–Si phase diagram

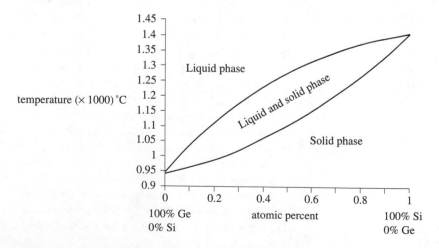

**27.** Fig. P1.3 shows the phase diagram for two materials A and B that do not form a solid solution (they are immiscible). From a molten mixture of the two and a composition of 30 atomic percent A and 70 atomic percent B, the entire system is cooled to a lower temperature.

   **(a)** At what temperature will solidification occur?

   **(b)** Which of the two will first solidify?

   **(c)** Right before the point of total solidification, what will be the composition of the liquid?

**Fig. P1.3**
Phase diagram for
A and B

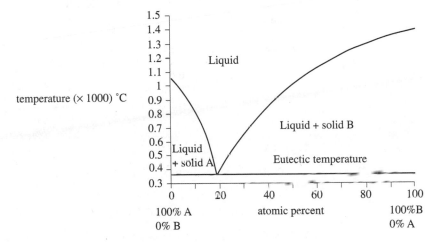

**28.** From the Al–Si phase diagram shown in Fig. P1.4 determine the solubility of Si in Al at 500 °C (the normally annealing temperature for an Al–Si contact). Estimate the depth of the average Si pit formed by the out-diffusion of Si into Al if the pits are cylindrical and occupy 1% of the Si surface. Assume an Al thickness of 2 $\mu$m and that it is saturated with Si.

**Fig. P1.4**
Al–Si phase diagram

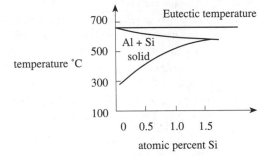

**29.** Fig. P1.5 shows the phase diagram for a given metal silicide ($M_xSi_y$). In the areas labelled A–D, identify the following:

   **(a)** liquid phase for both the metal and Si;

   **(b)** liquid phase plus Si;

**(c)** liquid phase plus silicide; and

**(d)** solid phase for both the silicide and Si. The five arrows in the figure identify the compositions of the silicides that exist. Label the arrows.

**Fig. P1.5**
Metal silicide phase
diagram

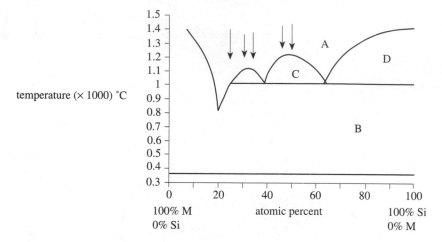

## 1.7   Techniques for Crystal Growth and Thin-Film Deposition

**30.** What are the main advantages of the Czochralski method in crystal growth? Name at least three.

**31.** In crystal growth, can you suggest which part of a crystal boule will be more likely to be subject to the presence of native defects? Why? If the dislocation density increases with the square of the radius of the boule, estimate the average dislocation density if the radius of the boule is $r_0$.

**32.** Determine the unit cell dimensions in amorphous $SiO_2$ if 1 μm of single-crystal Si produces 2.27 μm of $SiO_2$. The unit cell of $SiO_2$ has an fcc structure (similar to Si), and there are 8 $Si^{4+}$ ions and 16 $O^{2-}$ ions.

**33.** State the chemical reaction or reactions used to form $Si_3N_4$ using the CVD technique.

**34.** Suggest why there is a maximum film thickness during epitaxial growth in the presence of lattice mismatch between the epitaxial film and the substrate (see Table 1.7). Plot the maximum film thickness as a function of the lattice mismatch. Could you deduce an empirical relationship?

**35.** If the MBE technique is used to deposit an InP thin film with a thickness of 5 nm, determine the approximate number of atomic layers present.

## 1.8   Crystal Imperfections

**36.** In a given solid, if there are 10 defects per million atoms at 900 °C, suggest what the ratio of the defect density to the density of atomic sites at room temperature (300 K) will be. Assume these defects have an activation energy of 1.0 eV.

**37.** Suggest how the activation energy of defects in a metal can be related to its melting point.

38. Assuming that a typical dislocation site in Si there are 500 missing atoms, compute the density of dislocations in Si if these dislocations are due to segregation of point defects initially formed at 900 °C. The activation energy of the defects is 2.0 eV and $K_d = 5.5$. The mass density of Si is 2.33 Mg/m$^3$ and Si has an atomic weight of 28.09 amu (1 amu = $1.66 \times 10^{-27}$ kg).

39. If the resistivity of single crystal Cu is 0.018 $\mu\Omega\cdot$m and that of the thin film is 0.027 $\mu\Omega\cdot$m, suggest what is the barrier height $\Phi_B$ in the thin film at 300 K. The ratio of the two resistivities is $\exp(-\Phi_B/(kT))$, where $k$ is the Boltzmann constant and $T$ is the absolute temperature. Based on Table 1.9, suggest which of the listed metal thin films has the highest barrier height.

40. Name two metals (other than Au) that will form solid solutions with Ag. Verify that they obey the Hume-Rothlery rules.

41. According to Table 1.10 Cr is a $p$-type dopant in GaAs with an ionization energy of 63 meV. Determine the fraction of Cr that will be ionized at room temperature (300 K). Will Zn or Si be a better dopant in GaAs?

## 1.9 Diffusion in Solids

42. If the diffusion profile of impurities in a solid follows an exponentially decaying function, i.e., $c_i = c_0\exp(-x/\lambda)$, where $\lambda$ is a constant. Determine the impurity flux as a function of $x$.

43. If the diffusion constant of impurities in a solid is increased by a factor of 2, what will be the fractional increase in impurity density inside the solid, assuming that the boundary conditions remain the same?

44. If the surface dopant density $c_s$ is a constant, how long does it take for diffusion to bring the dopant density to $0.75c_s$ at a distance 1 cm from the solid surface? The diffusion constant of the dopants is 0.0001 m$^2$/s and the initial dopant density in the solid is zero.

45. Suggest what are the effects of atomic size on the diffusion of impurities in a solid.

46. Assuming that an Al–Si contact has 10% of its area covered by Al spikes, what will be the increase in current density due to the Al spikes if the ratio of the resistivities between Al and Si is 10,000?

## 1.10 Physical Diagnostic Tools

47. In a diffraction experiment, the measured Bragg angle for first-order effect is 10°.
    (a) Determine the interplanar distance if the X-ray wavelength is 0.05 nm.
    (b) Assuming a cubic crystal structure, what is the lattice constant if diffraction is due to the (1 2 1) planes?

48. Suggest the key reasons that an electron microscope can perform better than an optical microscope.

49. During SEM measurement, what will be the emitted X-ray wavelengths if (a) Cu, (b) Ni, and (c) Au are present at the sample surface?

50. What are the advantages and the limitations of Rutherford back scattering (RBS) in the study of the chemical composition of solids?

51. During SEM measurement, if secondary electrons are emitted from a spherical surface, suggest what the differences in the emission intensity from the top surface and from the equatorial surface of the sphere will be.

**52.** During SEM measurement, it takes 10 s for the electron beam to sweep through the entire surface of a sample at a frequency of 1 kHz and a total of 50 current dips, each lasting 20 μs, are observed in the scan. Suggest what fraction of the sample surface may be defective.

# CHAPTER 2

# Electrical Properties of Solids

## 2.1  INTRODUCTION

One of the most fundamental questions in the study of the electrical properties of solids concerns why some solids are good current conductors and others are not. This is not a simple question to answer because there are many contributing factors, but we are sure that the answer does not come from a study of the crystal structures alone. Current conduction in a solid, as we are well aware, largely depends on the movement of the charges, namely, the free electrons and the holes (holes exist primarily in semiconductors). Inside a solid, electrons behave differently than in a vacuum (for instance, inside a vacuum tube). Generally speaking, the majority of the electrons in a solid are attached to the nuclei, although some (sometimes quite a few) are able to move about freely. These electrons, which we called free electrons in Chapter 1, are able to respond to an applied electric field and give rise to current flow. To determine whether a solid is a good current conductor or not, we need to know more about the free electrons.

For instance, we need to know how many there are and how well they can move through the solid. The interactions of the free electrons inside a solid can be quite complicated, especially in view of the large number of them present (for a good current conductor, electron density can be up to $10^{28}/m^3$). In fact, it is almost impossible to keep track simultaneously of the movement of all the free electrons and their interactions. Fortunately, statistical physics has developed a way to deal with this situation; the dynamics of the free electrons can be collectively represented in an energy band diagram, which provides the energy-momentum relationship. Indeed, the development of the energy band diagram and the study of the different conduction mechanisms in solids are the main themes of this chapter.

In the first few sections of this chapter, we examine the concept of free electrons and their properties. These include a study on the energy distribution of the electrons inside a solid and the Fermi-Dirac statistics. Next, we introduce the energy-momentum relationship of the free electrons and explain how it can evolve into an energy band diagram. We then highlight the differences between the energy band diagrams of a conductor, a semiconductor, and an insulator. In connection with these, we also introduce the concept of *holes,* which are considered to be the mirror images of the free electrons. Related phenomena such as the Hall effect, magnetoresistance effect, thermoelectric effect, velocity saturation effect, and negative-resistance effect are described. The second part of this chapter is devoted to a study of the different current-conduction mechanisms found in solids. We conclude the treatment with the *continuity equation,* which describes charge transport. Finally, in the last section of the chapter we move on to a study of the properties of insulators and explain the origin of dipoles, their interactions with electromagnetic fields, associated energy losses, and ac conduction in insulators.

## 2.2    ELECTRONS IN A SOLID

From basic physics, we know that in an isolated atom, electrons are bound to the nucleus by electrostatic forces, and it requires a substantial amount of energy for an electron to move, say, from an orbit to infinity in space (the potential at infinity is often treated as a reference potential). This energy is sometimes called the ionization energy, $E_i$. Its value can range from tens of electron volts (1 eV is the energy needed to move an electron across a potential difference of 1 V) to several hundred electron volts. The situation is quite different in a solid. Inside a solid, the electron "sees" a large number of atoms or ions, and it also experiences forces exerted by the other electrons.

Statistically, electrons inside a solid do not have the same energy; some have more energy than others. Those few electrons having more energy overcome the electrostatic forces of their own nuclei and are "freed" within the solid. These electrons are called *free electrons.* Free electrons are distinguished from valence electrons, which remain attached to the individual nucleus. The energy needed to create a free electron is called the **excitation energy,** $E_{\text{excite}}$; its value varies from solid to solid. Table 2.1 lists the typical values of $E_{\text{excite}}$ for different solids.

As we can see from the table, $E_{\text{excite}}$ is essentially zero for the metals but is quite large for the semiconductors and the insulators. Generally speaking, at room temperature (assumed to be 300 K), very few electrons have energy more than a fraction of an electron volt, which means that the majority of the electrons are bound to the nuclei. There are, however, a few free electrons. Let us for the moment compare $E_{\text{excite}}$ to $E_{\text{th}}$, the average thermal energy of an electron. By definition, the average **thermal energy** in electron volts is given by

$$E_{\text{th}} = kT \qquad \textbf{(I)}$$

Table 2.1 Typical values of $E_{excite}$ for different solids

|  | Solid | $E_{excite}$ (eV) |
|---|---|---|
| Metals | Al | ~0 |
|  | Cu | ~0 |
|  | Zn | ~0 |
| Semiconductors | Si | 1.12 |
|  | GaAs | 1.42 |
|  | CdS | 2.42 |
| Insulators | $SiO_2$ | 9 |
|  | $Si_3N_4$ | 5 |

where    $k$ (= $86.2 \times 10^{-6}$ eV/K, or $1.38 \times 10^{-23}$ J/K) is the Boltzmann constant, and

$T$ is the absolute temperature in K.

At $T = 300$ K, $E_{th} \approx 0.026$ eV.

By referring to Table 2.1, we can easily see why very few of the electrons in solids other than the metals are free. Nevertheless, since there are many atoms inside a solid, the number of free electrons in some solids can still be substantial (as an example, 1 $mm^3$ of Si has about $10^6$ free electrons). Table 2.2 lists the free **electron densities**—i.e., the number of free electrons per unit volume—in different solids at 300 K.

Consider the case when $E_{excite} = 0$. Since the free electron energy is now equal to $E_{th}$ (see Eq. (1)), we have

$$E_{th} = kT = m_0 v_{th}^2/2 \qquad (2)$$

where    $m_0$ (= $0.91 \times 10^{-30}$ kg) is the mass of an electron, and

$v_{th}$ is the **thermal velocity** in m/s.

Rearranging Eq. (2) gives

$$v_{th} = \sqrt{2kT/m_0} \qquad (3)$$

Table 2.2 Values of free electron densities in different solids at $T = 300$ K

|  | Solid | Free Electron Density (/$m^3$) |
|---|---|---|
| Metals | Al | $1.81 \times 10^{29}$ |
|  | Cu | $8.45 \times 10^{28}$ |
| Semiconductor | Si | $1.45 \times 10^{16}$ |
|  | GaAs | $2.10 \times 10^{12}$ |
| Insulators | $SiO_2$ | $< 10^8$ |
|  | $Si_3N_4$ | $< 10^8$ |

If we substitute the values of $m_0'$, $k$, and $T$ (= 300 K) into Eq. (3), we can show that $v_{th} \approx 1 \times 10^5$ m/s ($\approx 3.6 \times 10^5$ km/h). This is a very high velocity by conventional measure.

**Example 2.1**   Compute the thermal energy $E_{th}$ of free electrons in a solid at **(a)** 300 K and **(b)** at 77 K.

*Solution:*   From Eq. (1), $E_{th} = kT = 1.38 \times 10^{-23} \times T$ J/K.
  **(a)** At $T = 300$ K, $E_{th} = 1.38 \times 10^{-23} \times 300$ J $= 4.14 \times 10^{-21}$ J (or 0.0258 eV).
  **(b)** At $T = 77$ K, $E_{th} = 1.38 \times 10^{-23} \times 77$ J $= 1.06 \times 10^{-21}$ J (or 0.0066 eV).   §

**Example 2.2**   Compare the thermal energy $E_{th}$ of free electrons at 300 K with the value of the excitation energy for Si ($E_{excite} = 1.12$ eV).

*Solution:*   At 300 K, $E_{th} = 4.14 \times 10^{-21}$ J (see Example 2.1). From Table 2.1, $E_{excite}$ of Si = 1.12 eV = 1.12 eV $\times 1.6 \times 10^{-19}$ J/eV = $1.79 \times 10^{-19}$ J. $E_{excite}$ is, therefore, about 50 times $E_{th}$, and we can expect only a small fraction of the electrons in Si to be free.   §

In general, the thermal motion of free electrons is random (see Fig. 2.1) since there are no directional forces involved. In fact, random motion will not contribute to current flow, even though the free electrons themselves are negatively charged. A current will only result if the free electrons somehow move together in one direction. This movement happens when there is an applied voltage across the solid. The applied voltage establishes a potential gradient (electric field) for the free electrons, and because of their negative charge, the free electrons will move up the potential gradient (i.e., against the electric field). This is illustrated schematically in Fig. 2.2.

We have so far ignored the effect of the lattice atoms or ions. Inside a solid, free electrons do not actually move in straight lines, as shown in Fig. 2.1, because they also encounter diffraction (see Section 1.9). Diffraction by the different lattice planes can significantly alter the paths of free electrons, and it results in trajectories similar to the one shown in Fig. 2.3. Notwithstanding this effect, free electrons still move

Fig. 2.1
Random motion of free
electrons

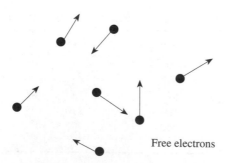

Free electrons

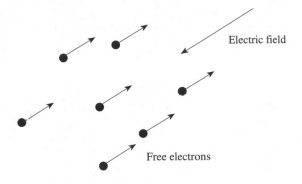

**Fig. 2.2**
Motion of free electrons under an applied electric field

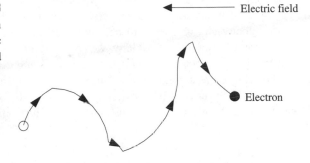

**Fig. 2.3**
Trajectory of an electron under an applied electric field

preferentially toward the positive potential and result in a current flow in the same direction as the electric field, i.e., from the positive-voltage terminal to the negative-voltage terminal.

In addition to diffraction by the lattice planes (which is elastic and does not contribute to energy loss), the motion of the free electrons is also affected by the periodic potential of the charged nuclei (if you assume the electrons to move like waves, you may consider this effect to be analogous to ocean waves propagating over an undulating seabed). To account for this background periodic potential, the free electrons are assigned an **effective mass,** $m_e^*$, in place of their physical mass, $m_0$. For most solids, the background potential assists in the motion of the free electrons, and so $m_e^*$ is smaller than $m_0$. Some solids are **anisotropic**—i.e., their lattice properties are dependent on the crystal orientation; in these cases, so are the values of $m_e^*$. For the purpose of calculations, averaged values of $m_e^*$ are frequently used. Table 2.3 lists the values of $m_e^*$ for free electrons in metals and in semiconductors (note that the effective mass of free electrons in insulators is seldom of interest because of their poor conduction properties).

The motion of a free electron in the presence of an applied voltage is also affected by **collisions** and **scattering** with impurities and lattice imperfections (for instance, a displaced atom). Direct collisions, in fact, rarely occur; most of the time, the impurities and the imperfections are electrically charged and long-range coulombic interactions (scattering) prevail. During collision and scattering, a free electron

Table 2.3    Values of $m_e^*/m_0$ in metals and in semiconductors

|  | Solid | $m_e^*/m_0$ |
|---|---|---|
| Metals | Al | 1.2 |
|  | Cu | 1.0 |
|  | Zn | 0.85 |
| Semiconductors | Si | 0.19 |
|  | GaAs | 0.063 |
|  | CdS | 0.2 |

loses its energy acquired through acceleration (under an electric field), and the *instantaneous* velocity is reduced to zero (even though acceleration will proceed once the collision or scattering is over). Over a long period of time, the free electron will behave as if it has an average velocity, which is called the **drift velocity, $v_{drift}$**. The drift velocity can be shown to be much smaller than $v_{th}$ and is nonzero only when a **current** is flowing. Pictorially, we can imagine the free electrons in a solid to be like little projectiles with an effective mass $m_e^*$ and a drift velocity $v_{drift}$ (when they take part in current conduction). This concept is illustrated in Fig. 2.4. During current conduction, energy is transferred, through collisions and scattering, from the electric field to the free electrons and eventually to the lattice in the form of heat.

**Example 2.3**    Distinguish between thermal velocity and drift velocity for a free electron.

*Solution:*    The thermal velocity of a free electron has no specific direction of motion, whereas drift velocity is in a direction opposite to the applied electric field. Thermal velocity is given approximately by $\sqrt{2kT/m_e^*}$, which, at 300 K, is roughly equal to $1 \times 10^5$ m/s. Drift velocity, on the other hand, is proportional to the electric field (see Section 2.5) and is zero when there is no current flowing.    §

**Example 2.4**    In a collisionless solid, compute the time required to achieve an electron velocity of $1 \times 10^5$ m/s from rest if the electric field is $\acute{E} = 1 \times 10^3$ V/m. Assume $m_e^* = m_0$.

Fig. 2.4
Electron velocity in an
electric field

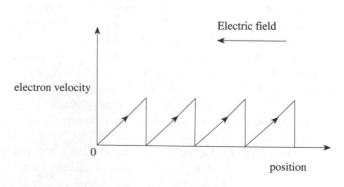

**Solution:**   According to physics, electrical force is $F = \Delta(mv)/\Delta t$, so $\Delta t = m\Delta v/F$, where $F = -q\acute{E}$, $q$ is the electron charge, and $\acute{E}$ is the electric field. Since $m = m_0 = 0.91 \times 10^{-30}$ kg, $\acute{E} = 1 \times 10^3$ V/m, and $\Delta v = 1 \times 10^5$ m/s. The required acceleration time is

$$
\begin{aligned}
\Delta t &= m_0 \Delta v/(q\acute{E}) \\
&= (0.91 \times 10^{-30} \text{ kg} \times 1 \times 10^5 \text{ m/s})/(1.6 \times 10^{-19} \text{ C} \times 1 \times 10^3 \text{ V/m}) \\
&= 0.57 \times 10^{-9} \text{ s} \quad \S
\end{aligned}
$$

# 2.3   ENERGY DISTRIBUTION OF ELECTRONS

We mentioned in the previous section that in a solid, the electron energies are not the same; some electrons have more energy than others (from this point onward, for simplicity, we shall refer to free electrons as electrons). The "natural" distribution of electrons inside a solid is closely related to the fact that electrons are indistinguishable from one another and that there can be at most one electron with the same energy and spin (**spin** is a parameter related to the magnetic properties of the electrons) in a physical location. This relationship is the Pauli exclusion principle; mathematically, it leads to an electron energy distribution known as the **Fermi-Dirac distribution function** $f(E)$, which is given by

$$ f(E) = 1/\{\exp[(E - E_F)/(kT)] + 1\} \tag{4} $$

where     $E$ is the energy of an electron in eV (1 eV $= 1.6 \times 10^{-19}$ J),

$E_F$ is a constant called the Fermi energy in eV,

$k$ is the Boltzmann constant in eV/K, and

$T$ is the absolute temperature in K.

Physically, $f(E)$ measures the probability (or chance) that electrons in the distribution will have an energy equal to $E$. In a collection of electrons, a large $f(E)$ therefore implies that most of the electrons have energy equal to $E$. This concept can be verified if we assume that $E$ ($= E_+$) is large and $(E_+ - E_F)/kT \gg 1$. Equation (4) then becomes

$$ f(E_+) \approx \exp[-(E_+ - E_F)/(kT)] \approx 0 \tag{5} $$

Equation (5) suggests that few of the electrons in the distribution have large energies (something that we should expect). On the other hand, when $E$ ($= E_-$) is small and $(E_- - E_F)/(kT) \ll 1$, Eq. (4) becomes

$$ f(E_-) \approx \exp(0) = 1 \tag{6} $$

A probability of 1 implies certainty, so we can expect the majority of the electrons in the distribution to have low energies.

To sum up, the Fermi-Dirac distribution function describes how a collection of electrons distribute themselves in terms of their energies, $E$. This is illustrated in Fig. 2.5, where $f(E)$ is plotted against $E$. As shown in the figure, $f(E)$ is a continuous function of $E$, and it has the limiting values of 0 or 1 (corresponding to zero probability or certainty), depending on whether $E$ is large or small. For most values of $E$, $f(E)$ falls between 0 and 1. In fact, the most rapid change in $f(E)$ occurs near $E = E_F$. Within $\pm 3kT$ from $E = E_F$, $f(E)$ actually changes from 0.047 to 0.952, and these values are very close to the asymptotic limiting values. Because of this significant change of $f(E)$ near $E_F$, $E_F$ is sometimes considered to be the demarcation energy between filled electron states and empty electron states. Physically, $E_F$ is related to the average internal energy of the electrons. At a finite temperature, $E_F$ is called the **Fermi level** (a term most commonly used in semiconductors). As we can tell from Eq. (4), in addition to the dependence on $E$ and $E_F$, $f(E)$ is also temperature dependent, as shown in Fig. 2.5. Indeed, the slope of the distribution function becomes less steep at a higher temperature because there are more electrons with higher energies.

**Example 2.5**    Compute the electron energy with respect to the Fermi level—i.e., $E - E_F$— at $T = 300$ K if **(a)** $f(E) = 0.1$, and **(b)** $f(E) = 1 \times 10^{-6}$.

*Solution:*    From Eq. (4), $E - E_F = kT \ln[1/f(E) - 1]$.
　　**(a)** When $f(E) = 0.1$, $E - E_F = 8.62 \times 10^{-5}$ eV/K $\times$ 300 K $\times$ $\ln(1/0.1 - 1) = 5.67 \times 10^{-2}$ eV.
　　**(b)** When $f(E) = 1 \times 10^{-6}$, $E - E_F = 8.62 \times 10^{-5}$ eV/K $\times$ 300 K $\times$ $\ln(1/1 \times 10^{-6} - 1) = 0.345$ eV.   §

**Example 2.6**    Verify that $f(E) = 0.047$ when $E - E_F = +3kT$ and $f(E) = 0.952$ when $E - E_F = -3kT$.

*Solution:*    Based on Eq. (4), when $E - E_F = +3kT$, $f(E) = 1/\{\exp[(3kT)/(kT)] + 1\} = 1/[\exp(3) + 1] = 0.047$. When $E - E_F = -3kT$, $f(E) = 1/\{\exp[(-3kT)/(kT)] + 1\} = 0.952$.   §

**Fig. 2.5**
Fermi-Dirac distributions

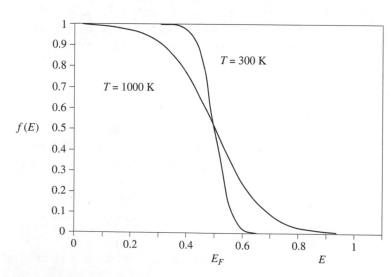

**Example 2.7**   If $f(E) = 0.1$ at $T = 300$ K, what is $f(E)$ at 1000 K? Assume $E_F$ is temperature independent.

**Solution:**   Based on Eq. (4), $E - E_F = kT(1/f(E) - 1)$. Since $f(E) = 0.1$, $E - E_F = 8.62 \times 10^{-5}$ eV/K $\times$ 300 K $\times \ln(1/0.1 - 1) = 0.057$ eV. At $T = 1000$ K, $f(E) = 1/(\exp(0.057$ eV$/(8.62 \times 10^{-5}$ eV/K $\times$ 1000 K)$) + 1) = 0.34$.   **§**

Note that as temperature rises, there will be more electrons with higher energies.

## 2.4   THE ENERGY BAND DIAGRAM

Current conduction in a solid involves changing the energies and the velocities of the electrons, and this change normally occurs in the presence of an applied electric field. As can be expected, the dynamics of electrons inside a solid differ from those in a vacuum because in the latter, electrons do not suffer from collisions or scattering. Inside a solid, collisions and scattering will prevail, and electrons normally attain an average drift velocity $v_{drift}$ in a given electric field $\acute{E}$. The electron energy $E$ and $v_{drift}$ together form the so-called state parameters. Thus, an electron under the influence of an applied voltage will move from a low-energy and low-velocity state to a high-energy and high-velocity state, as illustrated in Fig. 2.6. Generally speaking, in a solid there are many **electron states,** so the transitions between states simply depend on the occupancy (or availability) of the states. In some instances, intermediate states may also be involved. Furthermore, the distribution of electron states in a solid is also different than in a vacuum. In a vacuum, there is essentially an unlimited number of electron states, so there are no restrictions on the state values. The distribution of electron states in a solid, on the other hand, depends strictly on the crystal structure of the solid; there are strict rules governing the energy and the velocity an electron may acquire. In fact, as we shall see, not all of these electron states are available.

Let us examine the electron states in a solid in detail. We recall that they originate from the orbits of the atoms or ions. In the formation of the solid, the individual

**Fig. 2.6**

An electron moving from a low-energy, low-velocity state to a high-energy, high-velocity state

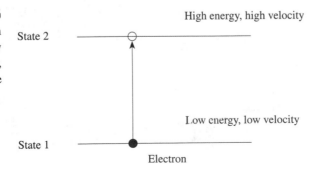

atoms or ions come close together to form bonds. Because of their proximity, electrons in the outer energy orbits far away from the nuclei will overlap spatially. If these outer orbits have the same energy (i.e., they are from similar energy levels), the residing electrons will violate the Pauli exclusion principle (which states that no two electrons with the same energy and spin can coexist together in the same location). In order to abide by the principle, the energies of the overlapping orbits will have to change. What actually happens is that the energies shift and deviate from one another by an incremental amount. In most solids, since there are many atoms and, hence, overlapping orbits, bands of closely spaced electron states are formed. These are called **energy bands.** An energy band is therefore a continuum of closely spaced electron states (or energy states). The term energy state is used here to denote electron states having different energies.

Fig. 2.7 shows the transformation from discrete energy states into energy bands in a hypothetical solid. In the figure, each orbit becomes a part of an energy band; depending on the interatomic (or interionic) separation, there may or may not be any spacings in between the energy bands. These spacings are called **energy gaps;** within the energy gaps, there are no electron states. As we shall see, the size of an energy gap has an important bearing on the conduction properties of a solid. Solids with a small interatomic (or interionic) separation often have fewer energy gaps because most of the energy bands will overlap to form wider bands. A wide band has more energy states and, hence, more electrons.

In order to quantify the density of electron states in a solid, we need to define an additional parameter, the effective density of states of the energy band. The **effective density of states** of a given energy band is a measure of the number of electron states per unit volume in the solid averaged over the entire energy band. Its value can be estimated if the energy band structure of the solid is known. For most energy bands, the value lies in the neighborhood of $1 \times 10^{24}$–$10^{25}$/m$^3$ (see Table 2.8 on page 107).

We can now examine microscopically current conduction in a solid using the energy band diagram. Since current conduction involves a change in the drift velocity (from zero) of the electrons, we also expect a corresponding increase in the electron energy. Physically, current conduction usually requires electrons from low-energy states to move to high-energy states within a given energy band; obviously, an

**Fig. 2.7**
Diagram showing atomic states and energy bands

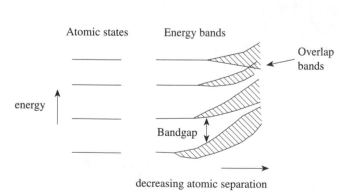

empty energy band will not qualify for current conduction since there are no electrons. A similar argument applies to a completely filled energy band since there are no vacant electron states for electrons to move to. In fact, it is only in the case of a partially filled energy band that the electrons will be able to conduct a current. Electrons in a partially filled energy band will be able to increase their energies by moving into the vacant electron states. Whether a partially filled energy band exists, however, depends on the energy band structure of the solid as well as the energy distribution of the electrons. Fig. 2.8 shows the energy band structure of Si. Since velocity and momentum are interrelated, the energy band structure is usually plotted with electron energy versus momentum. Because of symmetries, only the important crystallographic directions are considered.

The filling of the energy states in an energy band depends on the Fermi-Dirac distribution function $f(E)$ and, in particular, on the location of the Fermi level $E_F$. Since the Fermi level represents the average internal energy of the electrons, the closer $E_F$ is to an energy band, the more likely it is that the energy states in that energy band will be occupied. Temperature, of course, also plays an important role in the occupation of the energy states. As discussed in the previous section, $f(E)$ is temperature dependent, and the number of electrons with high energies increases with increasing temperature. An energy band that is empty at a low temperature can, therefore, become partially filled as the temperature increases. This fact explains why insulators are better conductors at a higher temperature.

As shown in Fig. 2.7, the energy bands in a solid can overlap to form a wider band as the interatomic (or interionic) separation decreases. This situation occurs most frequently in metals because of their close-packed crystal structures, and the net result is the formation of partially filled energy bands with large numbers of electrons. This situation is illustrated in Fig. 2.9. Metals are usually very good current conductors. Unlike metals, semiconductors and insulators have energy bands that do not overlap at low temperature, and the latter are either totally filled or totally empty. In fact, even for a semiconductor only at a moderate temperature is there appreciable current conduction. These solids are known to be poor current conductors. Normally, at a finite temperature the size of the energy gap will determine the number of electrons that are in the energy bands. At room temperature, some

**Fig 2.8**
Energy band diagram of a typical solid

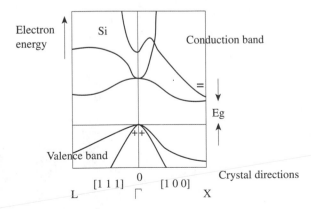

Fig. 2.9
Overlapping bands in a
metal

Band 1

Band 2

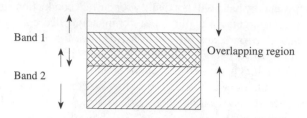

Overlapping region

of the electrons will be able to cross the energy gap, as shown in Fig. 2.10. This
number, however, is very small for an insulator. As a rule, the smaller the energy
gap, the greater the probability that electrons will be able to move across the gap.

In most semiconductors and insulators, the excitation of electrons from a filled
energy band to an empty energy band leaves behind vacant electron states. This
phenomenon is unique to semiconductors and insulators; the vacant electron states
are called **holes.** Holes are positively charged, and they act like mirror images of
electrons (in the sense that their motions are due to the motions of the electrons,
as in Section 2.7). Holes, as do electrons, take part in current conduction. In semiconductors and insulators, therefore, current conduction depends on the densities of the
holes and the electrons. Physically, holes and electrons move in opposite directions
under an applied voltage because they have different charges. To highlight this
difference and to account for the fact that electron and hole energies change differently, separate energy bands for electrons and holes are adopted. Thus, conduction
electrons are found primarily in **conduction bands,** whereas holes exist in **valence
bands.** Conduction bands and valence bands are separated by energy gaps.

**Example 2.8**     A hypothetical solid has $1 \times 10^{26}$ atoms, and each atom has 6 electrons. **(a)** Suggest
the minimum number of energy states needed to accommodate all the electrons
within a single energy band, and **(b)** if the lowest energy state in the energy band
is 0 eV and the energy states are separated by $0.5 \times 10^{-26}$ eV, compute the energy
of the most energetic electron.

Fig. 2.10
Energy bands in
semiconductors

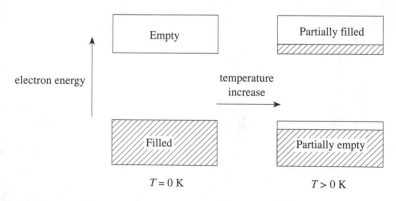

electron energy

temperature
increase

Empty

Partially filled

Filled

Partially empty

$T = 0$ K

$T > 0$ K

**Solution:**

(a) There are $6 \times 10^{26}$ electrons. Since each energy state can accommodate 2 electrons with different spins, there must be at least $3 \times 10^{26}$ energy states.

(b) If the energy states are separated by $0.5 \times 10^{-26}$ eV, the energy of the most energetic electron is $3 \times 10^{26} \times 0.5 \times 10^{-26}$ eV = 1.5 eV. **§**

**Example 2.9** What is the probability that an electron is present in the conduction band of **(a)** a diamond crystal (an insulator with an energy gap of 5.6 eV) and **(b)** a Si crystal (a semiconductor with an energy gap of 1.12 eV)? Assume $T = 300$ K (or $kT = 0.0259$ eV) and the intrinsic Fermi level is in the middle of the energy gap.

**Solution:** Given that $E_c - E_F = E_g/2$ (where $E_g$ is the energy gap).

(a) In pure diamond, $E_c - E_F = 5.6$ eV$/2 = 2.8$ eV. Therefore, $f(E_c) = 1/\{\exp[E_g/(2kT) + 1]\} = 1/[\exp(2.8/0.0259) + 1] = 1.12 \times 10^{-47}$.

(b) In pure Si, $E_c - E_F = 1.12$ eV$/2 = 0.56$ eV. Therefore, $f(E_c) = 1/[\exp(0.56/0.0259) + 1] = 4.07 \times 10^{-10}$. **§**

Note the significant difference in the probabilities of finding electrons in the conduction band of diamond (an insulator) and in the conduction band of Si (a semiconductor).

# 2.5 MATHEMATICAL FORMULATION OF THE CONDUCTION PROCESSES

Quantitatively, current conduction in a solid is measured in terms of current $I$, which is the net charge flow across the solid in unit time. Because of their negative charge, electrons will move against the applied electric field (i.e., toward the positive voltage terminal), which results in a current that is in the opposite direction. Holes, on the other hand, have positive charge, and they move in the same direction as the electric field. Current conduction therefore involves both electrons and holes (holes exist only in semiconductors). It is important to point out here that the valence electrons are attached to the nuclei and do not take part in the current conduction. Quantitatively, current is measured in amperes (A); 1 A is defined as the charge flow of 1 coulomb per second (C/s) (i.e., the flow of $6.25 \times 10^{18}$ electrons in 1 s). The ampere is a large unit for normal usage, and smaller units such as milliamperes (1 mA = $1 \times 10^{-3}$ A) and microamperes (1 μA = $1 \times 10^{-6}$ A) are more frequently used. For an applied voltage $V_a$ in volts, the **resistance** $R$ in ohms (Ω) is defined as

$$R = V_a/I \tag{7}$$

Equation (7) is called Ohm's law. Since $R$ is related to the current flow, it is expected that $R$ should also depend on the geometry of the solid. This is illustrated in Fig. 2.11 for a rectangular solid.

Generally speaking, we can separate out the geometric component of $R$ and write

$$R = \rho(L/A_{cs}) \tag{8}$$

Fig. 2.11
Physical dimensions of a
rectangular solid

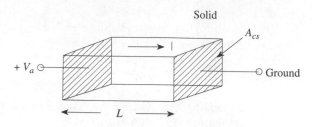

where    $\rho$ is the **resistivity** of the solid in $\Omega \cdot m$,

$L$ is the length in m, and

$A_{cs}$ is the cross-sectional area in $m^2$.

The parameter $\rho$ is independent of the geometry of the solid; typical values of $\rho$ for different solids are listed in Table 2.4. As observed, $\rho$ can vary by many orders of magnitude, indicating the significant differences in the conduction properties of the different solids.

The inverse of resistivity is called **conductivity, $\sigma$**; it is a measure of the effectiveness of the solid in current conduction. Conductivity in siemens/meter (S/m) depends on the electron density and **hole density** and is given by

$$\sigma = 1/\rho = nq\mu_n + pq\mu_p \qquad (9)$$

where    $q$ is the electron charge in C,

$n$ and $p$ are the respective densities of the electrons and the holes $/m^3$, and

$\mu_n$ and $\mu_p$ are the respective electron and hole **mobilities** in $m^2/V \cdot s$.

The mobilities are the parameters that govern the motion of the electrons and the holes inside a solid. Generally speaking, $\mu_n$ and $\mu_p$ can be further expressed in terms of the relaxation times, i.e.,

$$\mu_n = q\langle \tau_e \rangle / m_e^* \qquad (10)$$

$$\mu_p = q\langle \tau_h \rangle / m_h^*$$

where    $\langle \tau_e \rangle$ and $\langle \tau_h \rangle$ are the respective relaxation times for the electrons and the holes in s, and $m_e^*$ and $m_h^*$ are the respective effective masses in kg of the electrons and the holes.

Table 2.4    Typical values of the resistivities of a metal, a semiconductor, and an insulator

|  | Metal | Semiconductor | Insulator |
|---|---|---|---|
|  | Al | Si | $SiO_2$ |
| $\rho$ ($\Omega \cdot m$) | $2.7 \times 10^{-8}$ | $2.3 \times 10^3$ | $10^{12}$ |

We define the relaxation time as the average time lapse between successive collisions or scattering. If the relaxation times are long, both the electrons and the holes will be able to travel great distances before they scatter or suffer a collision. This, in turn, produces large mobilities. The reverse is also true if the relaxation times are short. We can now transform Eq. (7) to examine the microscopic properties of the electrons and the holes. For simplicity, we shall consider only electrons (a similar treatment can be applied to the holes). By definition of the current, we know

$$I = \Delta Q/\Delta t = nqA_{cs}\, \Delta x/\Delta t \qquad (11)$$

where   $\Delta Q$ $(= nqA_{cs}\, \Delta x)$ is the amount of charge in C that has moved through the distance $\Delta x$ (m) in $\Delta t$ (s),

$n$ is the electron density /m$^3$,

$q$ is the electron charge in **C**, and

$A_{cs}$ is the cross-sectional area in m$^2$.

Note that $A_{cs}\, \Delta x$ represents the spatial volume in cubic meters transversed by a single electron in $\Delta t$ seconds. When this quantity is multiplied by the charge density $(nq)$, the product is $\Delta Q$. We can also express the electric field $\acute{E}$ (in V/m) as

$$\acute{E} = V_a/L \qquad (12)$$

where   $V_a$ is the applied voltage in V, and

$L$ is the separation between the terminals in m.

Combined with Eqs. (11) and (12), Eq. (7) becomes

$$R = L\acute{E}/(nqv_{\text{drift}}A_{cs}) \qquad (13)$$

where we have replaced $\Delta x/\Delta t$ by $v_{\text{drift}}$, the drift velocity of the electrons in m/s. Equating Eqs. (8) and (13) now gives

$$\rho = \acute{E}/(nqv_{\text{drift}}) \qquad (14)$$

Since $1/\rho = 1/nq\mu_n$ (see Eq. (9) and remember that we have ignored the holes), Eq. (14) becomes

$$\mu_n = -v_{\text{drift}}/\acute{E} \qquad (15)$$

Equation (15) is the microscopic form of Ohm's law for a single electron. It links the drift velocity (which is related to the current) to the electric field (which is related to the applied voltage) through the electron mobility. A large $v_{\text{drift}}$ will result if there is either a large $\acute{E}$ or a large $\mu_n$. The negative sign of Eq. (15) simply ensures that the electron moves opposite to the direction of the electric field and hence to the current flow (in the case of a hole, the negative sign will be missing). Table 2.5 lists the values of the electron mobilities in different solids at 300 K.

In addition to collisions and scattering by impurities and crystal imperfections, electrons in a solid are also scattered by lattice vibrations. This scattering is called *lattice scattering,* and it involves the interactions between the electron waves and the lattice vibrational waves. During scattering, the accelerating electrons actually transfer a portion of their energies to the lattice vibrations and are themselves slowed down in the process (acceleration resumes once the scattering event is over). Over many successive scatterings, the electrons will acquire an average velocity that we called the drift velocity. Drift velocity due to lattice scattering is temperature dependent. As the temperature increases, the lattice vibrates more rigorously, and the frequency of scattering also increases (this ideally leads to a temperature dependence of $T^{-3/2}$ for $\mu_n$). In the absence of impurities and crystal imperfections, lattice scattering is the dominant scattering mechanism.

Collisions and scattering by impurities and crystal imperfections are called *impurity scattering;* the frequency increases with the densities of impurities and crystal imperfections. Ideally, the collision of an electron with a neutral impurity is independent of temperature because it depends only on the cross-sectional area of the impurity. This is, however, not true in the case of a charged impurity since the interaction will involve long-range coulombic forces, as illustrated in Fig. 2.12.

**Table 2.5**  Values of electron mobilities in different solids at 300 K

| Solid | Mobility (m²/V·s) |
|-------|-------------------|
| Al | 0.00435 |
| Cu | 0.00136 |
| Zn | 0.00081 |
| Si | 0.145 |
| GaAs | 0.850 |
| CdS | 0.034 |
| $SiO_2$ | ~0 |
| $Si_3N_4$ | ~0 |

**Fig. 2.12**
Different types of scattering in solids

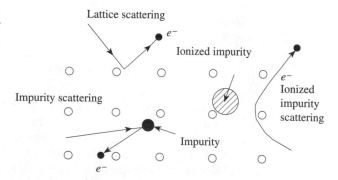

During charged impurity scattering, the scattering probability decreases rapidly with increasing temperature because the faster-moving electron, on the average, will spend less time near the impurity. This reduces the scattering probability and increases the electron mobility ($\mu_n$ in this case is proportional to $T^{3/2}$). Ionized impurity scattering therefore decreases as temperature increases.

When more than one type of scattering mechanism is involved, the combined **relaxation time** $\langle \tau_T \rangle$ (see Eq. (10)) is given by

$$\langle \tau_T \rangle = 1/\Sigma_i(1/\tau_i) \tag{16}$$

where    $\Sigma$ is the summation symbol,

$\tau_i$ is the individual relaxation time in s, and

$i$ identifies the scattering mechanism.

Equation (16) suggests that the shortest relaxation time will dominate, and the mobility of the electrons (Eq. (15)) is limited by the most effective scattering mechanism. Fig. 2.13 shows the temperature dependence of the resistivity of a metal. As observed, at low temperature, the resistivity is dominated by impurity scattering (this is known as the *Matthiessen rule*).

So far, we have considered only electrons and have ignored the holes. Holes are important in semiconductors; because they are the mirror images of the electrons, most of our discussions on conductivity, mobility, and scattering also apply to holes. The only difference is that unlike electrons, holes move in the same direction as the electric field, and they have a different effective mass.

**Example 2.10**    Compare the velocities of electrons moving in a vacuum and in Si.

*Solution:*   Since the effective mass of electrons in Si is 0.19 times $m_0$ (see Table 2.3), we can expect the drift velocity of electrons in Si to be about 5 times as large as its drift velocity in a vacuum (subject to a saturation value in Si).   §

**Example 2.11**    A copper (Cu) atom has 29 electrons. Compute the fraction of electrons in the solid if the conductivity of Cu is $1.7 \times 10^7$ S/m. Assume that the electron mobility is $12.6 \times 10^{-4}$ m$^2$/V·s and the atomic density of Cu is $2 \times 10^{29}$/m$^3$.

**Fig. 2.13**
Resistivity of a metal

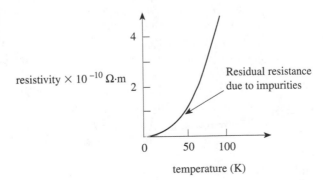

temperature (K)

*Solution:*   The free electron density $n$ in Cu is

$$\sigma/q\mu_n = 1.7 \times 10^7 \text{ S/m}/(1.6 \times 10^{-19} \text{ C} \times 0.00126 \text{ m}^2/\text{V}\cdot\text{s}) = 0.84 \times 10^{29}/\text{m}^3$$

The total number of electrons present in Cu per unit volume is $n_T = 2 \times 10^{29} \times 29 = 5.8 \times 10^{30}$. Thus, the fraction of free electrons in Cu is $n/n_T = 0.84 \times 10^{29}/5.8 \times 10^{30} = 0.014$.   §

**Example 2.12**   A wire made of an Al alloy is 1 mm in diameter and 1 m in length. **(a)** If there is a current of 10 mA flowing and the voltage across the wire is 0.432 mV, determine the conductivity of the wire. **(b)** If the electron mobility is $3 \times 10^{-3}$ m²/V·s, what is the electron density in the Al alloy? **(c)** What is the drift velocity of the electrons?

*Solution:*
   **(a)** Based on Eq. (7), $R = V_a/I = 0.432 \text{ V}/10 \text{ A} = 0.0432 \text{ }\Omega$. This gives a conductivity $\sigma$ of $L/(A_{cs}R) = 1 \text{ m}/[3.14 \times (0.5 \times 10^{-3} \text{ m})^2 \times 0.0432 \text{ }\Omega] = 29.5 \times 10^6$ S/m.
   **(b)** Based on Eq. (9),

$$n = \sigma/(q\mu_n) = 29.5 \times 10^6 \text{ S/m}/(1.6 \times 10^{-19} \text{ C} \times 3 \times 10^{-3} \text{ m}^2/\text{V}\cdot\text{s})$$
$$= 6.14 \times 10^{28} /\text{m}^3.$$

   **(c)** Based on Eq. (15), $v_{\text{drift}} = \mu_n \acute{E} = \mu_n V/L = 3 \times 10^{-3} \text{ m}^2/\text{V}\cdot\text{s} \times 0.432 \times 10^{-3} \text{ V}/0.01 \text{ A} = 1.3 \times 10^{-3}$ m/s.   §

**Example 2.13**   A solid is known to have two different types of scattering mechanisms. If the electron relaxation times are 0.8 ms and 1.2 ms, respectively, what will be the combined relaxation time?

*Solution:*   Based on Eq. (16), $\langle \tau_T \rangle = 1/\Sigma_i(1/\tau_i) = 1/(1/0.8 \text{ ms} + 1/1.2 \text{ ms}) = 0.48$ ms.   §

Note that the combined relaxation time is less than the individual relaxation time.

## 2.6   CONDUCTORS

Solids are divided into conductors, semiconductors, and insulators. **Conductors,** which are predominantly metals, have partially filled energy bands; there are many electrons (on the order of $1 \times 10^{28}/\text{m}^3$) and vacant (electron) states in these energy bands. With the large number of electrons, the electron density of a metal is virtually independent of temperature, and any change in conductivity is due solely to a change in the electron mobility (see Eq. (10)). Because of this, the temperature dependence of the conductivity of a metal can sometimes be expressed as

$$\sigma = \sigma_0/[1 + \alpha_T(T - T_0)] \tag{17}$$

where     $\sigma_0$ is the conductivity at 300 K in S/m,
         $\alpha_T$ is a temperature coefficient /K, and
         $T_0 = 300$ K.

Equation (17) suggests that the conductivity of a metal goes down with increasing temperature if $\alpha_T$ is positive. This is what happens with lattice scattering (see Section 2.5). The effect of a small quantity of impurities added to a metal also changes its conductivity. For a reasonably small addition, the conductivity is given by

$$\sigma = \sigma_i/(1 + \beta\kappa) \tag{18}$$

where    $\sigma_i$ is the intrinsic conductivity of the metal in S/m,

        $\beta$ is a constant, and

        $\kappa$ is the fraction of impurities added.

As the impurity density goes up, the conductivity decreases. This result agrees with what we observe during impurity scattering. Note that Eqs. (17) and (18) are actually the "linearized" forms of Eq. (9) after taking into account the first-order temperature dependence of $\mu_n$. Table 2.6 lists the typical values of $\sigma_0$ and $\alpha_T$ for metals and the values of $\beta$ (for Cu) in the presence of different impurities.

**Example 2.14**    From the periodic table (see Appendix A), suggest which groups of elements or compounds are most likely to be (a) metals and (b) semiconductors or insulators.

*Solution:*
   (a) The Group I and Group II elements are often metals because of their close-packed crystal structures.
   (b) The Group IV elements have covalent bonds, and they often form semiconductors or insulators. Most Group I–Group VII compounds and Group II–Group VI compounds have ionic bonds, and they are insulators.   §

**Example 2.15**    Calculate the conductivity of gold (Au) at 200 °C if its resistivity at 20 °C is 24.4 × $10^{-9}$ $\Omega$·m.

*Solution:*    From Table 2.6, $\alpha_T$ for Au is 0.0034/°C. Based on Eq. (17),

$$\begin{aligned}\sigma &= \sigma_0/[1 + \sigma_T(T - T_0)] \\ &= (1/24.4 \times 10^{-9}\ \Omega\text{·m})/[1 + 0.0034/\text{K} \times (200\ \text{K} - 20\ \text{K})] = 25.4 \times 10^6\ \text{S/m.}\ \S\end{aligned}$$

**Example 2.16**    From Table 2.6, estimate the resistivity of Cu if there is 0.1 wt% of chromium (Cr) present.

**Table 2.6**    Values of $\sigma_0$ and $\alpha_T$ for some metals and the values of $\beta$ for Cu in the presence of common impurities

| Metal | $\sigma_0$ (S/m) | $\alpha_T$ (/K) | Impurity | $\beta$ (/wt%) |
|-------|------------------|-----------------|----------|----------------|
| Fe | 1.03 × $10^{-11}$ | 0.00651 | | |
| Au | 4.09 × $10^{-11}$ | 0.00340 | | |
| Cu | 5.80 × $10^{-11}$ | 0.00393 | Fe | 5.41 |
| | | | Cr | 2.47 |
| | | | Al | 0.88 |

(*Source:* J. K. Stanley, *Electrical and Magnetic Properties of Metals*, (Materials Park, Ohio: ASM International, 1963.) Reprinted by permission of ASM International).

*Solution:*    Based on Eq. (18), $\sigma = \sigma_i/(1 + \beta\kappa)$. For impure Cu, $\sigma = 1/\rho = (1/5.8 \times 10^{-11})$ S/m $\times (1 + 0.1 \times 2.47) = 21.9 \times 10^{-9}$ $\Omega$·m.    §

## Hall Effect

The **Hall effect** is the most common method for determining the electron density in a metal (or a semiconductor). During Hall measurement, a rectangular bar of the solid and a source of known magnetic induction are required. Fig. 2.14 shows a Hall measurement setup. As shown, a voltage $V_a$, when applied along the length of the bar (in the $x$ direction), generates a current $I$. If the magnetic induction **B** (in T) is in the $z$ direction, the electrons that flow opposite to the direction of $I$ are deflected in the $-y$ direction. Accordingly, the Lorentz force (in N) is given by

$$\mathbf{F_L} = -q(\acute{\mathbf{E}} + \mathbf{v_{drift}} \times \mathbf{B}) \tag{19}$$

where    $q$ is the electron charge in C,

$\acute{\mathbf{E}}$ is the electric field in V/m,

$\mathbf{v_{drift}}$ is the drift velocity in m/s, and

$\times$ is the notation for cross product.

The accumulation of electrons on one side of the bar results in the exposure of the lattice ions (or the ionized dopants, in the case of a semiconductor) on the opposite side, as illustrated in Fig. 2.15. As suggested by Eq. (19), the quantity of electrons involved is directly proportional to the magnetic induction and the current. This charge separation produces a transverse voltage known as the Hall voltage, $V_H$. As $V_H$ builds up, it also exerts an electrostatic force $\mathbf{F_H}$ that counteracts the Lorentz force $\mathbf{F_L}$ (Eq. (19)) and reduces further charge flow in the $y$ direction. $\mathbf{F_H}$ (in N) is given by

$$|\mathbf{F_H}| = qV_H/d_w \tag{20}$$

**Fig. 2.14**
Hall measurement setup

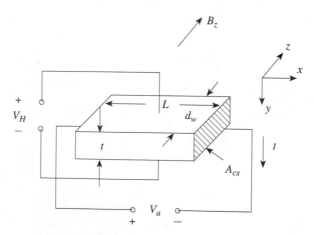

Fig. 2.15

Hall effect showing charge accumulation

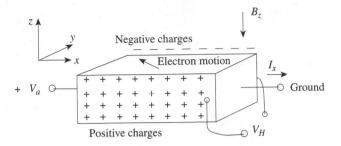

where $V_H$ is the Hall voltage in V, and

$d_w$ is the width of the bar in m.

In the steady state, $\mathbf{F_L} = -\mathbf{F_H}$, or Eqs. (19) and (20) are equal and opposite (note that $\acute{\mathbf{E}} = 0$). Using Eq. (12), i.e., $I = nqA_{cs}\,|v_{drift}|$, it can be shown that

$$V_H = R_H IB/t_h \tag{21}$$

where $R_H\ (= 1/(nq))$ is the Hall constant in C·m$^3$, and

$t_h\ (=A_{cs}/d_w)$ is the thickness of the bar in m.

If $V_H, I, B,$ and $t_h$ are known, $R_H$ (and hence the electron density, $n$) can be determined. Furthermore, using the sign notation appearing in Fig. 2.15, it can be shown that

$$R_H = -1/(nq) \tag{22}$$

A negative $V_H$ implies the presence of electrons (the reverse—i.e., a positive $V_H$ in a semiconductor—is indicative of the dominance of holes).

Hall measurement can also be used to determine electron mobility if the dimensions of the bar are also known. Using Eqs. (13), (15), (22), and (7), the electron mobility becomes

$$\mu_n = R_H LI/(V_a A_{cs}) \tag{23}$$

where $L$ is the length of the metal bar in m, and

$A_{cs}$ is the cross-sectional area in m$^2$.

Hall effect is sometimes used to determine an unknown magnetic induction. The procedure is essentially the reverse of the one described here, and the electron density of the bar has to be known.

**Example 2.17** Determine the Hall coefficient $R_H$ if the current through a solid is 1 mA, the width of the solid is 0.005 m, the cross-sectional area is $1 \times 10^{-6}$ m$^2$, the Hall voltage is 2.5 mV, and the magnetic induction is 0.1 T.

***Solution:*** Based on Eq. (21), $R_H = tV_H/(IB) = A_{cs}V_H/(IBd_w) = 1 \times 10^{-6}$ m$^2 \times$ $2.5 \times 10^{-3}$ V/$(1 \times 10^{-3}$ A $\times 0.1$ T $\times 5 \times 10^{-3}$ m$) = 5 \times 10^{-3}$ m$^3$/C. §

## Magnetoresistance Effect

The **magnetoresistance** effect is the increase in the resistance of a solid in the presence of an applied magnetic induction. Fig. 2.16 shows the setup for magnetoresistance measurement. The Lorentz force (Eq. (19)) in the presence of a vertical magnetic induction $\mathbf{B}_z$ can be decomposed into the following two components:

$$F_x = q\mathbf{v}_x/\mu_n = q\acute{\mathbf{E}}_x + |\mathbf{v}_y \times \mathbf{B}_z| \tag{24}$$

$$F_y = q\mathbf{v}_y/\mu_n = q\acute{\mathbf{E}}_y - |\mathbf{v}_x \times \mathbf{B}_z| \tag{25}$$

where   $\mathbf{v}_x$ and $\mathbf{v}_y$ are the respective velocities of the electrons in the $x$ direction and in the $y$ direction in m/s,

$\acute{\mathbf{E}}_x$ and $\acute{\mathbf{E}}_y$ are the respective electric fields in the $x$ direction and in the $y$ direction in V/m, and

$\mu_n$ is the electron mobility in m$^2$/V·s.

If $L$, the length of the solid, is much shorter than $d_w$, its width, $\acute{\mathbf{E}}_y$, is essentially zero, and Eqs. (24) and (25) can be combined to solve for $v_x$:

$$\mathbf{v}_x = \mu_n\acute{\mathbf{E}}_x/(1 + \mu_n^2\mathbf{B}_z^2) \tag{26}$$

The *apparent mobility* $\mu'$ $(= v_x/\acute{E}_x)$ in the presence of $\mathbf{B}_z$ becomes

$$\mu' = \mu_n/(1 + \mu_n^2\mathbf{B}_z^2) \tag{27}$$

Since $\rho = 1/(nq\mu_n)$, we can also write

$$\rho' = \rho(1 + \mu_n^2\mathbf{B}_z^2) \tag{28}$$

where   $\rho'$ $(= 1/(nq\mu'))$ is the *apparent resistivity* in $\Omega$·m.

Fig. 2.16
Setup for
magnetoresistance
measurement

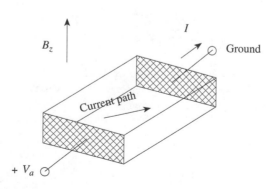

Equation (28) therefore suggests that the magnetic induction $\mathbf{B}_z$ increases the resistivity of the solid. By measuring $\rho'$ and $\rho$ in the presence and in the absence of the known magnetic induction $\mathbf{B}_z$, we can find the electron mobility. Conversely, if the magnetic field is not known but $\mu_n$ is given, $\mathbf{B}_z$ can be determined. The magnetoresistance effect can, therefore, be used as a simple technique to determine an unknown magnetic induction. The condition that $d_w \gg L$ can be met by using a structure called a **Corbino disk,** as shown in Fig. 2.17.

**Example 2.18**   If the fractional increase in the resistivity of Cu due to magnetoresistance effect is 1%, determine the magnetic induction.

*Solution:*   Based on Eq. (28), $\rho'/\rho = 1.01 = 1 + \mu_n^2 B_z^2$ and $\mathbf{B}_z = \sqrt{0.01}/\mu_n = 0.1/\mu_n$. From Table 2.5, $\mu_n = 0.00126$ m$^2$/V·s, so, $\mathbf{B}_z = 0.101/0.00126$ T = 79.8 T.   §

## Thermoelectric Effect

In addition to taking part in current flow, electrons in a solid also respond to a temperature gradient. This response is sometimes called *thermoelectric effect*. A metal wire will develop a voltage across it if there is a temperature gradient. The voltage is known as **thermoelectric voltage,** $\Delta V$. $\Delta V$ in volts is given by

$$\Delta V = K_s \Delta T \tag{29}$$

where   $K_s$ is the Seebeck coefficient in V/K, and

$\Delta T$ is the temperature difference in K.

A **thermocouple** is a temperature-measuring device based on the thermoelectric effect. It consists of two different metal wire junctions (A/B and B/A) connected in an open loop (i.e., A/B/A). One of the junctions is placed in a cold reference $T_c$

Fig. 2.17
Corbino disk with a large
$d_w/L$

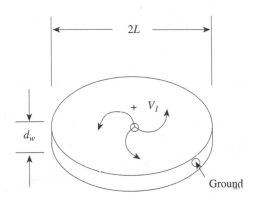

(such as an ice bath at 0 °C), and the other junction is placed at the location where the temperature $T_h$ is to be measured. Fig. 2.18 shows a typical setup.

As shown, the voltage $\Delta V_{AA}$ in volts developed across the loop A/B/A due to the temperature difference at the two junctions is given by

$$\Delta V_{AA} = K_{sA}(T_h - T_c) - K_{sB}(T_h - T_c) \tag{30}$$

$$= (K_{sA} - K_{sB})(T_h - T_c)$$

where $K_{sA}$ and $K_{sB}$ are the Seebeck coefficients for metal A and metal B, respectively, in V/K.

Since $T_c = 0$ °C, $\Delta V_{AA}$ is proportional to $T_h$. By measuring $\Delta V_{AA}$ and knowing $K_{sA} - K_{sB}$, the temperature of the hot junction can be determined. Table 2.7 lists the values of $K_{sA} - K_{sB}$ for some common thermocouples.

Although the thermoelectric effect is most frequently used for temperature measurements, it can also be used for refrigeration by reversing the thermoelectric process, i.e., by passing a current through a thermocouple to cause cooling in one of the junctions. Using a sufficiently large number of thermocouples (connected in parallel), solid-state refrigeration of tens of degrees in a small system can often be achieved.

**Example 2.19**  A chromel-alumel thermocouple is used to monitor the temperature of a hot furnace. If the voltage output from the thermocouple is 10 mV and the cold junction is an ice bath ($T_c = 0$ °C), calculate the furnace temperature.

**Fig. 2.18**
Thermocouple
measurement

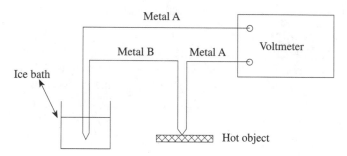

**Table 2.7**  Seebeck coefficients for the common thermocouples (the cold junction is at 0 °C)

| Metal A | Metal B | $K_{sA} - K_{sB}$ (mV) |
| --- | --- | --- |
| Chromel | Alumel | 0.040 |
| Platinum | Platinum/rhodium | 0.006 |
| Copper | Constantin | 0.039 |

***Solution:***   Based on Eq. (30), $(T_h - T_c) = \Delta V_{AA}/(K_{sA} - K_{sB})$. Since $\Delta V_{AA} =$ 10 mV and (from Table 2.7) $K_{sA} - K_{sB} = 0.040$ mV/°C, $T_h =$ 10 mV/0.040 mV/°C = 250 °C.   **§**

## Other Types of Conductors

Metals and their alloys are frequently used for interconnections and contact formation. These include metals such as Al, Cu, Au, Pt, W, Mo, and Hf and their alloys. Most of these solids are chosen for their high conductivities ($\sim 10^7$ S/m) and the fact that they can be easily deposited on substrates by vapor-phase-deposition techniques. The Al-Au alloy, for example, is a popular contact material for Si based microcircuits because of its compatibility with Si and $SiO_2$, its low cost, and its stability. More recently, as finer (submicrometer) linewidths have been needed for interconnections in integrated circuits (ICs), refractory metals such as Mo and W have been the choice because they can form much thinner lines and withstand higher temperatures. Other conductors, such as Ti-W alloys and Au-Ge alloys, are preferred contact materials for the Group III and Group V compound semiconductors and devices. Metal silicides, compounds of Si and metals, are also becoming more popular because of their high conductivities and their resistance to atmospheric contamination. Silicides can be formed by direct deposition of a metal onto Si, even though they require heat treatment at a few hundred degrees Celsius. During formation, however, some of the Si atoms in the substrate will be consumed (a thickness roughly equal to the thickness of the metal layer). Currently, the most popular silicides include $PtSi$, $TiSi_2$, $TaSi_2$, $WSi_2$, and $NiSi_2$.

# 2.7   SEMICONDUCTORS

A pure **semiconductor** is an insulator at low temperature, and its conductivity increases with increasing temperature due to thermal excitation of the electrons from the valence bands to the conduction bands. (Remember the temperature dependence of the Fermi-Dirac distribution function). The more common semiconductors, including Si, Ge, and GaAs, have a diamond or zinc-blende crystal structure, i.e., with four nearest neighbors (see Section 1.3). These solids can have a wide range of conductivities if dopants are added. The conductivity of a pure semiconductor is directly related to the number of electrons in the conduction bands and holes in the valence bands (see Eq. (9)). At room temperature and without dopants, only a moderate number of electrons and holes are present. Holes are the vacant electron states in the valence bands, and they move as a result of the capture of electrons. For instance, an electron originated from an atom B when captured by a hole residing at an atom A is effectively the same as the hole moving from atom A to atom B. This situation is illustrated in Fig. 2.19. Different from electrons, holes move in the same direction as the electric field and the current. For simplicity, electrons and holes are often called (charged) **carriers.** In a semiconductor, the total conductivity is therefore given by the sum of the carrier conductivities (see Eq. (9)).

    As mentioned earlier, the electron and hole densities depend on the occupancy of the electron states in the energy bands and the Fermi-Dirac distribution function

Fig. 2.19
Movement of a hole

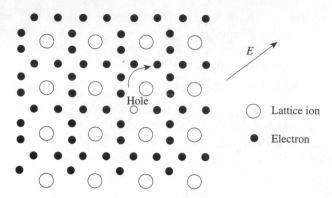

(Eq. (4)). Since $f(E)$ is the probability that an electron will have an energy equal to $E$, its complement, $1 - f(E)$, gives the probability for a hole with energy $E$. It is given by

$$1 - f(E) = \exp[(E - E_F)/(kT)]/\{\exp[(E - E_F)/(kT)] + 1\} \qquad \textbf{(31)}$$

As in the case of electrons, if we know $E_F$ and $T$, the probability of a hole in the valence band can be determined. Fig. 2.20 shows the energy band diagram of a semiconductor and the effect of the Fermi-Dirac distribution function on the electron states.

**Example 2.20**    In intrinsic (pure) Si at 300 K, $n = p = 1.45 \times 10^{16}/\text{m}^3$, $\mu_n = 0.15 \text{ m}^2/\text{V·s}$, and $\mu_p = 0.045 \text{ m}^2/\text{V·s}$, determine the conductivity of intrinsic Si.

*Solution:*    From Eq. (9), $\sigma = nq\mu_n + pq\mu_p = 1.45 \times 10^{16}/\text{m}^3 \times 1.6 \times 10^{-19}\text{C} \times (0.15 \text{ m}^2/\text{V·s} + 0.045 \text{ m}^2/\text{V·s}) = 0.45 \times 10^{-3} \text{ S/m}.$   **§**

## Intrinsic Semiconductors

It is important to distinguish between an intrinsic semiconductor and an extrinsic semiconductor. An **intrinsic semiconductor** is a pure semiconductor, and there are

Fig. 2.20
Energy band diagram of a
semiconductor

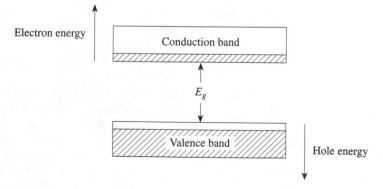

no detectable impurities present. Because of this, every electron generated in an intrinsic semiconductor will be accompanied by the creation of a hole; i.e., $n = p$. It is therefore possible to define an **intrinsic Fermi level** $E_{Fi}$ for such a semiconductor. If we assume that the effective densities of states in the conduction band and in the valence band are the same (see Table 2.8 for the actual values), then the probability of finding an electron in the conduction band (Eq. (4)) must be the same as the probability of finding a hole in the valence band (Eq. (32)). This leads to

$$f(E_c) = 1/\{\exp[(E_c - E_F)/(kT)] + 1\} = 1 - f(E_v) \tag{32}$$

$$= \exp[(E_v - E_F)/(kT)]/\{\exp[(E_v - E_F)/(kT)] + 1\}$$

In arriving at Eq. (32), we have evaluated Eq. (4) at the conduction band edge—i.e., $E = E_c$—and Eq. (31) at the valence band edge—i.e., $E = E_v$. Rewriting $E_F$ as $E_{Fi}$ for an intrinsic semiconductor, Eq. (32) gives

$$E_{Fi} = (E_c - E_v)/2 = E_v + E_g/2 \tag{33}$$

where   $E_g$ is the energy gap in eV.

With the assumptions mentioned earlier, the intrinsic Fermi level will be in the middle of the energy gap. This placement is in basic agreement with the fact that $f(E)$ is approximately symmetrical at $E = E_{Fi}$ (see Fig. 2.5).

To compute the carrier densities $n$ and $p$ (/m$^3$) in an intrinsic semiconductor, we assume that the effective density of states (number of electron states per unit volume) in the conduction band is $N_c$ (/m$^3$). Then $n$ is given by the product of $N_c$ and the probability function that the electron states in the conduction band ($E = E_c$) are occupied, i.e.,

$$n = N_c f(E_c) \approx N_c \exp[-(E_c - E_F)/(kT)] \tag{34}$$

In arriving at Eq. (34), we used Eq. (4) and assumed that $E_c - E_F >> kT$. This approximation is generally good for most semiconductors with a reasonably large $E_g$ and at a low temperature. A similar expression for the hole density $p$ is given by

$$p = N_v(1 - f(E_v)) = N_v \exp[-(E_F - E_v)/(kT)] \tag{35}$$

where   $N_v$ is the effective density of states in the valence band /m$^3$.

In arriving at Eq. (35), we used Eq. (31) and assumed that $E = E_v$ and $E_F - E_v >> kT$. Combining Eqs. (34) and (35), the product of $n$ and $p$ (or $np$) becomes

$$np = N_c N_v \exp(-E_g/kT) \tag{36}$$

where   $E_g = E_c - E_v$.

Equation (36) is one of the most important relationships observed in semiconductors. First of all, the product $np$ is shown to be independent of the position of the Fermi level, and so it is also independent of such changes due to the presence of dopants (thus, it is valid in extrinsic semiconductors). Secondly, since $N_c$, $N_v$, and $E_g$ are constants at a given temperature, the product $np$ will also be constant at a given temperature. Equation (36) therefore represents a unique relationship between the values of $n$ and $p$. Physically, any increase in the density of one of the carriers will result in a decrease in the density of the other. Such a self-regulating mechanism is a consequence of the *law of mass action* that governs the creation and annihilation of electrons and holes. An excess in one of the carriers will automatically increase the annihilation rate until a balance is achieved. Equation (36) also provides the temperature dependence of $pn$, as illustrated in Fig. 2.21. Although $E_g$ is often assumed to be temperature independent, in actual fact, it varies somewhat with temperature. In Si, for instance, the temperature dependence of $E_g$ obeys the following empirical relationship:

$$E_g(T) = 1.17 - (4.73 \times 10^{-4})T^2/(T + 636) \tag{37}$$

where  $E_g(T)$ is the energy gap in eV, and

$T$ is the absolute temperature in K.

The constants appearing in Eq. (37) were measured experimentally. The value of $E_g$ for Si at 300 K has been evaluated to be 1.12 eV (see Table 2.8). In connection with the temperature dependence of $n$ and $p$, we note that both $N_c$ and $N_v$ are also temperature dependent. Ideally, $N_c$ and $N_v$ vary as $T^{3/2}$. Table 2.8 summarizes the physical properties of the important semiconductors at 300 K.

The conductivity of an intrinsic semiconductor (see Eq. (9)) also depends on the carrier mobilities. For most semiconductors, the electron mobility is higher than the hole mobility by a factor of 2 or more. Carrier mobilities in intrinsic semiconductors are determined by lattice scattering; the latter is more important at high temperatures, when the lattice vibrations intensify. In addition, carrier mobilities also depend on

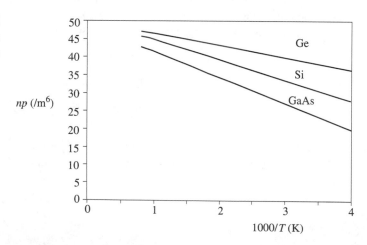

**Fig. 2.21**
*np* versus inverse temperature

**Table 2.8**  Physical properties of the important semiconductors at 300 K ($\epsilon_o$ is the permittivity of vacuum)

| Semiconductor | $E_g$ (eV) | $\epsilon_s$ ($\times \epsilon_0$) | $\mu_n$ ($\times 10^{-1}$ m²/V·s) | $\mu_p$ | $N_c$ ($\times 10^{25}$/m³) | $N_v$ | $n_i$ ($\times 10^{16}$/m³) |
|---|---|---|---|---|---|---|---|
| Si | 1.12 | 11.9 | 1.45 | 0.45 | 2.8 | 1.04 | 1.45 |
| Ge | 0.66 | 16 | 3.90 | 1.90 | 1.04 | 0.6 | $2.4 \times 10^3$ |
| GaP | 2.26 | 10 | 0.30 | 0.15 | | | |
| GaAs | 1.42 | 13.1 | 8.50 | 0.40 | 0.047 | 0.7 | $1.8 \times 10^{-4}$ |
| InP | 1.35 | 12.1 | 4.00 | 0.60 | | | |
| InSb | 0.18 | 18 | 100 | 1.70 | | | |

the effective masses. Higher carrier mobilities are found in semiconductors with a small effective mass, as in the case of the Group III–Group V compound semiconductors, such as GaAs and InP, where the electron mobilities are 3 to 5 times higher than the corresponding values in Si. Even higher carrier mobilities can be found in the small energy gap semiconductors, such as InSb (see Table 2.8). Fig. 2.22 shows the temperature dependence of the carrier mobilities in an intrinsic semiconductor.

From Eq. (36) and the fact that $p = n$, we can define an intrinsic carrier density $n_i$ (/m³) by

$$n_i = p = n = \sqrt{N_c N_v} \exp[-E_g/(2kT)] \tag{38}$$

where  $n_i$ depends exponentially on $E_g$ and $T$.

Combining Eqs. (9) and (38), the temperature dependence of the conductivity for an intrinsic semiconductor has the simple form

$$\ln(\sigma/\sigma_0') = -E_g/2kT \tag{39}$$

where  $\sigma_0'$ is a conductivity parameter in S/m and is assumed to be temperature independent.

**Fig. 2.22**
Temperature dependence
of carrier mobilities

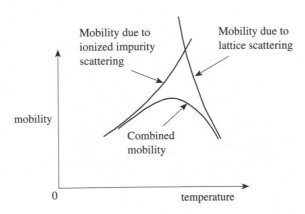

Table 2.9    Theoretical values of $\sigma_0'$ for the common semiconductors

| Semiconductor | $\sigma_0'$ (S/m) |
|---|---|
| Si | 1.81 |
| Ge | 0.37 |
| GaAs | 3.82 |

Table 2.9 lists the theoretical values of $\sigma_0'$ for the common semiconductors.

**Example 2.21**    Compute the value of the energy gap of Si at 400 K.

*Solution:*    Based on Eq. (37), $E_g(T) = 1.17 - (4.73 \times 10^{-4})T^2/(T + 636)$ eV. At 400 K, $E_g(400 \text{ K}) = 1.17 - 4.73 \times 10^{-4} \times 400^2/(400 + 636)$ eV = 1.10 eV.    §

**Example 2.22**    At 300 K, compute the ratio of the density of electrons in intrinsic Si to density of Si atoms, given that the (weight) density of Si is 2.33 g/cm$^3$ and the atomic mass is 28.09 amu.

*Solution:*    The atomic density of Si is $N_{Si}$ = density of Si $\times N_{AG}$/atomic mass = 2.33 g/cm$^3$ $\times 1 \times 10^6$ cm$^3$/m$^3$ $\times 0.6023 \times 10^{24}$/mol/28.09 g = $5 \times 10^{28}$/m$^3$. From Table 2.8, $n_i = 1.4 \times 10^{16}$/m$^3$, so, $n_i/N_{Si} = 1.4 \times 10^{16}/5 \times 10^{28} = 2.8 \times 10^{-13}$.    §

Note that there are only 2.8 electrons per $10^{13}$ Si atoms (this number is very small).

**Example 2.23**    An intrinsic semiconductor has a conductivity of 250 S/m at 20 °C and 1100 S/m at 100 °C. What is the value of the energy gap $E_g$?

*Solution:*    Based on Eq. (39), $\ln(\sigma_1/\sigma_2) = E_g/2kT_2 - E_g/2kT_1$. Solving for $E_g$ gives $E_g = 2k \ln(\sigma_1/\sigma_2)/(T_1^{-1} - 1/T_2^{-1}) = 2 \times 86.2 \times 10^{-6}$ eV/K $\times$ $\ln(1100$ S/m/250 S/m)/(1/293 K $-$ 1/373 K) = 0.349 eV.    §

**Example 2.24**    Calculate the intrinsic conductivity of GaAs at 50 °C (= 323 K).

*Solution:*    For intrinsic GaAs, $n = p = n_i$, or $\sigma = n_i q(\mu_n + \mu_p)$. From Table 2.8, at 300 K, $\sigma = 1.8 \times 10^{12}$/m$^3$ $\times 1.6 \times 10^{-19}$ C $\times$ (0.85 m$^2$/V·s + 0.4 m$^2$/V·s) = $3.6 \times 10^{-7}$ S/m. Since $\sigma = \sigma_0 \exp(-E_g/2kT)$, at 50 °C, $\sigma = 3.6 \times 10^{-7}$ S/m $\times \exp[1.42 \times (1/300$ K $-$ 1/323 K)/(2 $\times$ 8.62 $\times$ $10^{-5}$ eV/K)] = $2.9 \times 10^{-6}$ S/m.    §

**Example 2.25**    In intrinsic CdTe, what fraction of the current is carried by electrons? Assume that in CdTe, $\mu_n = 0.07$ m$^2$/V·s, and $\mu_p = 0.007$ m$^2$/V·s.

*Solution:*    Since $n = p$, Eq. (9) gives $I_n/(I_n + I_p) = \mu_n/(\mu_n + \mu_p)$, where $I_n$ is the electron current and $I_p$ is the hole current. This ratio becomes 0.07/(0.07 + 0.007) = 0.909.    §

## Extrinsic Semiconductors

**Extrinsic semiconductors** contain dopants, and their properties are different from intrinsic semiconductors. The primary effect of adding dopants to a semiconductor is to change the electron and hole densities. These changes usually make the semiconductor more conducting. Dopant atoms are similar to the semiconductor atoms except for a difference in the valence (i.e., the number of electrons in the outermost energy orbit). Dopants become a part of the semiconductor lattice through substitution of the host atoms or ions since they are quite similar in size. In the energy band diagram, the dopant states are located slightly away from the band edges, as illustrated in Fig. 2.23.

Because of the difference in valence, a dopant atom, when it becomes ionized, either gives away or accepts an electron from the semiconductor. For instance, a Group V atom, when added to a Group IV semiconductor, provides an extra electron and becomes positively charged. It is called a **donor;** a donor increases the electron density. When a Group III atom is added to a Group IV semiconductor, however, it will receive an electron and become negatively charged. It is called an **acceptor;** an acceptor increases the hole density. In either case, because of Eq. (36), any increase in the electron density due to the donors will reduce the hole density, and vice versa. Thus, for a donor density equal to $N_D$ (/m$^3$), the electron and hole densities are given by

$$n = N_D \tag{40}$$

$$p = n_i^2/N_D$$

where    $n_i$ is the intrinsic carrier density (see Eq. (38)).

A similar set of equations result when the acceptor density is $N_A$ (/m$^3$):

$$p = N_A \tag{41}$$

$$n = n_i^2/N_A$$

In an extrinsic semiconductor, the dominant carrier (the one that exists in a larger quantity) is called the *majority carrier.* The minority carrier has a lower density. In

**Fig. 2.23**
Location of donor and
acceptor energy states

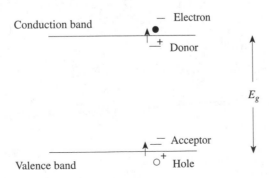

fact, if electrons dominate, the semiconductor is called an **n-type semiconductor;** if holes dominate, it is called a **p-type semiconductor.** These n-type and p-type semiconductors can interchange when a proper amount of dopants is added (for instance, it is possible to make an n-type semiconductor p-type by adding a sufficient amount of acceptors). Changes in the electron and hole densities also directly affect the internal energy of the electrons and hence the position of the Fermi level, $E_F$ (or the Fermi energy). In an n-type semiconductor, $E_F$ is closer to the conduction band edge, whereas in a p-type semiconductor, it is near the valence band edge. The overall effect is to maintain the validity of the Fermi-Dirac distribution function (Eq. (4)). Based on Eqs. (34) and (35), the Fermi levels in an n-type semiconductor and in a p-type semiconductor are given, respectively, by

$$E_{Fn} = E_c - kT \ln(N_c/N_D)$$

$$F_{Fp} = E_v + kT \ln(N_v/N_A) \tag{42}$$

Note that as $N_D$ (or $N_A$) approaches $N_c$ (or $N_v$), the second term in Eq. (42) approaches zero, and the Fermi levels will be at the band edges. A plot of the location of the Fermi levels in intrinsic n-type and p-type semiconductors is shown in Fig. 2.24.

In addition to changing the electron and hole densities, dopants also affect the carrier mobilities. Highly doped semiconductors suffer from impurity scattering and can result in substantially lower mobilities. Fig. 2.25 shows how the carrier mobilities vary with the dopant density in Si and GaAs. As stated in Section 2.5, ionized impurity scattering in semiconductors is more important at low temperature.

**Example 2.26**   An extrinsic Si semiconductor contains 100 parts per billion of Al by weight. What is the mole % of Al in Si?

***Solution:***   Let us assume that in 100 g of extrinsic Si, the weight of Al is $100 \times (100/1 \times 10^9)$ g $= 1 \times 10^{-5}$ g. The number of moles of Al is weight of Al/atomic weight of Al $= 1 \times 10^{-5}$ g/26.98 g/mol $= 3.71 \times 10^{-7}$ mol. The number of moles of Si is weight of Si/atomic weight of Si $= 100$ g/28.09 g/mol $= 3.56$ mol. Therefore, the mol % Al in Si is $3.71 \times 10^{-7}$ mol $\times 100\%/3.56$ mol $= 1.04 \times 10^{-5}\%$.   §

**Example 2.27**   In a p-type semiconductor, the Fermi level is 0.1 eV above its intrinsic value ($= E_g/2$). If the energy gap is 1.11 eV, what is the probability that an electron is found in the conduction band at 25 °C ($T = 298$ K)?

Fig. 2.24
Different positions of the
Fermi level

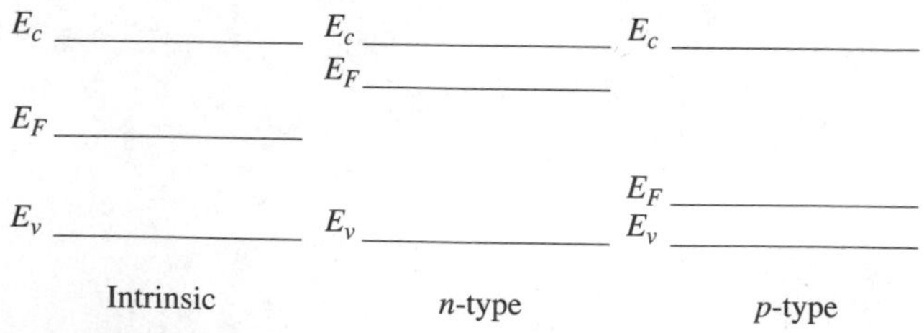

|  | Intrinsic | n-type | p-type |

$E_c$ ———  $E_c$ ———  $E_c$ ———

$E_F$ ——— (n-type)

$E_F$ ———

$E_F$ ——— (p-type)  $E_v$ ———

$E_v$ ———  $E_v$ ———

Fig. 2.25

Carrier mobilities versus dopant density

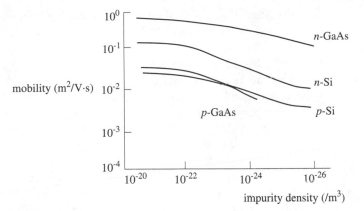

(*Source:* W. E. Beadle, J. C. C. Tsai, and R. D. Plummer, eds., *Quick Reference Manual for Silicon Integrated Circuit Technology* (New York: J. Wiley & Sons, Inc.) © 1985 by Bell Laboratories Inc.; H. C. Casey, Jr. and M. B. Panish, *Heterostructure Lasers—Part B, Materials and Operating Characteristics*. (New York: Academic Press). © 1978 by Bell Laboratories Inc. Reprinted with permission of Lucent Technologies Bell Laboratories.)

*Solution:* The Fermi level is $E_F = E_c - E_{Fi} + 0.1$ eV $= E_c - 1.11/2 + 0.1$ eV $= E_c - 0.45$ eV. Using Eq. (4), $f(E) = 1/\{\exp[(E - E_F)/(kT)] + 1\}$. Since $E_c - E_F = 0.45$ eV and $T = 298$ K, $f(E_c) = 1/\{\exp[0.45$ eV$/(86.2 \times 10^{-6}$ eV/K $\times 298$ K$)] + 1\} = 2.2 \times 10^{-8}$. §

**Example 2.28**   The extrinsic conductivity of a phosphorous-doped (*n*-type) Ge sample at 300 K is 60 S/m. Calculate the P concentration in parts per billion by weight if the P atoms are all ionized. Assume $\mu_n = 0.364$ m²/V·s.

*Solution:*   At 300 K,

$$N_D = n = \sigma/q\mu_n = 60 \text{ S/m}/(1.6 \times 10^{-19} \text{ C} \times 0.364 \text{ m}^2/\text{V·s})$$
$$= 1.03 \times 10^{21}/\text{m}^3$$

The % weight of P in Ge is $N_D \times$ atomic weight of P$/(N_{AG} \times$ density of Ge$) = 1.03 \times 10^{21}/\text{m}^3 \times 30.97$ g/mol$/(0.6023 \times 10^{24}/\text{mol} \times 5.32 \times 10^6$ g/m³$) = 9.96 \times 10^{-9}$, or 9.96 ppb.   §

**Example 2.29**   The energy of a photon is equal to $hc/\lambda_L$, where $h$ ($= 0.662 \times 10^{-34}$ J·s) is Planck's constant, $c$ ($= 3 \times 10^8$ m/s) is the speed of light, and $\lambda_L$ is the wavelength. **(a)** Calculate the wavelength of light necessary to promote an electron from the valence band to the conduction band in Si, and **(b)** repeat the calculations for promotion of an electron from an As donor in Si (the energy level of As is 0.049 eV below the conduction band edge) to the conduction band.

*Solution:*
   **(a)** From Table 2.8, for Si, $E_g = 1.12$ eV. The required wavelength is $\lambda_L = hc/E_g = 0.663 \times 10^{-33}$ J·s $\times 3 \times 10^8$ m/s$/(1.12$ eV $\times 0.16 \times 10^{-18}$ V/eV$) = 1.11 \times 10^{-6}$ m $= 1.11$ μm.

**(b)** Since $E_{donor} = 0.049$ eV, $\lambda_L = 0.663 \times 10^{-33}$ J·s $\times 3 \times 10^8$ m/s/($0.049$ eV $\times 0.16 \times 10^{-18}$ V/eV) $= 2.54 \times 10^{-5}$ m $= 25.4$ μm.    §

Note that roughly one-twentieth of the photon energy (or light with a much longer wavelength) is required to promote an electron from the As donor level to the conduction band as from the valence band.

There are many different ways that dopants can be added to a semiconductor. The simplest way is to add the dopants to the melt during the crystal growth, so the dopants become a part of the "background" impurity. The uniformity of the dopants in this case is reasonably good, but the technique lacks the flexibility of selective doping (i.e., to add dopants only to selective parts in the semiconductor). Dopants are more frequently added to a semiconductor by diffusion. During this process, the dopants are initially deposited on the semiconductor surface in the form of an oxide. Through a rapid high-temperature diffusion process, a highly doped surface layer is initially formed (in what is called predeposition or predep). These dopants are then redistributed (in a process called drive-in) into the semiconductor at a temperature of 900 °C or higher. The high temperature is needed to make the diffusion time reasonably short (see Section 1.9). Even at a high temperature, the drive-in time can be anywhere between a few minutes to several hours, depending on the required dopant profile.

Thermal diffusion is very suitable for selective doping (through windows formed on a protective coating deposited on top of the semiconductor) and for large-scale processing of semiconductors since many thin slices of wafers can be stacked together and processed in the same furnace (batch-mode operation). The dopant profile obtained by thermal diffusion is nonuniform and has a much higher dopant density near the surface, as shown in Fig. 2.26.

One way to produce a relatively uniform dopant profile in a semiconductor is to use an ion beam technique called *ion implantation*. In this case, the dopants are first ionized by thermal vaporization in a chamber and then fed to an accelerator (with a voltage difference of hundreds of kilovolts) to form an energetic ion beam. The ion beam is then allowed to bombard the semiconductor placed in a target chamber (end station) to a thickness proportional to the ion energy. The size of the ion beam is usually quite small (only a few square millimeters), and often the beam is allowed to scan across the semiconductor surface many times to achieve a better (dopant) uniformity. A schematic of the setup for ion implantation is shown in Fig.

**Fig. 2.26**
Diffusion and ion
implantation profiles

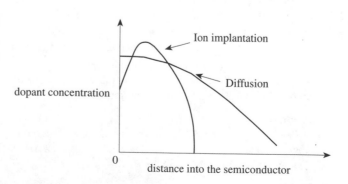

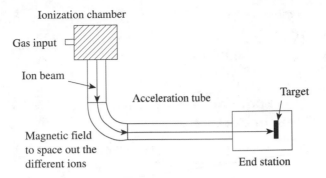

Fig. 2.27
Setup for ion implantation

2.27. High voltage and high vacuum in the chambers are required. In addition, post-implantation heat treatment is also needed to activate the dopants.

## Temperature Effect

At room temperature, the majority carrier density of an extrinsic semiconductor is often the same as the dopant density (if $n_i << N_D$ or $N_A$) since the energy required to ionize the dopants is quite small. As temperature increases, $n_i$ also increases, whereas $N_D$ and $N_A$ remain unchanged [this is the result of the exponential temperature dependence of $n_i$ (see Eq. (38))], and at a sufficiently high temperature, the intrinsic carrier density will dominate (i.e., $n_i >> N_D$ or $N_A$). This result causes $n = p = n_i$. As temperature decreases, however, some of the dopants may not have enough (thermal) energy to ionize, which can result in a reduction in the majority carrier density; i.e., the carriers are "frozen" in the dopants. The effect is called **freeze-out.** Even at a low temperature, the relationship between the majority carrier density and the minority carrier density (Eq. (36)) always holds. Fig. 2.28 shows how the electron density of an $n$-type semiconductor varies with temperature. At very low temperature, a semiconductor is similar to an insulator.

**Example 2.30**

An $n$-type semiconductor has donor density $N_D = 1 \times 10^{22}/m^3$. At what temperature will the intrinsic carrier density equal to $N_D$? We are given that $N_c = 2.46 \times 10^{25}/m^3$, $N_v = 1 \times 10^{25}/m^3$, and $E_g = 1.1$ eV, and they are assumed to be temperature independent.

Fig. 2.28
Log of electron concentration versus temperature

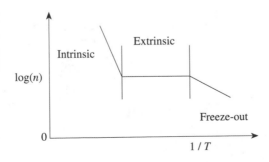

*Solution:*   From Eq. (38), $n_i = N_D = \sqrt{N_c N_v} \exp(-E_g/2kT)$, which gives

$$\begin{aligned}
T &= E_g/[2k \ln(\sqrt{N_c N_v}/N_D) \\
&= 1.1 \text{ eV}/[2 \times 1.38 \times 10^{-23} \text{ J/K} \times \ln(\sqrt{2.46 \times 10^{25}/\text{m}^3 \times 1 \times 10^{25}/\text{m}^3}/(1 \times 10^{22}/\text{m}^3)] \\
&= 867 \text{ K}. \quad \S
\end{aligned}$$

## Thin-Film and Amorphous Semiconductors

Thin-film semiconductors are becoming more important because of their low production cost and the possibility of depositing high-quality thin films on different types of substrates. Thin-film semiconductors can be deposited by evaporation, sputtering, and chemical-vapor deposition (CVD) (see Section 1.7). The quality of these films varies considerably from sample to sample, although the electrical properties are primarily determined by the trap states and defects present at the grain boundaries. Most of the trap states are located near the middle of the energy gap; because of their high density, they have the effect of **pinning** (fixing) the Fermi level. Pinning creates a band structure, as shown in Fig. 2.29.

The barrier height at the grain boundaries has been found to depend on the dopant density. At a low density, the barrier height remains low since very few of the (charged) carriers will be trapped at the grain boundaries. Increasing the dopant density increases the number of trapped carriers, and the barrier height increases proportionally. Finally, when all the trap states are filled, the Fermi level at the grain boundaries is free to move up or down the energy gap, and the barrier height is reduced to zero since the charges of the trapped carriers are now fully compensated for by the charges on the dopant ions. In this case, the grain boundaries are no different from the grains themselves. Normally, the density of the trap states at the grain boundaries ranges from $10^{14}$ to $10^{16}/\text{m}^2$. Because of the presence of an intergranular barrier at a low dopant density, the conductivity of a thin-film semiconductor is given by

$$\sigma = \sigma_{go} \exp[-\Phi_B/(kT)] \tag{43}$$

where   $\sigma_{go}$ is the conductivity of the grains in S/m, and

$\Phi_B$ is the barrier height in eV.

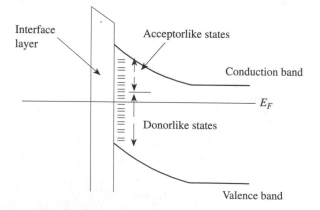

**Fig. 2.29**
Surface states at grain boundaries

Similarly, the carrier mobility of a thin-film semiconductor can be expressed as

$$\mu = \mu_0 \exp[-\Phi_B/(kT)] \tag{44}$$

where    $\mu_0$ is the grain mobility in $m^2/V\cdot s$.

Fig. 2.30 shows how the carrier mobility of a polycrystalline Si thin film varies with doping.

Grain boundaries are the locations where gas absorption usually takes place. Absorption of $H_2$, for instance, has been observed to lower the barrier height of polycrystalline Si, which is a very important factor in the formation of low-resistivity Si thin films. Defects are also expected to be present in significant quantities near the grain boundaries because of the techniques used in thin-film deposition. These defects are known to form electron and hole states deep inside the energy gap and in some semiconductors have been found to extend the range of light absorption to values much smaller than the energy gap. In GaP, for instance, the threshold energy for light absorption can be reduced to as low as 0.8 eV (compared with a value of 2.27 eV observed in single crystal). Defects also reduce the lifetime of carriers, which is known to affect the performance of thin-film optical devices.

Semiconductor thin films with no observable grain size are called *amorphous thin films*. They are widely used in low-cost display devices and solar cells. High-quality amorphous-Si ($\alpha$-Si) thin films can be formed by glow discharge using $SiH_4$ as the source; thin plates of $\alpha$-Si as large as 1 $m^2$ have been reported. These films are labeled as $\alpha$-Si:H because they are rich in $H_2$. $H_2$ serves the purpose of removing the dangling bonds in $\alpha$-Si (see Section 1.8) and results in a relative low density of defect states. $\alpha$-Si:H has a (direct) energy gap of 1.7 eV, and the values of the carrier mobilities are around $10^{-3}$ $m^2/V\cdot s$.

Fig. 2.31 shows the distribution of the defect states in the energy gap of undoped $\alpha$-Si. These defect states are known to extend continuously across the energy gap and have peak densities near the band edges. As in the case of a polycrystalline semiconductor, the location of the Fermi level in an intrinsic amorphous thin film is pinned to the middle of the energy gap (in fact, slightly closer to the conduction band edge). Considerable amounts of dopants are needed to make $\alpha$-Si conducting because of the presence of the large quantities of defect states. In an active device

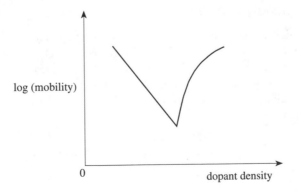

**Fig. 2.30**
Log of mobility versus
dopant density

log (mobility)

0                                                        dopant density

Fig. 2.31
Distribution of defect
states in α-Si

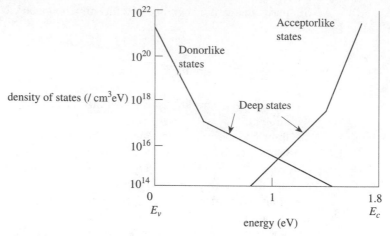

(*Source:* W. B. Jackson, R. J. Nemanich, and N. M. Amer, "Energy dependence of the carrier mobility-lifetime product in hydrogenated amorphous silicon," *Phy. Rev. B 27,* no. 8, p. 4860. Reprinted by permission of the American Physical Society.)

made with α-Si, the Fermi level can sometimes be displaced through carrier injection from a nearby region of high-dopant density. This is one possible mechanism for increasing the conductivity of α-Si during device operation.

## 2.8 OTHER CONDUCTION MECHANISMS

In Section 2.5, we studied the drift velocity of the electrons and showed how charge flow in the presence of an electric field can give rise to current conduction in solids. In this section, we consider other types of current-conduction mechanisms, namely, diffusion current, generation current, and recombination current. These currents exist primarily in semiconductors; under varying physical conditions, each can give rise to significant current flow.

### Diffusion Current

Previously, we showed how a solid under an applied voltage will produce a current (Eq. (7)). Current also flows in the presence of a carrier density gradient (with no applied voltage). The migration of electrons and holes from high density to low density is the manifestation of the natural process called *diffusion* (see Section 1.9). Diffusion of the (charged) carriers obeys Fick's first law, and in one dimension is written as

$$\phi_n = -D_n \, dn/dx \qquad\qquad (45)$$

$$\phi_p = -D_p \, dp/dx$$

where   $\phi_n$ and $\phi_p$ are the electron flux and hole flux in $m^2/s$, respectively,

$D_n$ and $D_p$ are the electron diffusivity and hole diffusivity in $m^2/s$, respectively, and

$dn/dx$ and $dp/dx$ are the carrier density gradients $/m^4$.

According to Eq. (45), the motion of the electrons or holes under a carrier density gradient is independent of the sign of the charge, and both electrons and holes will move along the carrier density gradient. The resulting electron and hole currents, however, are in opposite directions, as illustrated in Fig. 2.32.

The total diffusion current $I_{\text{diff}}$ is, therefore, the sum of the individual diffusion currents and is given by

$$I_{\text{diff}} = qA_{cs}(-\phi_p + \phi_n) = -qD_pA_{cs}\,dp/dx + qD_nA_{cs}\,dn/dx \qquad (46)$$

If $dp/dx = dn/dx$, the electrons and the holes move in the same direction, and the total diffusion current depends on the difference between $D_p$ and $D_n$. Normally, diffusion currents exist only in semiconductors but not in metals because it is difficult to create charge separation in a highly conducting medium. There are many important semiconductor devices that make use of the diffusion currents; one outstanding example is the p-n junction (see Chapter 3), where the electron and hole densities are allowed to vary appreciably from the n-region to the p-region.

**Example 2.31**   A hypothetical n-type semiconductor has a uniform donor density gradient $\Delta N_D/\Delta x$ along its long axis. Use this to show that the electron mobility is proportional to the electron diffusivity $D_n$. Assume that the donors are fully ionized.

*Solution:*   Initially, both diffusion current and drift current are present in the semiconductor. These are given by $I_{\text{diff}} = qD_nA_{cs}\,\Delta n/\Delta x$ and $I_{\text{drift}} = -qn\mu_n\acute{E}A_{cs}$, respectively. In the steady state, there is no current passing through the semiconductor, i.e., $I_{\text{diff}} = -I_{\text{drift}}$, or $\mu_n = D_n\,\Delta n/\Delta x/(n\acute{E})$. Based on Eq. (34), $n = N_c\exp[-(E_c - E_F)/kT]$. Since $\Delta E_c/\Delta x = q\acute{E}$ and $\Delta E_F/\Delta x = 0$ (the average internal energy of the electrons has to be the same throughout), this leads to $\Delta n/\Delta x = -qn\acute{E}/(kT)$, or $\mu_n = qD_n/kT$.   **§**

**Fig. 2.32**
Diffusion of electrons and holes

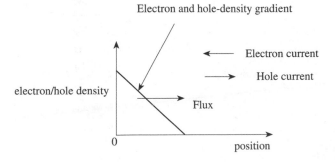

Electron and hole-density gradient

electron/hole density

Electron current

Hole current

Flux

0

position

This relationship is known as the **Einstein relationship,** and it relates carrier mobility to the diffusivity. A similar expression exists for holes.

## Generation and Recombination Currents

Generation and recombination currents exist in semiconductors. **Carrier generation** may be viewed as the creation of electrons and holes, and **carrier recombination** is their annihilation (i.e., the recovery of the valence electrons by the host atoms). This is illustrated schematically in Fig. 2.33. In both cases, exchange of energy is involved. In general, at equilibrium the generation rates for electrons and holes are exactly equal to their recombination rates, and there are no net changes in the carrier densities. Such a dynamic balance exists only as long as both the electron density $n$ and the hole density $p$ remain constant.

Let us consider a region in a semiconductor where the electrons and the holes are depleted, i.e., $n \ll n_0$, $p \ll p_0$ (the subscript 0 stands for the equilibrium values). In this case, the thermal generation rates will exceed the recombination rates, and additional electrons and holes will be created. If somehow these carriers are removed from the semiconductor before they recombine (for instance, they are allowed to pass out of the semiconductor through an external circuit), a current called the generation current, $I_g$, will result. $I_g$ (in A) is given by

$$I_g = qn_i WA_{cs}/\tau_g \tag{47}$$

where    $q$ is the electron charge in C,

$n_i$ is the intrinsic carrier density of the semiconductor /m$^3$,

$W$ is the width of the depletion region in m,

$A_{cs}$ is the cross-sectional area of the depletion region in m$^2$, and

$\tau_g$ is the **generation lifetime** in s.

According to Eq. (47), the generation current depends on both the volume of the depletion region (i.e., $WA_{cs}$) and the generation lifetime, $\tau_g$. In the energy band

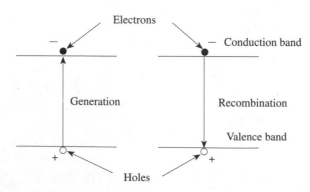

**Fig. 2.33**
Generation and recombination process

diagram, generation may be viewed as electrons being thermally excited from the valence band to the conduction band (with the simultaneous creation of holes in the valence band). In some semiconductors, electrons and holes are also excited from impurities (besides band-to-band excitation) even though the number of (charged) carriers so created will be less. The generation lifetime, $\tau_g$, in Eq. (47) can be considered as a measure of the average time needed for the electrons and holes to be formed, and it depends on the amount of energy required. For instance, $\tau_g$ is shorter in a semiconductor with a smaller energy gap (this, in fact, is sometimes qualitatively linked to the uncertainty principle: $\Delta E\ \Delta t \approx h'$, where h' is the Planck constant/$2\pi$).

Recombination of electrons and holes occurs naturally in a semiconductor and is a checking mechanism against excessive carrier generation. If the excess carriers originate from the conduction band or the valence band, the process is called band-to-band recombination and is the prevalent recombination mechanism found in intrinsic semiconductors. Electrons and holes also recombine via impurities. In this case, the majority carrier is first trapped at the impurity site, and recombination takes place when a minority carrier is captured. Fig. 2.34 shows the trapping of an electron and its subsequent recombination when a hole is also captured.

As expected, the recombination rate is dependent on the capture lifetime of the hole (the minority carrier). Because of the smaller energy steps involved in recombination via impurities, this type of recombination mechanism will normally prevail over band-to-band recombination. In fact, the most effective recombination centers are located near the middle of the energy gap since the energy required to capture either type of carrier is about the same. Recombination of carriers from the different regions in a semiconductor can also give rise to current flow, as in the case of carriers that are injected into the semiconductor from an external circuit. This process is known as recombination current. The equation for recombination current is similar in form to that of the generation current (Eq. (47)) and is given by

$$I_r = -q\ \Delta n_{\min} A_{cs}/\tau_r \tag{48}$$

where    $\Delta n_{\min}$ is the excess minority carrier density /m$^3$, and

   $\tau_r$ is the **recombination lifetime** of the minority carriers in s.

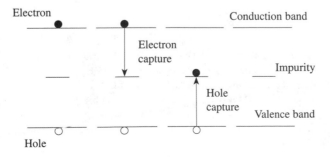

**Fig. 2.34**
Electron and hole
recombination via an
impurity

As expected, $\tau_r$ is the average time needed for the minority carriers to be captured (and recombine).

For band-to-band recombination, $\tau_r$ is inversely proportional to the dopant density $N_{\text{dopant}}$ (i.e., $N_D$ or $N_A$); the same is true when recombination is via impurities, provided that $N_{\text{dopant}}$ is replaced by the impurity density $N_I$. In the extreme case, when $n$ and $p$ are both large ($>>N_I$), as in high injection, recombination will occur independent of the density of the recombination centers but through direct recombination of the carriers. The excess energy generated in the recombination process is most frequently dissipated as heat and is sometimes called *Auger recombination.* In each Auger recombination, two of the majority carriers are included and $\tau_r$ is proportional to $1/N^2_{\text{dopant}}$. Fig. 2.35 shows how the carrier lifetimes vary with $N_I$ (or $N_{\text{dopant}}$) in Si.

Recombination also occurs near a semiconductor surface and is characterized by the **surface recombination velocity** $S_v$ (in m/s). Surface recombination lowers the surface carrier density, and $S_v$ is defined as

$$S_v = I_{\text{srec}}/(q\,\Delta n_s A_{cs}) \qquad (49)$$

where    $I_{\text{srec}}$ is the surface recombination current in A,

$\Delta n_s$ is the excess (minority) carrier density at the semiconductor surface /m³, and

$A_{cs}$ is the cross-sectional area in m².

**Fig. 2.35**
Carrier lifetimes versus
dopant density

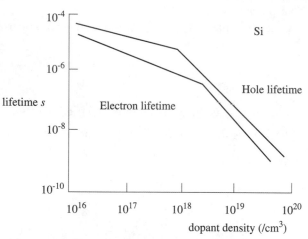

(*Source:* Reprinted from J. G. Fossum, R. P. Mertens, D. S. Lee, and J. F. Nus, "Carrier recombination and lifetime in highly doped silicon," *Solid-State Electronics* 26 (1983): 570, and M. S. Tyagi and R. Van Overstraeten, "Minority carrier recombination in heavily doped silicon," *Solid-State Electronics* 26 (1983): 587, with kind permission from Elsevier Science Ltd., The Boulevard, Langford Lane, Kidlington OX5 1GB, UK.)

$S_v$ is proportional to the density of surface recombination centers $N_{st}$ (/m$^2$) and is given by

$$S_v = \sigma' v_{th} N_{st} \qquad (50)$$

where    $\sigma'$ is the capture cross section of the recombination centers in m$^2$, and

$v_{th}$ is the thermal velocity of the carriers in m/s (see Section 2.2).

In summary, the overall current flow in a semiconductor is the sum of the drift currents, the diffusion currents, the generation currents, and the recombination currents. Under different physical conditions, each of these currents may prevail. Except for the drift currents, which depend on the strength of the electric field, the currents occur naturally under specific conditions and become significant only at nonequilibrium.

**Example 2.32**    Compute the generation current per unit volume in GaAs if the generation lifetime is 10 μs. Assume $n_i = 1.8 \times 10^{12}$/m$^3$.

*Solution:*    Based on Eq. (47), $I_g$ (per unit volume) $= q n_i / \tau_g = 1.6 \times 10^{-19}$ C $\times$ $1.8 \times 10^{12}$/m$^3$/$(1 \times 10^{-5})$ s $= 2.88 \times 10^{-2}$ A/m$^3 = 28.8$ mA/m$^3$.    §

**Example 2.33**    Compute the surface recombination current if the surface (minority) carrier density in a $n$-type Si sample is increased by a factor of $10^4$ (through carrier injection). Assume $N_D = 1 \times 10^{22}$/m$^3$, $n_i^2 = 2.10 \times 10^{32}$/m$^6$, $S_v = 1$ m/s, and $A_{cs} = 1 \times 10^{-6}$ m$^2$.

*Solution:*    As given, $\Delta n_s = 10^4 n_i^2 / N_D = 1 \times 10^4 \times (1.45 \times 10^{16}$/m$^3)^2 / 1 \times 10^{22}$/m$^3 = 2.10 \times 10^{14}$/m$^3$. Based on Eq. (49), $I_{srec} = q S_v \Delta n_s A_{cs} = 1.6 \times 10^{-19}$ C $\times 1$ m/s $\times 2.10 \times 10^{14}$/m$^3 \times 1 \times 10^{-6}$ m$^2 = 3.36 \times 10^{-11}$ A $= 33.6$ pA.    §

## 2.9  VELOCITY SATURATION AND NEGATIVE-RESISTANCE EFFECT

Equation (15), which suggests a linear relationship between the drift velocity and the electric field ($\acute{E} = V_a / L$) for the electrons, is true only when the applied voltage is small. In most semiconductors, the electron velocity actually saturates once the electric field exceeds a value of around $1 \times 10^6$ V/m, as shown in Fig. 2.36.

When $\acute{E}$ is large, both $\langle \tau_e \rangle$ and $m_e^*$ become dependent on $\acute{E}$; these electrons are called *hot* electrons. **Hot electrons** are electrons that have acquired more energy from the electric field than they can dissipate, so their motion no longer obeys the equations derived in Section 2.5. In addition, hot electrons also exhibit a lower mobility, which is reflected in the saturation of the drift velocity. An empirical equation for $v_{drift}$ (in m/s) taking into account **velocity saturation** is given by

$$v_{drift} = \mu_n \acute{E} / \sqrt{1 + (\mu_n \acute{E} / v_s)^2} \qquad (51)$$

Fig. 2.36
Electron velocities versus
electric field

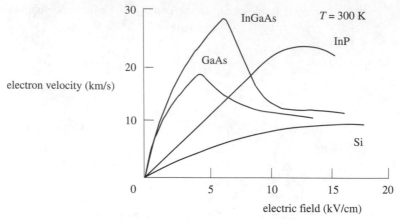

(Source: S. M. Sze, Semiconductor Device Physics and Technology. ©1985 by Bell
Laboratories Inc. Reprinted by permission of John Wiley & Sons, Inc.)

where    $\mu_n$ is the low-field electron mobility in $m^2/V\cdot s$,

$\acute{E}$ is the electric field in V/m, and

$v_s$ is the saturation velocity in m/s.

If $\acute{E}$ is small, $v_{\text{drift}} = \mu_n \acute{E}$. For a large $\acute{E}$, $v_{\text{drift}} = v_s$; the value of $v_s$ for most
semiconductors is around $1 \times 10^5$ m/s. As expected, $v_s$ also changes with tempera-
ture. The temperature dependence of $v_s$ in GaAs, for instance, is given by

$$v_s(T) = 2.4 \times 10^5/[1 + 0.8 \times \exp(T/600)] \qquad \textbf{(52)}$$

Using Eq. (52), it can be shown that indeed $v_s \approx 1.0 \times 10^5$ m/s at 300 K. In
Group III–Group V compound semiconductors such as GaAs, the electron mobility
also changes in a high electric field, but for an entirely different reason. In a high
electric field, those electrons normally residing in the low-field conduction band
will move into the high-field conduction band as their energy increases. This situation
is illustrated in the energy band diagram of GaAs shown in Fig. 2.37. The low-
field conduction band is located at the so-called $\Gamma$ valley (a notation used by
the crystallographers to identify the crystal orientations), whereas the high-field
conduction bands are at the $X$ and $L$ valleys. Because of the crystal structure, electron
mobilities in the different conduction bands are different. In fact, in the $\Gamma$ valley,
the electrons have a higher mobility, whereas in the $X$ and $L$ valleys, they have a
lower mobility. Thus, those electrons moving from the $\Gamma$ valley into the $X$ and $L$
valleys will exhibit a differential negative-resistance effect. This is reflected in the
electron velocities for GaAs, as shown in Fig. 2.36. The phenomenon is called the
Gunn effect; this effect is known to produce oscillations (the spontaneous building
up of (charged) carriers and current pulses in semiconductors). It is used extensively
in the development of microwave power devices.

Fig. 2.37

Energy band diagram for GaAs

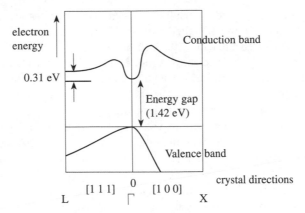

**Example 2.34**   Estimate the effective mobility of GaAs **(a)** when $\acute{E} = 100$ V/m and **(b)** when $\acute{E} = 1 \times 10^5$ V/m. Assume $\mu_n = 0.8$ m²/V·s and the saturation velocity is $1.0 \times 10^5$ m/s.

**Solution:**   Based on Eq. (51), $\mu_{\text{eff}} = v_{\text{drift}}/\acute{E} = \mu_n/[1 + (\mu_n\acute{E}/v_s)^2]^{1/2}$.
   **(a)** When $\acute{E} = 100$ V/m,

$$\mu_{\text{eff}} = 0.8 \text{ m}^2/\text{V·s}/[1 + (0.8 \text{ m}^2/\text{V·s} \times 100 \text{ V/m}/1.0 \times 10^5 \text{ m/s})^2] \text{ m}^2$$
$$\approx 0.8 \text{ m}^2/\text{V·s}$$

   **(b)** When $\acute{E} = 1 \times 10^6$ V/m,

$$\mu_{\text{eff}} = 0.8 \text{ m}^2/\text{V·s}/[1 + (0.8 \text{ m}^2/\text{V·s} \times 1 \times 10^5 \text{ V/m}/(1.0 \times 10^5 \text{ m/s})]^2$$
$$= 0.44 \text{ m}^2/\text{V·s} \quad \S$$

Note that there is reduction in the observed mobility by nearly a factor of 2.

**Example 2.35**   Suggest what happens to a current passing through a semiconductor as the electron mobility reduces to a lower value in a high electric field.

**Solution:**   According to Eq. (12), the current $I$ is proportional to $v_{\text{drift}}$. At low field, $I$ increases with increasing $v_{\text{drift}}$, and the resistance is a constant. As $v_{\text{drift}}$ saturates, $I$ no longer increases with the electric field, and the resistance goes up ($I$ also saturates). This changeover from low resistance to high resistance produces a differential negative-resistance effect.   $\S$

## 2.10   INSULATORS

Unlike conductors, insulators cannot pass a significant current since there are too few carriers. Instead, the response of an insulator to an applied voltage is through an effect known as *polarization*. **Polarization** occurs when positive and negative charges in the insulator are physically separated from each other. These separated charges are called dipoles. Dipoles can be the result of either an applied voltage or some form of stress-related structural deformation. Both permanent and induced dipoles are found in insulators. **Permanent dipoles** are fixed ion pairs, and they respond only to an applied voltage through changes in their orientation. Permanent dipoles often interact with each other to increase the effect of polarization, as in the case of a ferroelectric solid. **Induced dipoles,** on the other hand, are charge separations not normally found in an insulator. An example of induced polarization due to lattice displacement is shown in Fig. 2.38. In this case, the displacement of the lattice ions gives rise to a dipole moment.

Both dc (direct current) and ac (alternating current) polarization are possible in insulators. Dc polarization is due to static charge separation and is associated with such effects as piezoelectricity (stress-induced voltage) and pyroelectricity (heat-induced voltage). Ac polarization occurs when an alternating voltage is present and the array of dipoles in the insulator oscillates with the ac electric field. An ac current results from the backward-and-forward movement of the charges. In most insulators, there is a time delay in the ac current response, and some of the energy is stored in the polarization field. Ac polarization in insulators is not restricted to low-frequency radio waves but can also occur at optical frequencies ($\sim 10^{12}$–$10^{13}$ Hz). At these high frequencies, the interactions are better known as *optoelectronic effects;* there are numerous examples where high-frequency ac voltages are used to modulate the optical properties of solids or even to induce light emission (see Chapter 5).

## Dielectric Properties of an Insulator

The dielectric properties of an insulator are best described by the capacitance effect. In general, **capacitance** $C$ is measured in farads (F) and is defined as

$$C = QA_{cs}/V_a \qquad (53)$$

where    $Q$ is the charge density in C/m$^2$,

$A_{cs}$ is the area of the capacitor in m$^2$, and

$V_a$ is the voltage across the capacitor in V.

The amount of charge stored in a capacitor depends on the geometry of the **capacitor** and the dielectric properties of the insulator. The capacitance $C$ for a parallel-plate capacitor is given by

**Fig. 2.38**
Induced ionic polarization
due to applied field

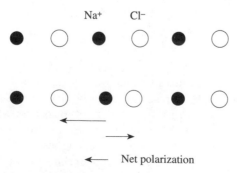

Na$^+$    Cl$^-$

$\longleftarrow$    Net polarization

$$C = \epsilon_i A_{cs}/d \tag{54}$$

where  $\epsilon_i$ is the permittivity of the insulator in F/m, and

  $d$ is the separation of the capacitor plates in m.

The **permittivity** of a solid reflects its ability to store charges and for an insulator, it is given by

$$\epsilon_i = \epsilon_0 \epsilon_r \tag{55}$$

where  $\epsilon_0$ ($- 8.85 \times 10^{-12}$ F/m) is the permittivity of vacuum, and

  $\epsilon_r$ is the **relative permittivity** or the **dielectric constant** of the insulator.

Since $\epsilon_r > 1$, Eq. (54) suggests that the capacitance of the parallel-plate capacitor increases with the presence of an insulator. Dielectric properties of an insulator are directly linked to the alignment of the dipoles. Schematically, a dipole may be viewed as a pair of positive and negative charges as shown in Fig. 2.39. If the dipoles respond to the electric field by changing their orientations or the separation, the insulator becomes polarized. As we shall see, the more intense the polarization is, the larger will be the relative permittivity of the insulator. The dipoles in an insulator are characterized by the **dipole moment** $p_d$ (in C·m), which measures the strength of the dipoles and is given by

$$p_d = \alpha' \acute{E} \tag{56}$$

where  $\alpha'$ is the polarizability in F·m$^2$, and

  $\acute{E}$ is the electric field in V/m.

Physically, **polarizability** may be viewed as a measure of the extent of the alignment of the dipoles. In a parallel-plate capacitor, the dipoles within the insulator actually nullify each other, except near the surface where the uncompensated dipoles are located. Near the surface, the dipole charges will offset a portion of the charges normally residing on the capacitor plates and allow additional charges to flow in from the voltage supply. This effectively increases the storage capacity of the parallel-

**Fig. 2.39**
An array of dipoles

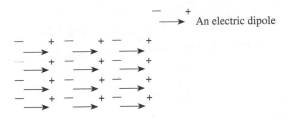

An electric dipole

plate capacitor, as illustrated in Fig. 2.40. Assuming that the excess surface charge density in the presence of an insulator is given by $\Delta Q$ (in $C/m^2$), then the capacitance can be written as

$$C = (Q_0 + \Delta Q)A_{cs}/V_a = C_0 A_{cs}(1 + \Delta Q/Q_0) \tag{57}$$

where    $Q_0$ is the surface charge density without the insulator in $C/m^2$, and

$C_0 \, (= Q_0 A_{cs}/V_a)$ is the free space capacitance of the parallel-plate capacitor in $F/m^2$.

As expected, $\Delta Q$ is proportional to the density of the dipoles, $N_{dipole}$ $(/m^3)$, and it is given by

$$\Delta Q = N_{dipole} p_d = N_{dipole} \alpha' \acute{E} \tag{58}$$

Using Eqs. (53) and (54) and the fact that $\acute{E} = V_a/d$, Eq. (58) can be written as

$$\Delta Q/Q_0 = N_{dipole}\alpha'/\epsilon_0 \tag{59}$$

When Eq. (59) is substituted into Eq. (57) and making use of Eq. (54), Eq. (55) becomes

$$\epsilon_i = \epsilon_0(1 + N_{dipole}\alpha'/\epsilon_0) \tag{60}$$

or

$$\epsilon_r = 1 + N_{dipole}\alpha'/\epsilon_0. \tag{61}$$

A more rigorous calculation will give

$$\epsilon_r = (1 + 2\, N_{dipole}\alpha'/3\epsilon_0)/(1 - N_{dipole}\alpha'/3\epsilon_0)$$

**Fig. 2.40**
A parallel-plate capacitor

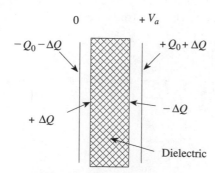

Equation (61) shows that $\epsilon_r$ does indeed increase with an increase in the dipole density. Insulators are therefore more effective in charge storage if they have either a high density of dipoles or if the dipoles readily align in the direction of the electric field. Table 2.10 lists the values of the relative permittivity, or dielectric constant, of common insulators.

In addition to the charge storage effect, a capacitor also conducts an ac current. Under an applied ac voltage, the dipoles in the capacitor oscillate periodically. This oscillation causes charging and discharging of the capacitor plates and results in an ac current flowing in the circuit. The ac current, $I_{ac}$, (in A) is given by

$$I_{ac} = \Delta Q A_{cs}/\Delta t \tag{62}$$

where  $\Delta Q$ is the incremental charge density that flows into and out of the capacitor plates due to the ac voltage in $C/m^2$,

$A_{cs}$ is the area of the parallel-plate capacitor in $m^2$, and

$\Delta t$ is the incremental time in s.

Using Eq. (53) (i.e., $\Delta V = \Delta Q A_{cs}/C$), we can write $I_{ac}$ as

$$I_{ac} = C\Delta V_a/\Delta t = C \, dV_a/dt \tag{63}$$

where  $dV_a/dt$ is the time rate of change of the applied voltage in V/s, and

$C$ is the capacitance in F.

If the applied voltage is sinusoidal, i.e., $V_a = V_0 \sin(\omega t)$, where $V_0$ is the peak voltage and $\omega$ is the angular frequency, then Eq. (63) will give a current of the form $I_{ac} = CV_0 \, \omega \cos(\omega t)$. This equation shows a time delay corresponding to a phase difference of $\pi/2$ rad between the current and the applied voltage, as illustrated in Fig. 2.41. Physically, the maximum ac current will occur if the voltage difference between the capacitor plates and the voltage supply is at its peak (or when the voltage lag is at its maximum).

**Example 2.36**  A parallel-plate capacitor has the following properties: $A_{cs} = 2 \times 10^{-3}$ $m^2$; $d = 1 \times 10^{-2}$ m; $V_a = 100$ V, and $\epsilon_r = 6$. (a) Find the capacitance. (b) If there are $1 \times 10^{15}$ dipoles/$m^3$ in the insulator, what is the polarizability? (c) Determine the amount of surface charge on the capacitor plates. (d) Repeat part (c) in the absence of the insulator. (e) What is the amount of polarization charge?

**Table 2.10**  Values of the relative permittivity, or dielectric constant, of common insulators

| Material | $\epsilon_r$ |
| --- | --- |
| Polymers | 1–2 |
| Water | 80 |
| Ceramics, glasses | 4–10 |
| Ferrites | $10^4$ |

Fig. 2.41
Current and voltage
relationship of a capacitor

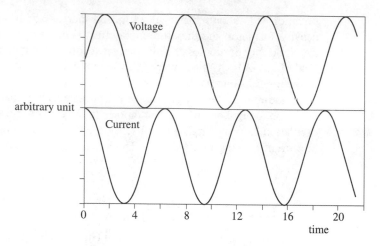

*Solution:*

(a) Based on Eq. (54), $C = \epsilon_i A_{cs}/d = 8.86 \times 10^{-12}$ F/m $\times 6 \times 2 \times 10^{-3}$ m$^2$/(1 $\times 10^{-2}$ m) = $1.06 \times 10^{-11}$ F.

(b) Based on Eq. (61), $\epsilon_r = 1 + \alpha' N_{dipole}/\epsilon_0 = 6$. Therefore, $\alpha' = (6 - 1) \times 8.85 \times 10^{-12}$ F/m/(1 $\times 10^{15}$/m$^3$) = $4.43 \times 10^{-26}$ F·m$^2$.

(c) $Q = CV_a = 1.06 \times 10^{-11}$ F $\times$ 100 V = $1.06 \times 10^{-9}$ C.

(d) $Q_0 = C_0 V = Q/\epsilon_r = 1.06 \times 10^{-9}$ C/6 = $0.176 \times 10^{-9}$ C.

(e) $\Delta Q = Q - Q_0 = (1.06 - 0.176) \times 10^{-9}$ C = $0.88 \times 10^{-9}$ C.  §

**Example 2.37**    If an insulator has $1 \times 10^{17}$ dipoles/m$^3$ and a polarizability of $1 \times 10^{-28}$ F·m$^2$, what is its relative permittivity?

*Solution:*    Based on Eq. (61), $\epsilon_r = 1 + \alpha' N_{dipole}/\epsilon_0 = 1 + 1 \times 10^{17}$/m$^3 \times 1 \times 10^{-28}$ F·m$^2$/8.86 $\times 10^{-12}$ F/m = 2.13.  §

Note that if we use the more rigorous calculations, then $\epsilon_r = 2.8$.

## Electronic Properties of Dipoles

We have shown that the dielectric properties of an insulator depend on the dipole density. In general, there are three types of dipoles found in insulators: (1) dipoles that are due to permanent charge separation of atoms or groups of atoms; (2) dipoles that arise from the polarization of the ions or the displacement of the ions; and (3) dipoles that exist when the valence electrons are polarized with respect to the nuclei. In insulators, one or more of these three types of polarization are present. Fig. 2.42 shows the frequency-dependence of the different types of polarization.

Permanent dipoles are responsible for what is known as *orientational polarization,* and they respond to dc or low-frequency electric fields by changing their orientations. The movement of these dipoles often involves frictional forces that can result in energy loss (due to damping) and heat dissipation. The effect of orientational polar-

Fig. 2.42
Polarizability versus
frequency of electric field

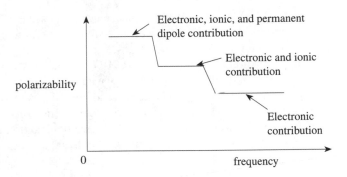

ization can be described by the following simple model. We assume two possible orientations of a dipole with internal energies given by $-\Delta E$ and $+\Delta E$ in joules. The dipole may be a polar molecule with two dipole states that correspond to parallel and antiparallel alignment with the applied electric field. If the probability that a dipole is parallel to the electric field is $\varpi$, then $1 - \varpi$ is the probability that the direction of the dipole is antiparallel. Based on the **Boltzmann distribution,** the ratio of these two probabilities will be exponentially dependent on the energy difference (i.e., $2\Delta E$), and

$$(1 - \varpi)/\varpi = \exp[-2\Delta E/(kT)] \tag{64}$$

Rearranging Eq. (64) gives

$$\varpi = 1/\{1 + \exp[-2\Delta E/(kT)]\} \tag{65}$$

Thus, the difference in the dipole moment $\Delta p_d$ (in C·m)—which is equal to the individual dipole moment, $p_d$, times the difference between the two probabilities—is given by

$$\Delta p_d = p_d[\varpi - (1 - \varpi)] = p_d(2\varpi - 1) \tag{66}$$

where    $p_d$ is the dipole moment in C·m.

Since $\Delta E = p_d\acute{E}$, where $\acute{E}$ is the electric field and assuming $\Delta E$ is small compared with $kT$, so that $\varpi = [1 + 2\Delta E/(kT)]/2$, it can be shown that

$$\Delta p_d = 2p_d\Delta E/kT = 2p_d^2\acute{E}/kT \tag{67}$$

Combining Eq. (67) with Eq. (56), the polarizability $\alpha'$ (in F·m$^2$) is given by

$$\alpha' = 2p_d^2/kT \tag{68}$$

In this model, we have shown that the orientational polarizability is proportional to the square of the dipole moment and is inversely proportional to temperature.

**Orientational dipoles,** therefore, become progressively randomized at a higher temperature. As an example, we examine the polarization of water molecules ($H_2O$), which is shown in Fig. 2.43.

In the natural state, the O atom in $H_2O$ is negatively charged with respect to the H atoms, and the two H atoms are separated at an angle of 105°. Without any electric field, the water molecules are randomly placed, and there is no polarization. However, in the presence of an electric field $\acute{E}_x$ (in V/m), the water molecules will align with the electric field to give a nonzero polarization along their midline. The energy $E(\Theta)$ (in J) of a single water molecule aligned at an angle $\Theta$ to the electric field $\acute{E}_x$ (in V/m) is given by

$$E(\Theta) = -p_0\acute{E}_x\cos(\Theta) \tag{69}$$

where  $p_0$ is the dipole moment of the water molecule in C·m.

According to Eq. (69), the lowest energy occurs when $\Theta = 0$, or when all the water molecules are aligned parallel to the electric field. Natural randomization, however, invalidates this, and instead the orientation of the water molecules follows the Boltzmann distribution, which in this case is given by

$$N_{\text{water}} = A_1\exp[-E(\Theta)/kT] \tag{70}$$

where  $N_{\text{water}}$ is the fraction of water molecules making an angle $\Theta$ with $\acute{E}$,

$A_1$ is a constant,

$k$ is the Boltzmann constant in J/K, and

$T$ is the absolute temperature in K.

Based on Eq. (70), the average dipole moment, $p_{\text{av}}$, is given by

$$p_{\text{av}} = \int_0^\pi [N_{\text{water}}p_0] \, d\Theta/\pi \approx p_0^2\acute{E}_x/(3kT) \tag{71}$$

Equation (71) suggests that after polarization, not all the water molecules will be aligned to the electric field. In fact, only a fraction equal to $p_0\acute{E}_x/(3kT)$ will do so.

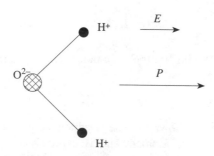

**Fig. 2.43**
Polarization of a water molecule

**Example 2.38**   From the simple two-state dipole model, compute the electric field necessary to align 90% of the dipoles if the dipole moment is $1.6 \times 10^{-28}$ C·m. Assume $T = 300$ K.

***Solution:***   Since $\varpi = 0.9$, $(1 - \varpi)/\varpi = \frac{1}{9}$. Based on Eq. (65), $\Delta E = \ln[\varpi/(1 - \varpi)]kT/2 = \ln(9) \times 1.38 \times 10^{-23}$ J/K $\times$ 300 K/2 $= 4.55 \times 10^{-21}$ J. The required electric field $\acute{E} = \Delta E/p_d = 4.55 \times 10^{-21}$ J/$1.6 \times 10^{-28}$ C·m $= 2.84 \times 10^{7}$ V/m.   **§**

**Example 2.39**   Compute the polarizability of water molecules if $p_0 = 1 \times 10^{-29}$ C·m, $\acute{E} = 1 \times 10^{5}$ V/m and $T = 300$ K.

***Solution:***   Based on Eq. (56), $\alpha' = p_{av}/\acute{E}$. According to Eq. (67), $\alpha' \approx p_0^2/3(kT) = (1 \times 10^{-29}$ C·m$)^2/(3 \times 1.38 \times 10^{-23}$ J/K $\times$ 300 K) $= 8.05 \times 10^{-39}$ F·m².   **§**

Positive and negative ions in a solid will move against each other, giving rise to **ionic polarization.** Among the different ionic solids, a class called **ferroelectrics** exhibit some very interesting properties. In particular, the induced polarization does not vanish even after the removal of the electric field. The effect is known as **hysteresis.** To understand hysteresis in ferroelectrics, we need to look at their crystal structures. The dipoles of ferroelectrics, unlike those of the ordinary ionic solids, are confined to regions called **ferroelectric domains.** Within each domain, the dipoles are always aligned in the same direction in order to minimize the overall energy of the solid. In the natural state, the domains are randomly oriented, and there is no observable polarization. The domains will, however, reorient in response to an electric field and give rise to an induced polarization similar to the case of the isolated dipoles. The main difference is that a part of this induced polarization will remain even after the electric field is removed. This result gives rise to the hysteresis effect, which is illustrated in Fig. 2.44.

As observed, remnant polarization can be removed if the direction of the electric field is reversed. Such a field-induced polarization effect is sometimes called **para-electricity.** The principle behind paraelectricity has been used in the design of

**Fig. 2.44**
Polarization curve for a ferroelectric solid

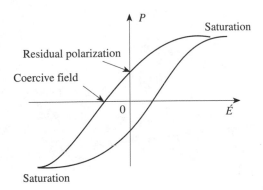

light switches that can control the passage of polarized light through an otherwise transparent solid. $BaTiO_3$ is an example; its crystal structure is shown in Fig. 2.45.

Another property of the ferroelectrics is that although the domains are relatively stable at low temperatures, they will disintegrate abruptly (due to a phase transformation) when the temperature exceeds a critical value known as the **Curie temperature** $T_c$. **Piezoelectricity** is another form of structural polarization and is related to the development of a voltage across an ionic insulator in the presence of applied pressure. Microscopically, piezoelectricity is the result of a change in the dipole moment when the crystal lattice is distorted. The process is reversible and is widely used in the design of pressure sensors (see Section 6.6). A typical piezoelectric material used for pressure sensing is quartz; its crystal structure is shown in Fig. 2.46.

**Example 2.40**     Defining the dipole moment as the product of the dipole charge times the separation, compute the dipole moment for the tetragonal $BaTiO_3$ unit cell shown in Fig. 2.45.

***Solution:***     From the unit cell structure, there is the equivalence of 3 $O^{2-}$ ions, 1 $Ti^{4+}$ ion, and 1 $Ba^{2+}$ ion. The dipole moment related to the shift in the $Ti^{4+}$ ion is $p_{Ti4+} = 4 \times 1.6 \times 10^{-19}$ C $\times 0.006 \times 10^{-9}$ m $= 3.84 \times 10^{-30}$ C·m. The dipole moment related to the shift in the $O^{2-}$ ion is $p_{O2-} = 2 \times 1.6 \times 10^{-19}$ C $\times (2 \times 0.006 \times 10^{-9}$ m $+ 0.009 \times 10^{-9}$ m$) = 6.72 \times 10^{-30}$ C·m. Since the $Ba^{2+}$ is considered to be stationary, the total dipole moment is $p_{Ti4+} + p_{O2-} = (3.84 + 6.72) \times 10^{-30}$ C·m $= 1.056 \times 10^{-29}$ C·m.     **§**

Fig. 2.45   Polarization of $BaTiO_3$ crystal

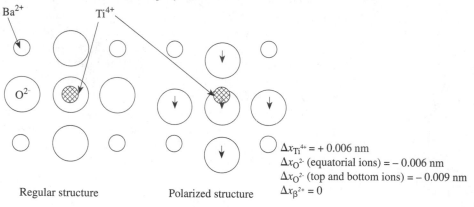

$\Delta x_{Ti^{4+}} = + 0.006$ nm
$\Delta x_{O^{2-}}$ (equatorial ions) $= - 0.006$ nm
$\Delta x_{O^{2-}}$ (top and bottom ions) $= - 0.009$ nm
$\Delta x_{B^{2+}} = 0$

Regular structure          Polarized structure

Fig. 2.46
Quartz crystal under stress

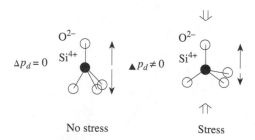

No stress          Stress

**Electronic polarization** occurs when the valence electrons are polarized with respect to the nuclei. It occurs at a much higher frequency (near optical frequencies) because of the lighter electron mass. Electronic polarization is present in all solids, including metals, semiconductors, and insulators. Let us examine the simple case of an H atom with a singly charged nucleus surrounded by an electron cloud. Assuming a finite displacement $\Delta x$ of the electron cloud in the presence of an electric field $\acute{E}$ (in V/m), Gauss law—which states that the charge enclosed in a volume, as is given by the total outward electric flux $D$ (in C/m$^2$)—can be written as

$$D = -q\Delta x^3/(d_0^3 4\pi\,\Delta x^2) \tag{72}$$

where $q$ is the electron charge in C,

   $\Delta x$ is the charge separation in m, and

   $d_0$ is the radius of the electron cloud in m.

The force $F$ (in N) resulting from Eq. (72) is then given by

$$F = qD/\epsilon_0 = -q^2\Delta x/(4\pi\epsilon_0 d_0^3) \tag{73}$$

where $\epsilon_0$ is the vacuum permittivity in F/m.

Since $F = q\acute{E}$, we can write

$$\acute{E} = F/q = -q\Delta x/(4\pi\epsilon_0 d_0^3) \tag{74}$$

Combining Eq. (74) with Eq. (56) and using the fact that $p = q\,\Delta x$, the polarizability of the H atom (in F·m$^2$) becomes

$$\alpha' = 4\pi\epsilon_0 d_0^3 \tag{75}$$

This calculation illustrates how the electronic polarizability can be determined. More accurate calculations will reveal that a correction factor is needed if a collection of H atoms is being considered.

So far, we have not considered the ac effect due to a time-varying electric field. In general, an ac electric field will generate time-dependent variations in the displacement of the dipoles. If we assume $\acute{E} = \acute{E}_0 \exp(j\omega t)$, where $\acute{E}_0$ is the peak amplitude in volts per meter, the expression representing the oscillatory motion of a dipole will be given by

$$m_i\, dx^2/dt^2 = -B_1\, dx/dt - C_1 x - q\acute{E}_0 \exp(j\omega t) \tag{76}$$

where $m_i$ is the mass of the lighter ion in kg,

   $x$ is the time-dependent displacement in m,

   $j\,(= \sqrt{-1})$ is an imaginary number,

   $\omega$ is the angular frequency in rad/s,

   $t$ is time in s, and

   $B_1$ and $C_1$ are constants in kg/s and kg/s$^2$, respectively.

The solution to Eq. (76) has the form

$$x = (q/m_i)\acute{E}_0\exp(j\omega t)/(\omega_0^2 - \omega^2 + j\omega B_1/m_i) \tag{77}$$

where $\omega_0 = \sqrt{C_1/m_i}$.

Since, by definition, polarizability ($\alpha'$) is given by $qx/[\acute{E}_0\exp(j\omega t)]$, it follows that

$$\alpha' = q^2/[m_i(\omega_0^2 - \omega^2 + j\omega B_1/m_i)] = \alpha_R + j\alpha_{Im} \tag{78}$$

where $\alpha_R$ and $\alpha_{Im}$ (in $F \cdot m^2$) are the real part and the imaginary part of $\alpha'$, respectively.

According to Eq. (78), at high frequency (i.e., when the $\omega^2$ term dominates over the $\omega$ term) the polarizability will peak at $\omega = \omega_0$; this represents **resonant absorption** in the solid. Fig. 2.47 shows how $\alpha_R$ and $\alpha_{Im}$ vary with the frequency of the ac electric field. At low frequency ($\omega \approx 0$), $\alpha'$ approaches $q^2/(m_i\omega_0^2)$. The time-dependent polarizability also translates into a frequency-dependent dielectric constant $\epsilon_r(\omega)$, which can be written as

$$\epsilon_r(\omega) = \epsilon_r(\infty) + [\epsilon_r(0) - \epsilon_r(\infty)]/(1 - \omega^2/\omega_t^2) \tag{79}$$

where $\epsilon_r(\infty)$ is the dielectric constant when $\omega$ approaches $\infty$,

$\epsilon_r(0)$ is the low-frequency dielectric constant, and

$\omega_t$ is the characteristic frequency of the dipoles in rad/s.

Equation (79) suggests that the dielectric constant of an ionic solid varies between $\epsilon_r(0)$ and $\epsilon_r(\infty)$ and has a singular value at $\omega = \omega_t$ when there is resonant absorption (in a real solid, damping also exists, and $\epsilon_r$ remains finite).

**Example 2.41**    A nonionic solid with no permanent dipole moment has $2 \times 10^{28}$ atoms/m$^3$. If the relative permittivity $\epsilon_r$ is 2.8, compute the frequency at maximum absorption.

**Fig. 2.47**
Frequency dependence of polarization

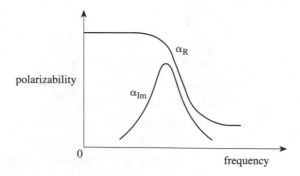

***Solution:*** Based on Eq. (61), $\epsilon_r = 1 + \alpha' N_{\text{dipole}}/\epsilon_0$. This gives $\alpha' = \epsilon_0(\epsilon_r - 1)/N_{\text{dipole}} = 8.85 \times 10^{-12}$ F/m $\times (2.8 - 1)/(2 \times 10^{28}/\text{m}^3) = 7.9 \times 10^{-40}$ F·m². Using the fact that $\alpha' = q^2/(m_i\omega_0^2)$, it can be shown that $\omega_0 = 1.6 \times 10^{-19}$ C$/\sqrt{7.9 \times 10^{40} \text{ F·m}^2 \times 0.91 \times 10^{-30} \text{ kg}} = 6 \times 10^{15}$ rad/s. **§**

## Energy Loss in Insulators

Charge separation across a capacitor is a form of energy storage; an ideal capacitor should retain the charges indefinitely. The most direct energy loss in a capacitor is due to leakage current. Leakage current may flow either along the edges of the insulator or directly through the insulator itself. Other than leakage current, current flow in an insulator is primarily due to ion hopping—i.e., ions physically moving from one lattice site to another.

Because of their heavier masses, ions are relatively immobile, and the ion current is quite small (usually of the order of picoamperes). Sometimes, current in an insulator involves electron transfer between ions. An outstanding example is the charge transfer between two ions of different valence states. For example, when two valence electrons in a $Ti^{2+}$ ion move to a $Ti^{4+}$ ion at a different site, this is effectively the same as the movement of the $Ti^{2+}$ ion to the site of the $Ti^{4+}$ ion. Ion hopping often requires an activation energy, and the energy loss during hopping will appear as heat. The mechanisms involved in surface leakage current are usually more complex; it is well known the dangling bonds and surface contaminants can substantially increase the leakage current. Furthermore, leakage current results in energy loss, and the associated heating effect can also give rise to dielectric breakdown in insulators.

A dielectric capacitor cannot hold too much charge, and when the internal electric field exceeds a critical value, breakdown occurs. Breakdown generates a large current in the insulator, and the latter can be seriously damaged. Breakdown often involves a carrier multiplication effect, which happens when the (few) carriers in the insulator have acquired sufficient energy from the electric field to start to ionize the other atoms. If the excess energy is large enough to break the energy bonds, additional electrons and holes are produced along the way. This process, sometimes called *avalanche breakdown,* or **carrier multiplication,** is illustrated in Fig. 2.48. Breakdown is usually accompanied by substantial heating, which again further increases the current through the insulator. Once initiated, breakdown will proceed indefinitely unless some form of current-limiting mechanisms, such as a fuse or a circuit breaker, is activated. Damage caused by breakdown is usually irreversible.

Ac loss in an insulator in quite complex and involves the activation and relaxation

**Fig. 2.48**
Electron multiplication
during breakdown

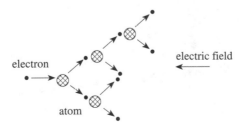

of the dipoles. In general, ac loss appears as an imaginary term in the permittivity of the solid (Eq. (78)) and any damping will give rise to power dissipation in the form of heating. Classically, the energy involved in ac loss is viewed as a frictional effect and is expressed in terms of a **loss factor,** $\tan(\delta)$, which is given by

$$\tan(\delta) = \text{energy loss per cycle}/(2\pi \times \text{maximum energy stored}) \qquad \textbf{(80)}$$

Table 2.11 lists the values of $\tan(\delta)$ in different insulators.

**Example 2.42**   The equivalent circuit for a lossy capacitor is shown in Fig. E2.1. Calculate the loss factor. Assume $V_a = V_0\sin(\omega t)$.

**Fig. E2.1**

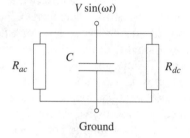

$V \sin(\omega t)$

$R_{ac}$    $C$    $R_{dc}$

Ground

*Solution:*   The energy stored per cycle in the capacitor is $E_{\text{stored}} = \int_{\text{cycle}} [V_a i]\, dt = \int_{\text{cycle}} [V_0^2 \omega C \sin(\omega t)\cos(\omega t)]\, dt = -V_0^2 C \cos(2\omega t)/4$. The peak value of $E|_{\text{stored}} = CV_0^2/2$. Energy loss due to the resistors is given by $E' = V_0^2/R_{ac} \int_{\text{cycle}} [\sin(\omega t)\sin(\omega t)]\, dt = V_0^2 \pi/(\omega R_{ac})$. Thus, the loss factor is $\tan(\delta) = V_0^2 \pi/(\omega R_{ac}) \times 2\pi CV_0^2/2 = 1/(\omega R_{ac})$.   **§**

**Example 2.43**   Silicon nitride is a good insulator. For a film thickness of 100 nm, the average breakdown voltage is 30 V. Calculate the breakdown field.

*Solution:*   The breakdown electric field is $\acute{E}_{\text{br}} = 30\text{ V}/(1 \times 10^{-7}\text{ m}) = 3 \times 10^8$ V/m.   **§**

**Table 2.11**   Values of $\tan(\delta)$ in different insulators

| Materials | $\tan(\delta)$ |
|---|---|
| Polymers | $10^{-4}$ |
| Water | 0.1 |
| Ceramics, glasses | $10^{-3}$ |
| Ferrites | $10^{-2}$ |

## GLOSSARY

| | |
|---|---|
| **excitation energy** | the energy needed to release an electron in a solid. |
| **thermal energy** | the energy possessed by electrons at a finite temperature. |
| **electron density** | the number of electrons per unit volume. |
| **thermal velocity** | the velocity of the electrons, assuming that their kinetic energy is equal to the thermal energy. |
| **effective mass** | a parameter equivalent to the mass of an electron or a hole in a solid. |
| **anisotropic solid** | a solid whose properties are dependent on the crystal orientation. |
| **collision** | the impact on a lattice atom or ion by an electron or a hole. |
| **scattering** | the transfer of energy to a lattice atom or ion by an electron or a hole. |
| **drift velocity** | average velocity of an electron or a hole under the influence of an applied voltage. |
| **current** | the charge flow in a unit time. |
| **spin** | an entity used to define the magnetic properties of an electron. |
| **Fermi-Dirac distribution function** | the energy distribution function that applies to electrons in a solid. |
| **Fermi level** | the average (internal) energy of a collection of electrons. |
| **electron state** | the values of the energy and momentum of an electron. |
| **energy band** | a continuum of electron or hole states. |
| **energy gap** | the spacing separating two energy bands. |
| **effective density of states** | the (average) number of electron or hole states per unit volume in an energy band. |
| **hole** | a vacant state in an energy band. |
| **conduction band** | the energy band where the electrons reside. |
| **valence band** | the energy band where the holes reside. |
| **resistance** | a parameter that measures the current conduction in a solid. |
| **resistivity** | a parameter that characterizes the conduction properties of a solid. |
| **conductivity** | the inverse of resistivity. |
| **hole density** | the number of holes per unit volume. |
| **mobility** | a parameter that measures the ability of an electron or a hole to conduct a current. |
| **relaxation time** | the average time lapse between two successive electronic scattering or collision events. |
| **conductor** | a solid that allows a current to pass through easily. |
| **Hall effect** | a method of measuring the carrier densities in metals and in semiconductors. |
| **magnetoresistance** | the apparent increase in the resistance of a solid due to the presence of a transverse magnetic induction. |
| **Corbino disk** | a solid shaped into a special geometry for measuring magnetoresistance. |
| **thermoelectric voltage** | the voltage developed across a solid due to a temperature gradient. |
| **thermocouple** | a device used for temperature measurement. |
| **semiconductor** | a solid that has moderate conduction properties. |
| **carrier** | either an electron or a hole. |
| **intrinsic semiconductor** | a pure semiconductor with few or no impurities. |
| **intrinsic Fermi level** | the Fermi level of an intrinsic semiconductor. |
| **extrinsic semiconductor** | a semiconductor containing dopants. |
| **donor** | a dopant that furnishes an electron. |
| **acceptor** | a dopant that furnishes a hole. |
| **$n$-type semiconductor** | a semiconductor whose electrical properties are dominated by electrons. |
| **$p$-type semiconductor** | a semiconductor whose electrical properties are dominated by holes. |
| **freeze-out** | the recovery of the carriers at a low temperature by the dopant ions. |
| **pinning** | the immobilization of the Fermi level by surface or impurity states. |
| **Einstein relationship** | the equation linking the diffusivity of carriers to their mobility and the temperature. |
| **carrier generation** | the creation of electrons and holes in a semiconductor. |

| | |
|---|---|
| carrier recombination | the removal of electrons and holes in a semiconductor. |
| generation lifetime | the average time required to create an electron or a hole. |
| recombination lifetime | the average time required to annihilate an electron or a hole. |
| surface recombination velocity | a parameter that measures the rate of surface recombination of carriers in a solid. |
| hot electron | an electron that has acquired more energy from the electric field than it can dissipate. |
| velocity saturation | the leveling off of the drift velocity in a high electric field. |
| polarization | the separation of positive and negative charges in a solid. |
| permanent dipoles | dipoles that are always present in a solid. |
| induced dipoles | dipoles that are created in the presence of an electric field or some applied stress. |
| capacitance | a parameter that measures the charge storage effect in a solid. |
| capacitor | a device that stores charges. |
| permittivity | a parameter that measures the ability of an insulator to store charges. |
| relative permittivity, or dielectric constant | a parameter that measures charge polarization in a solid. |
| dipole moment | a parameter that measures the strength of a dipole. |
| polarizability | the proportionality constant between the dipole moment and the electric field. |
| Boltzmann distribution | an approximate energy distribution function of the electrons (and also the dipoles) in a solid. |
| orientational dipole | a dipole whose polarization depends on the dipole orientation. |
| ionic polarization | polarization due to positive and negative ions polarized with respect to each other. |
| ferroelectrics | solids whose induced polarization does not vanish even after the removal of the electric field. |
| hysteresis | the effect of nonreversible polarization. |
| ferroelectric domains | regions within a solid that have uniform polarization. |
| paraelectricity | the change in the electrical properties in a solid due to an applied voltage. |
| Curie temperature | the temperature above which ferroelectric domains will vanish. |
| piezoelectricity | induced polarization in a solid when it is under mechanical stress. |
| electronic polarization | polarization in a solid when the valence electrons are polarized with respect to the nuclei. |
| resonant absorption | absorption when the incident energy is exactly equal to the transition energy of the electron or dipole states. |
| carrier multiplication | a state that occurs when carriers in a solid have acquired sufficient energy from the electric field to ionize other atoms, thereby producing additional carriers. |
| loss factor | a parameter that measures the energy loss in an insulator. |

## REFERENCES

Grovenor, C. R. M. *Microelectronic Materials*. Bristol and Philadelphia: Adam Hilgar, 1989.

Kittel, C. *Introduction to Solid State Physics*. New York: John Wiley & Sons, Inc., 1967.

Mayer, J. W. and S. S. Lau. *Electronic Materials Science for Integrated Circuits in Si and GaAs*. New York: Macmillan Publishing Company, 1990.

Omar, M. A. *Elementary Solid State Physics*. Reading, Mass.: Addison-Wesley Publishing Company, 1975.

Sze, S. M. *Semiconductor Devices, Physics and Technology*. New York: John Wiley & Sons, Inc., 1985.

## EXERCISES

### 2.2  Electrons in a Solid

1. If the density of electrons in a solid depends exclusively on the excitation energy, compute the ratio of electron densities in Si and CdS at 300 K using the values of $E_{excite}$

in Table 2.1. Assume that the electron density depends exponentially on the excitation energy and $kT$ at 300 K $\approx$ 0.026 eV.

2. The energy gap of InSb is 0.18 eV. Suggest the temperature at which the thermal energy will be the same as the value of the energy gap.

3. An electron has a rest mass $m_0 = 9.1 \times 10^{-31}$ kg. What will be the effective mass of the electron if it is moving (a) in a vacuum, and (b) in the conduction band of a solid with an energy versus momentum relationship given by $E = (h'k')^2/(1.4m_0)$, where $h'k'$ is the momentum (or crystal momentum).

4. Suggest what the effective mass of electrons in a solid will be if there are two overlapping conduction bands with effective masses of $0.9m_0$ and $0.7m_0$, respectively. Assume the electron densities in the conduction bands are the same.

5. Determine the drift velocity of electrons in a solid if the average kinetic energy of the electrons (over and above the thermal energy) is 10 meV. Assume $m_e^* = m_0$.

## 2.3   Energy Distribution of Electrons

6. Compute the minimum value of $(E - E_F)/(kT)$ if Eq. (5) is to have an error of no more than 10%. Repeat the computation for Eq. (6).

7. Compute the location of the Fermi level below the conduction band if there is a 50% chance that electrons are found in the conduction band. ($T = 300$ K.)

8. If the probability of the Fermi-Dirac function $f(E_1)$ is $1 \times 10^{-5}$ at 300 K, what is its value at 600 K? Assume $E_1 - E_F \gg kT$.

9. The Fermi-Dirac distribution gives the electron distribution at temperature $T$. It depends on a parameter called the Fermi level, $E_F$, which varies for different solids.

   (a) At 300 K, calculate the probability of occupancy when $E - E_F = 0.5$ eV. Repeat the calculations when $E - E_F = 0.2$ eV.

   (b) Repeat the results of part (a) at $T = 77$ K, the temperature of liquid nitrogen. Observe the difference.

10. In a particular metal, $E_F = 10$ eV. Compute the probability of finding an electron with energy between $E_F$ and $E_F + 3kT$. ($T = 300$ K.)

## 2.4   The Energy Band Diagram

11. Assume that you have particles labeled by their energy $E$ and momentum $k'$. Both $E$ and $k'$ can vary from 0 to infinity, and $E$ and $k'$ are related by the free electron relationship: $E = (h'k')^2/2m_0$, where $m_0$ is the electron mass and $h'$ is the Planck's constant.
   (a) What is the density of states function (number of states per unit energy per unit volume) when $E = 1.0$ eV? (b) What is the density of states (number of states per unit volume) for 1.0 eV $< E <$ 1.1 eV?

12. Two energy bands are separated by a distance of 0.5 eV. Suggest the electron density in the upper energy band if the Fermi level is located at the top of the lower band and the temperature is 300 K. Assume that the densities of states in the upper band, $N_c$, and in the lower band, $N_v$, are $1 \times 10^{24}/m^2$. What will be the hole density in the lower band?

13. Plot the interatomic distances of Ge, Si, GaAs, CdS, and GaN versus their energy gaps. Can you draw any conclusions from the plot?

14. Suggest how it is possible to transform the energy band diagram ($E$ versus $k'$) into a diagram such as is shown in Fig. 2.20.

15. If the minimum energy difference between two electron states is $h'$, Planck's constant (in J), compute the number of electrons present in a filled energy band with an energy range of 1.2 eV. Remember that each state can accommodate two electrons with different spins.

## 2.5  Mathematical Formulation of the Conduction Processes

16. A metal bar will pass a current of 1 mA through it if the applied voltage is 1 V. What is its resistance? If the length of the bar is 1 cm, what is the resistivity per unit area?

17. Two metal cubes with dimensions 1 cm × 1 cm × 1 cm are placed end to end. If one cube is made with Cu and the other is made with Ag, what will be the current passing through the cubes if the total voltage across the cubes is 1 V? What happens to the current if the dimensions of the Cu cube are reduced to 0.5 cm × 0.5 cm × 0.5 cm. The conductivities of Cu and Ag are $5.88 \times 10^7$ S/m and $6.21 \times 10^7$ S/m, respectively.

18. Compute the electron mobility if the effective mass of the electrons is $0.2m_0$ and the typical electron suffers a collision every nanosecond.

19. A solid carries a current of 10 mA/cm². Compute the drift velocity if the electron density in the solid is $1 \times 10^{20}$/m³.

20. Assume that the thermal electron (with $E_{th} = kT$) is spherical and has a radius of 0.05 nm. (a) How long does it take for the average electron to experience a collision if the density of the scattering centers is $1 \times 10^{20}$/m³? ($T = 300$ K.) (b) Repeat the calculations if $T = 450$ K. (c) What happens to the collision frequency?

21. In a solid, conductivity is due to two groups of electrons with identical densities. If one group has an effective mass of $0.1m_0$ and another group an effective mass of $0.3m_0$, suggest what the effective mass of the electrons determined from conductivity measurement will be.

22. (a) Compute the mobility of Ag if $\sigma = 6.21 \times 10^7$ S/m and $n = 5.85 \times 10^{28}$/m³.
    (b) Compute its relaxation time if $m_e^* = m_0$.
    (c) What will be the drift velocity of the electrons if the electric field is 10 V/m?

## 2.6  Conductors

23. Suggest what the temperature dependence of the electron mobility is according to Eq. (17).

24. Suggest what the impurity dependence of the electron mobility is according to Eq. (18).

25. Compute the Hall constant of a solid if the conductivity of the solid is 0.01 S/m and the carrier mobility is 0.01 m²/V·s. If the Hall setup is used to measure an unknown magnetic induction based on magnetoresistance effect, what will be the unknown magnetic induction if the resistivity in the presence of the magnetic induction is increased by 10%?

26. Show why the Corbino disk is a good approximation for $d_w \gg L$, i.e., when the sample width is much greater than its length. Estimate the percent error.

27. For the thermocouples listed in Table 2.7, suggest which thermocouple is more suitable for high-temperature measurements.

## 2.7  Semiconductors

28. What is the mobility ratio between electrons and holes if the conductivity of an intrinsic semiconductor is 0.001 S/m and the carrier densities are $1 \times 10^{17}$/m³? Assume $\mu_n = 0.04$ m²/V·s.

29. Compare the intrinsic carrier densities of Si and Ge. Suggest why the intrinsic carrier density is larger when the energy gap of the semiconductor is smaller.

30. Plot the temperature dependence of the energy gap in Si. What is the fractional decrease in the energy gap as the temperature increases from 300 K to 600 K? Suggest the reason why there is an energy gap shrinkage with temperature, whereas the lattice constant increases with temperature.

31. Using Eq. (39), verify the values of $\sigma_0'$ appearing in Table 2.9.

32. For an $n$-type GaAs sample with a donor density of $1 \times 10^{18}/m^3$, suggest at what temperature the sample will become intrinsic. Ignore the temperature dependence of the energy gap, $N_c$ and $N_v$.

33. Compute the hole density in GaAs at 300 K if the donor density is $1 \times 10^{24}/m^3$. $n_i$ (GaAs) $= 1.8 \times 10^{12}/m^3$.

34. Based on Fig. 2.25, suggest an empirical equation that fits the carrier mobilities for GaAs at different impurity concentrations.

35. Suggest at what temperature freeze-out will occur in an $n$-type Si sample if we define freeze-out as occurring when only 1% of the donors are ionized. Assume the donor level to be located 0.05 eV below the conduction band edge and $N_D = 1 \times 10^{24}/m^3$. What is the intrinsic carrier density at the freeze-out temperature? Ignore the temperature dependence of $E_g$, $N_c$, and $N_v$.

36. If the density of the trap states at the Si surface is $10^{14}/m^2 \cdot eV$, determine what fraction of carriers will go into these states as the Fermi level moves from midgap toward the conduction band edge.

37. If single-crystal GaAs has a conductivity of $\sigma = 0.1$ S/m, what will be the conductivity of granular GaAs at $T = 300$ K if the intergranular barrier height is 0.2 eV? Repeat the calculations for $T = 200$ K. Assume $\sigma$ is independent of temperature.

38. An amorphous solid may be considered as one that is 50% filled by voids. Suggest what the conductivity of amorphous Si is if the rest of the solid is filled with intrinsic Si.

## 2.8  Other Conduction Mechanisms

39. Compute the electron flux in a solid if the electron density gradient is $1 \times 10^{23}/m^4$ and the electron diffusivity is $1 \times 10^{-4}$ m$^2$/s. If the hole diffusivity is $1 \times 10^{-5}$ m$^2$/s, suggest the value of a hole density gradient that generates a similar current.

40. Compute the current density passing into a Si sample if the surface electron and hole densities are $1 \times 10^{18}/m^3$ and the thickness of the sample is 200 μm. Assume that the carrier densities at the lower surface is negligible compared with those at the front. $D_p = 1.2 \times 10^{-3}$ m$^2$/s and $D_n = 3.8 \times 10^{-3}$ m$^2$/s.

41. If a semiconductor sample has a dopant density gradient of $d(N_D)/dx$, compute the internal diffusion current due to the dopant density gradient if all the dopants are ionized. Show that the electrostatic force due to the ionized dopants will generate an opposite current to offset the diffusion current, so that the overall current in the sample is zero.

42. Using the Einstein relationship, compute the electron mobility in a semiconductor sample if the electron diffusivity is $1 \times 10^2$ m$^2$/V·s and $T = 300$ K (or $kT/q = 0.026$ eV).

43. If carrier generation at a semiconductor interface with a cross-sectional area of $1 \times 10^{-6}$ m$^2$ produces a current of 1 mA, estimate the width of the region involved in the generation process. The generation lifetime is 1 ns and $n_i = 1.45 \times 10^{16}/m^3$.

44. At an interface where recombination occurs, compute the recombination current if the excess carrier density is $1 \times 10^{12}/m^3$, the cross-sectional area is 1 mm $\times$ 1 mm, and the recombination lifetime is 1 μs.

**45.** Using Fig. 2.35, arrive at an empirical relationship for the hole lifetimes in Si for **(a)** band-to-band recombination (or recombination through dopants), and **(b)** Auger recombination.

**46.** If the surface diffusion current density of a semiconductor sample is 0.1 A/m$^2$ and the surface excess carrier density is $1 \times 10^{20}$/m$^3$, compute the surface recombination velocity. If the density of the surface recombination center is $1 \times 10^{21}$/m$^3$, estimate the capture cross section of the recombination centers.

## 2.9  Velocity Saturation and Negative-Resistance Effect

**47.** Using Eq. (51), give an expression for $\acute{E}$ in terms $\mu_0$ and $v_s$ when the low-field carrier mobility $\mu_0$ is reduced by 50%.

**48.** Using Eq. (52), compute the saturation velocity of GaAs at 500 K. Give a brief explanation for the temperature dependence of the saturation velocity.

**49.** Explain the differences in the relationship between the electron velocity and the electric field in Si and GaAs, as shown in Fig. 2.36. Do you expect the saturation velocity of most semiconductors to approach a value of around $1 \times 10^5$ m/s? Give a reason why this may be so.

**50.** Based on the energy band diagram of GaAs, as shown in Fig. 2.37, suggest the value of the effective temperature of an electron that moves from the $\Gamma$ valley to the $L$ valley.

**51.** Qualitatively, suggest how the negative resistance effect may be used to induce signal amplification in a semiconductor.

**52.** Although velocity saturation is a well-established phenomenon in semiconductors, determine whether there are instances when velocity saturation may be avoided. (Consider the case when a very high electric field is applied for a time duration much less than the relaxation time of the carriers.)

**53.** Although it is possible to induce the negative-resistance effect through carrier transition from a high-mobility conduction band valley to a low-mobility conduction band valley, as in the case of GaAs, determine whether it is possible to induce negative-resistance effect through the release of holes (optically) from trap states in an $n$-type semiconductor. Explain your arguments.

## 2.10  Insulators

**54.** Determine the capacitance of a parallel-plate capacitor if the voltage observed across the capacitor is 2.5 V after the capacitor is charged for 1 ms with a current of 100 $\mu$A. If the area of the capacitor plate is 1 mm $\times$ 1 mm and the separation is 50 $\mu$m, determine the relative permittivity of the insulator.

**55.** If there are $1 \times 10^{24}$ dipoles in a piece of dielectric with dimensions of 5 mm $\times$ 5 mm $\times$ 50 $\mu$m and the relative permittivity is 5, what will be the polarizability?

**56.** Attempt to explain the differences in the dielectric constants of the different solids listed in Table 2.10.

**57.** Will an ac voltage lag behind the ac current in a capacitor or the other way around? Briefly explain your argument.

**58.** Using Eq. (65), determine the ratio of the dipoles in the direction of the electric field over those that are randomly oriented if the energy difference between the states of the dipole molecules is 0.5 eV and $T = 250$ K.

**59.** Is it possible to reduce Eq. (71) to Eq. (68), the polarizability of a bistable molecule?

**60.** Determine the dipole density in $BaTiO_3$.

**61.** From Fig. 2.46, determine the dipole moment in a quartz tetrahedron if its height is reduced by 10% under an applied pressure. The bond length of $Si^{4+}-O^{2-}$ is 0.171 nm.

**62.** Determine the polarizability in a H atom if its atomic radius is used for the value of $d_0$ appearing in Eq. (75). Using a similar approach, compute the polarizability of He.

**63.** Sketch the magnitude of the frequency response of an electronic oscillator as given by Eq. (77). If $C_1$ is 10 kg/s$^2$, what is the resonant frequency?

**64.** Estimate the value of $\tan(\delta)$ for $SiO_2$ at 1 MHz, assuming that the resistivity is $1 \times 10^9$ $\Omega \cdot m$ and the capacitance is 1 pF. Also, $\omega = (2\pi)(\text{frequency})$.

# p-n Junctions and Related Devices

## 3.1 INTRODUCTION

In the previous chapter, we learned that the electronic properties of semiconductors are quite different from those found in the other solids because of their unique energy band structures and the presence of both free electrons and holes. Free electrons and holes are responsible for the formation of n-type and p-type semiconductors, from which a p-n junction is made. Functionally, a p-n junction acts like an electronic switch; it has a wide range of applications, including those in switching circuits, in signal generation and detection, as a power (light) source, and even in energy conversion. Notwithstanding these important uses, a p-n junction is also the most common structure found in other more complicated semiconductor devices, such as transistors and microwave oscillators. In addition, there are other reasons why semiconductors are so important. For instance, many semiconductors—such as Si in the form of silicates—are found abundantly on Earth, and they can be purified easily using chemical techniques to form high-quality single crystals. Single-crystal semiconductors usually have predictable physical and electronic properties, from which reliable devices can be made. In addition to the p-n junctions, other related devices, such as the Schottky (metal-semiconductor) junction, also provide rectification. These devices are called diodes, a term borrowed from the two-terminal vacuum-tube device. Transistors are p-n junction devices of the next level of complexity. Generally speaking, transistors can be divided into the junction transistors, an example being bipolar junction transistors (BJT), and field-effect transistors (FET). The BJT is used primarily in signal amplification and switching, although it is an equally good light detector and power device. There are many different types of FETs; the most important one is the metal-oxide-semiconductor field-effect transistor, or MOSFET, which is widely used because of its structural simplicity and low power consumption (as in the complementary structure). Over the years, many new BJTs and FETs have been developed. Recent emphasis has been on heterostructure versions of these devices, employing layers of different semiconductor materials instead of one uniform substrate.

In this chapter, we start with a description of the operation of *p-n* junctions, including their structures and physical properties. These included the current-voltage (*I-V*) characteristics as well as the equivalent-circuit models. A similar treatment is given to Schottky junctions and ohmic contact. The second part of this chapter is devoted to transistors. Different types of transistor structures are examined, as are their principles of operation, materials considerations, and performance evaluation The final section reviews new transistor structures and their potential advantages over conventional transistors.

## 3.2    CONCEPT OF A *p-n* JUNCTION

A ***p-n* junction** is also called a semiconductor diode, or simply a **diode.** The word *diode* is the extension of a name given to a rectifying vacuum tube. A *p-n* junction may be formed by many techniques, including alloying a *p*-type semiconductor and an *n*-type semiconductor together or locally reversing the carrier type of an extrinsic semiconductor through the addition of a suitable dopant. One requirement in the formation of a *p-n* junction is that electrons (this term is used in place of the term *free electrons*) and holes must be free to move across the interface when the junction is formed. A typical *p-n* junction is shown in Fig. 3.1. A properly prepared *p-n* junction must have **rectification** properties; i.e., current flows easily in one direction but not in the opposite direction. Such properties of the *p-n* junction are present regardless of the doping levels and the type of semiconductors used. Even so, some semiconductors nevertheless have better rectification properties than others. We shall begin the study of a *p-n* junction by examining its energy band diagram. The concept of energy bands was introduced in Chapter 2.

### Energy Band Diagram for a *p-n* Junction

In the formation of a *p-n* junction, because of the difference in the carrier densities on the two sides, we expect that there will be migration of the carriers across the interface; i.e., carriers will flow from the high-density side to the low-density side.

**Fig. 3.1**
An alloyed *p-n* junction

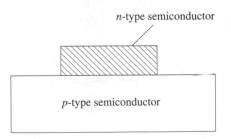

*n*-type semiconductor

*p*-type semiconductor

For simplicity, we shall separate the *p-n* junction into three different regions: The region near the interface where the carriers have been depleted due to diffusion is called the **junction** region; the remaining two regions of the *p-n* junction, where few or no changes in the carrier densities have occurred, are called the ***p*-side** and the ***n*-side.** Fig. 3.2 is a schematic diagram of the *p-n* junction showing the three regions. As expected, the junction region is responsible for the rectification properties.

In a doped semiconductor, in addition to the presence of the electrons and holes, there are also donor ions and acceptor ions. These ions are not mobile (at least not at room temperature), and they are called **stationary charges** to distinguish them from the electrons and holes (charged carriers), which are mobile. In the junction region where the rectification properties arise, we have both stationary charges and the (charged) carriers. The carriers will diffuse across the junction interface, leaving behind the stationary charges. The stationary charges form a bipolar (oppositely charged) layer made up of donor ions and acceptor ions. The region is sometimes called the **space-charge region.** In the space-charge region, there is a very strong internal electric field because of the small separation (of the order of a fraction of a micrometer) between the ions. The direction of the electric field points from the positively charged donor ions toward the negatively charged acceptor ions—i.e., from the *n*-side toward the *p*-side. The electric field strength depends on the amount of ions exposed, which effectively controls the extent of carrier flow across the interface. Thus, in the junction region, there are two opposite forces acting on the carriers. There is the *diffusion force,* which drives the carriers across the interface due to the differences in the carrier densities, and the *electrostatic force,* which is due to the space charge that tries to retain them. At equilibrium, when the two forces exactly balance each other, there is no net flow of carriers. At this point, the space-charge region is the same as the junction region. It is also called the *depletion region* because of the absence of the carriers.

Much information about a *p-n* junction can be extracted from a study of the energy band diagram, which is shown in Fig. 3.3. The energy band diagram is drawn for the electrons, and the vertical axis shows the electron potential energy (in absolute sense, this is negative potential energy). Before a *p-n* junction is formed, the two sides of the *p-n* junction have their own separate energy band diagrams and Fermi levels, as illustrated in Fig. 3.4. Once they are alloyed together, thermodynamics dictate that the Fermi level (which represents the average internal energy of all the

**Fig. 3.2**
A *p-n* junction showing the three major regions

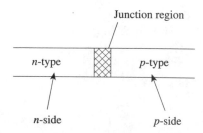

**Fig. 3.3**

Energy band diagram of a
*p-n* junction

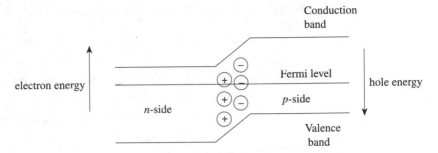

**Fig. 3.4**

Alignment of the Fermi
levels in a *p-n* junction

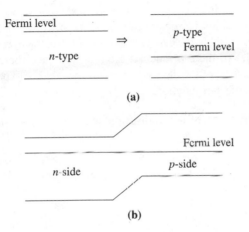

electrons) must be uniform across the entire *p-n* junction. In order to achieve this, the energy bands have to be adjusted, which results in band bending in the junction region.

Looking at it in another way, band bending in a *p-n* junction is related to the presence of space charge. Space charge gives rise to an internal electric field, which translates into a potential gradient, or band bending, as it is called. Physically, in a *p-n* junction, the electrons moving from the *n*-side to the *p*-side encounter a potential barrier generated by the space charge or the dopant ions. A similar situation exists for the holes when they move from the *p*-side to the *n*-side. Because holes are positively charged, they reside in the valence band, and their potential energy increases downward in the energy band diagram. In Fig. 3.4, we see that the energy bands shift upward from left to right and have a positive slope (assuming the $+x$ direction is from left to right). Since the electric field is equal to the negative of the potential gradient, this implies that the electric field also points from left to right, in basic agreement with the fact that the positive ions are located to the left of the *p-n* junction.

The amount of band bending across a *p-n* junction is called the **built-in potential**

**(energy)** $\Phi_{bi}$. It is usually given in electron volts. The built-in potential $\Phi_{bi}$ is given by

$$\Phi_{hi} = kT \ln(N_A N_D / n_i^2) \tag{1}$$

where    $k$ (=$8.62 \times 10^{-5}$ eV/K) is the Boltzmann constant,

   $T$ is the absolute temperature in K.

   $n_i$ is the intrinsic carrier density of the semiconductor /m$^3$,

   $N_A$ is the acceptor density /m$^3$, and

   $N_D$ is the donor density /m$^3$.

In fact, $\Phi_{bi}$ stands for the shift in the energy bands needed to align the Fermi levels. Its value in most *p-n* junctions is less than 1 eV. It is also true that $\Phi_{bi}$ in a *p-n* junction cannot be measured externally since $\Phi_{bi}$ is balanced internally by the diffusion potential of the carriers. At equilibrium, the average energies of the carriers on both sides of the *p-n* junction are the same, and there is no net current flow in any cross section of the device. Therefore, we do not expect to measure any voltage drop across the *p-n* junction.

**Example 3.1**    Compute the value of the built-in potential, $\Phi_{bi}$, of a *p-n* junction at 300 K for Si if $N_A = 1 \times 10^{24}$/m$^3$, $N_D = 1 \times 10^{21}$/m$^3$, $n_i = 1.45 \times 10^{16}$/m$^3$ and $kT \approx 0.026$ eV.

*Solution:*    Based on Eq. (1), $\Phi_{bi} = kT \ln(N_A N_D / n_i^2) = 0.026$ eV $\times$ $\ln(1 \times 10^{24}$/m$^3 \times 1 \times 10^{21}$/m$^3/(1.45 \times 10^{16}$/m$^3)^2)$ eV $= 0.75$ eV.    §

**Example 3.2**    Compute $n_i$ of a semiconductor at 400 K given that $E_g = 1.1$ eV and $n_i(300$ K$) = 1.45 \times 10^{16}$/m$^3$.

*Solution:*    From Chapter 2, Eq. (38),

$$
\begin{aligned}
n_i(400 \text{ K}) &= n_i(300 \text{ K})\exp(-Eg/2k400)/\exp(-Eg/2k300) = 1.45 \times 10^{16}/\text{m}^3 \\
&\quad \times \exp(-1.1 \text{ eV} \times 1.6 \times 10^{-19} \text{ C}/(2 \times 1.38 \times 10^{-23} \text{ J/K} \times 400 \text{ K})) \\
&\quad \times \exp(-1.1 \text{ eV} \times 1.6 \times 10^{-19} \text{ C}/(2 \times 1.38 \times 10^{-23} \text{ J/K} \times 300 \text{ K}))/\text{m}^3 \\
&= 3 \times 10^{19}/\text{m}^3 \quad §
\end{aligned}
$$

## The *I-V* Characteristics of a *p-n* Junction

Let us first examine the current-voltage (*I-V*) **characteristics of a *p-n* junction** as measured experimentally. This can easily be done by supplying a fixed voltage $V_a$ (not too big) in volts across the *p-n* junction and recording the current flow $I$ in amperes for different values of $V_a$. Fig. 3.5 shows a typical *I-V* curve.

As observed, the characteristics are nonlinear. If, however, the same characteristics are plotted semilogarithmically as shown in the inset, the curve for $\ln(I)$ versus $V_a$ will become a straight line under a large forward-bias voltage—i.e., when the positive voltage is applied to the *p*-side of the junction. Based on this observation, we can (empirically) develop the following current-voltage relationship for the forward-bias *p-n* junction:

$$I = I_0 \exp(V_a / V_t) \tag{2}$$

Fig. 3.5
*I-V* characteristics of a
*p-n* junction

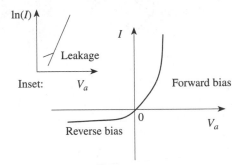

where $V_t$ is a constant in V and $I_0$ is a constant in A.

Equation (2) can be used to compute the forward-bias resistance of the *p-n* junction. This is called the **dynamic resistance,** $R_{\mathrm{dyn}}$. The dynamic resistance, $R_{\mathrm{dyn}}$ ($\Omega$), is given by

$$R_{\mathrm{dyn}} = d(V_a)/dI = V_t/I \qquad (3)$$

Combining Eqs. (2) and (3), we find that $R_{\mathrm{dyn}}$ decreases exponentially with increasing bias voltage. Physically, we can measure the dynamic resistance of a *p-n* junction at different bias points and plot it against the bias voltage, as shown in Fig. 3.6. For a sufficiently large forward bias ($> 0.5$ V), the resistance will be very small (say, $\sim$ 1–2 $\Omega$, a value often overshadowed by the series resistance of the voltage-supply circuit). In many applications, $R_{\mathrm{dyn}}$ is considered to be negligible, and the forward-bias *p-n* junction behaves like a **short circuit,** or a device with almost no resistance. In the notation of a switching circuit, a short circuit is the ON state of a switch.

Also shown in Fig. 3.5 is the reverse current of the *p-n* junction, which is very small and relatively independent of the reverse-bias voltage. The reverse-bias *p-n*

Fig. 3.6
Dynamic resistance of a
*p-n* junction

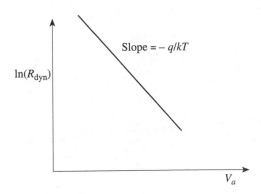

junction behaves like an open circuit and is the OFF state of a switch. To include the reverse current in the *I-V* characteristics, Eq. (2) becomes

$$I = I_0 \exp(V_a/V_t) - I_r \tag{4}$$

where    $I_r$ is the reverse current (or the *leakage* current) in A.

If $I_r$ happens to be identical to $I_0$, Eq. (4) becomes

$$I = I_0[\exp(V_a/V_t) - 1] \tag{5}$$

Equation (5) is called the **ideal diode equation,** and $I_0$ is the **saturation current.** The saturation current may be viewed as the value of $I$ when $V_a$ approaches $-\infty$. For most *p-n* junctions, $I_r$ is not the same as $I_0$.

**Example 3.3**   Compute the dynamic resistance of an ideal *p-n* junction when (a) $V_a = 0.1$ V, and (b) $V_a = 0.6$ V, if $I_0 = 1$ nA. Assume $V_t = 0.026$ V.

*Solution:*   Based on Eqs. (3) and (5), $R_{\mathrm{dyn}} = kT/(qI)$.
   **(a)** When $V_a = 0.1$ V, $I = 1 \times 10^{-9} \times \exp(0.1 \text{ V}/0.026 \text{ V})$ A $= 4.7 \times 10^{-8}$ A. Thus, $R_{\mathrm{dyn}} = 0.026$ V$/(4.7 \times 10^{-8}$ A$) = 5.5 \times 10^5$ $\Omega = 550$ k$\Omega$.
   **(b)** When $V_a = 0.6$ V, $I = 1 \times 10^{-9} \times \exp(0.6 \text{ V}/0.026 \text{ V})$ A $= 1.05$ A. Thus, $R_{\mathrm{dyn}} = 0.026/1.05 = 0.024$ $\Omega$.  §

**Example 3.4**   Obtain an expression for the forward-bias dynamic resistance of a *p-n* junction assuming $V_a$ is very small.

*Solution:*   Under forward bias, based on Eq. (4), $I \approx I_0(1 + V_a/V_t)$ when $V_a$ is small. This gives $d(V_a)/dI \approx V_t/I_0$.  §

   Let us try to deduce the parameters appearing in Eq. (4). We assume for simplicity a $p^+$-$n$ junction with the *p*-side more heavily doped. This situation gives a much larger hole current across the *p-n* junction and allows us to ignore the electron current for the time being. A $p^+$-$n$ junction is called a **one-sided step junction.** Fig. 3.7 is a schematic diagram showing the hole **diffusion current,** $I_p$, flowing into the *n*-side in a $p^+$-$n$ junction.

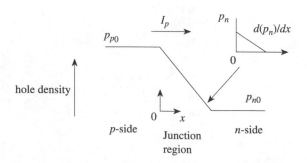

**Fig. 3.7**
Hole current through a
$p^+$-$n$ junction

Mathematically, this current (A) is given by

$$I_p(x) = qD_p[d(p_n(x))/dx]A_{cs} \tag{6}$$

where   $q$ is the electron charge in C,

$D_p$ is the hole diffusivity in m²/s,

$A_{cs}$ is the cross section of the *p-n* junction in m², and

$d(p_n(x))/dx$ is the hole density gradient in the *n*-side in /m⁴.

In Fig. 3.7 we have chosen $x = 0$ to be at the edge of the junction region near the *n*-side; *x* increases from left to right. In Eq. (6), we can see that the only variable is the hole density gradient $d(p_n(x))/dx$. Assuming that the hole density in the *n*-side decreases exponentially with position as in the diffusion of free carriers with fixed boundary conditions (see Section 1.9), we can write

$$p_n(x) = p'\exp(-x/L_p) + p_{n0} \tag{7}$$

where   $p'$ is the excess hole density at $x = 0$ in /m³,

$p_{n0}$ ($= n_i^2/N_D$) is the equilibrium hole density in the *n*-side in /m³, and

$L_p$ is the **diffusion length** of the holes in m.

Differentiating Eq. (7) with respect to position *x* gives

$$d(p_n(x))/dx = -(p_n(x) - p_{n0})/L_p \tag{8}$$

Substituting Eq. (8) into Eq. (6) gives

$$I_p(x) = qD_p(p_n(x) - p_{n0})A_{cs}/L_p \tag{9}$$

Equation (9) shows that the hole diffusion current, $I_p$, depends on the excess hole density, $p'$, through $p_n(x)$. Physically, as *x* increases, the hole diffusion current decreases and is progressively replaced by the electron current, $I_n(x)$. The total current in any one cross section of the *p-n* junction remains the same. The maximum hole current exists at $x = 0$ when $p_n(x) = p' + p_{n0}$. To evaluate $p'$, we assume that the forward-bias voltage produces a quasi–Fermi level for the holes in the *n*-side, which is shifted from the equilibrium Fermi level by the same amount as the applied voltage $V_a$. Schematically, this is illustrated in Fig. 3.8.

**Fig. 3.8**
A forward-bias *p-n*
junction

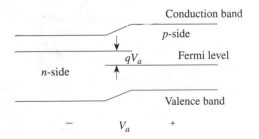

Since the hole density varies exponentially with the shift in the quasi–Fermi level, we have

$$p' = p_{n0}[\exp(qV_a/kT) - 1] \tag{10}$$

where $(E - E_F)/kT \gg 1$ is assumed.

Combined with Eqs. (7) and (10), Eq. (9) becomes

$$I_p = [qD_p n_i^2 A_{cs}/(N_D L_p)][\exp(qV_a/kT) - 1] \tag{11}$$

Equation (11) is identical to Eq. (5) if we set $V_t = kT/q$ and $I_0 = qD_p n_i^2 A_{cs}/(N_D L_p)$.

We can now remove the condition of a $p^+$-$n$ junction and generalize the $I$-$V$ characteristics of a $p$-$n$ junction to include the electron current, $I_n$, in addition to the hole current, $I_p$. The total current $I$ for a forward-bias $p$-$n$ junction is given by their superposition:

$$I = I_p + I_n = I_0[\exp(V_a/V_t) - 1] \tag{12}$$

where  $I_0 = qD_p n_i^2 A_{cs}/(N_D L_p) + qD_n n_i^2 A_{cs}/(N_A L_n)$,

$D_n$ is the electron diffusivity in $m^2/s$,

$N_A$ is the acceptor density in $m^3$, and

$L_n$ is the diffusion length of the electrons in m.

So far, we have only modeled the forward-bias $I$-$V$ characteristics. Under reverse bias, the potential difference across the $p$-$n$ junction increases. This causes the energy bands in the $n$-side to shift downward with respect to those in the $p$-side, so the carrier flow across the $p$-$n$ junction is suppressed. Current flow in a reverse-bias $p$-$n$ junction is, in fact, dominated by the thermal generation of carriers. In the depletion region, the thermally generated carriers are swept across the junction by the internal electric field. They will become a part of the **reverse-bias current.** Since the thermal energy available to the carriers at room temperature is quite small ($\approx 0.026$ eV) compared with the value of the energy gap (1–3 eV), very few carriers are thermally generated at room temperature, and the reverse-bias current is often negligible. It can be shown that the **thermal generation current,** $I_G$, is given by

$$I_G = qn_i W_j A_{cs}/(2\tau_0) \tag{13}$$

where  $W_j$ is the width of the junction region in m, and

$\tau_0$ is the generation lifetime of the carriers in s.

Current due to thermal generation of carriers is sensitive to temperature changes through its dependence on $n_i$ (see Section 2.7). In fact, $I_G$ may be reduced by cooling the $p$-$n$ junction.

To include the generation current in a $p$-$n$ junction, we simply add the term $I_G$ in Eq. (13) to Eq. (12). The only other current component that we have not discussed so far is the recombination current. **Recombination current** is a consequence of carrier recombination in the junction region under forward bias, and it removes some of the carriers that would otherwise become a part of the diffusion currents (see Eq. (6)). Comparatively speaking, recombination current is quite small; at a small bias, it is often overshadowed by the **leakage current** in the *p-n* junction. Assuming that the most effective recombination centers are always located near the middle of the energy gap (so that the energies involved in the capture of the carriers are roughly the same and not limited by one of the two capture processes), we can show that the recombination current will give rise to a correction factor appearing in the exponent of the equation for the *I-V* characteristics (see Eq. (12)). The modified equation becomes

$$I = I_0 \exp[(V_a/\eta_I V_t) - 1] - I_G \tag{14}$$

where    $\eta_I$ is the **ideality factor,** which takes into account the recombination effect.

Equation (14) is plotted graphically in Fig. 3.9 for different values of $\eta_I$. (Equation (14) is only an approximation near $V_a = 0$ since $I_G$ is the steady-state reverse-bias current when $V_a$ is large and negative.)

For an ideal *p-n* junction, $\eta_I = 1$. When recombination and other leakage effects dominate, $\eta_I$ varies between 2 and 4. We have now developed the complete *I-V* characteristics of the *p-n* junction, which in general agree with experimental results. Table 3.1 (p. 154) lists the parameters useful in the calculations of the *I-V* characteristics of *p-n* junctions for the common semiconductors at 300 K.

In the design of a *p-n* junction, different materials parameters will have to be chosen; their selection often depends on the specific application. For instance, for use as an electronic switch, a *p-n* junction will benefit from having a small ideality factor and a small saturation current. These features will generate sharp turn-on and turn-off characteristics. To enhance these features, the diffusion current should dominate the forward-bias *I-V* characteristics, and $\eta_I$ should be close to unity. Both generation current and recombination current, which we want to eliminate, depend on the density of the impurity centers present in the junction region; they can be minimized by using a *defect-free* sample. A reduction in the reverse-bias current is

**Fig. 3.9**

*I-V* characteristics of a *p-n* junction for different values of $\eta_I$

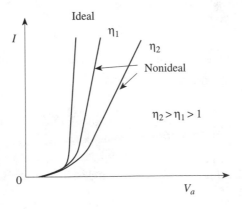

Table 3.1
Parameters related to the
*I-V* characteristics of *p-n*
junctions for common
semiconductors at $T = 300$ K

|  | Si | Ge | GaAs |
|---|---|---|---|
| $n_i$ $(10^{16}/m^3)$ | 1.45 | $2.4 \times 10^3$ | $1.79 \times 10^{-4}$ |
| $\mu_n$ $(m^2/V \cdot s)$ | 0.15 | 0.39 | 0.85 |
| $\mu_p$ $(m^2/V \cdot s)$ | 0.045 | 0.19 | 0.04 |
| $E_g$ (eV) | 1.12 | 0.66 | 1.42 |
| $D_n$ $(10^{-3}\ m^2/s)$ | 3.6 | 10.1 | 22 |
| $D_p$ $(10^{-3}\ m^2/s)$ | 1.22 | 4.92 | 1.03 |
| $\tau_p$ (ms) | 2.5 | 1 | $10^{-5}$ |
| $\tau_n$ (ms) | 2.5 | 1 | $10^{-5}$ |
| $\tau_0$ (ns) |  | 500 |  |

achieved if a wide-energy-gap semiconductor is used, since $n_i$ appearing in Eq. (13) is exponentially dependent on $-E_g$.

The selection of the semiconductors is another important issue to consider. Most semiconductors will form *p-n* junctions provided that they can be made into *p*-type and *n*-type semiconductors. In these cases, the *p-n* junctions are called **homojunctions. Heterojunctions** are formed by two different kinds of semiconductors. Compared with homojunctions, heterojunctions usually have more defects in the junction region due to the lattice mismatch (see Section 1.8), and this has been a critical problem in the design of heterojunction devices. Frequently, in a heterojunction diode, a part of the input current will be lost through carrier trapping and recombination near the junction interface. Some semiconductors, such as CdS, are unsuitable for making *p-n* junctions because they cannot be doped into *p*-type or *n*-type semiconductors due to self-compensation in the presence of native defects.

The control of dopants in these semiconductors is also a serious problem. *P-n* junctions that are made of wide-energy-gap semiconductors, such as ZnO and SiC, suffer from the presence of internal defects. These defects are responsible for the nonideal characteristics, excessive thermal noise, and large current transients caused by carrier trapping. A few semiconductors, such as ZnS, have very low carrier mobilities and are not suitable for making *p-n* junctions because of the high internal resistance. Surface defects in semiconductors give rise to leakage current, and in planar devices, they can significantly degrade the device performance, particularly at a low bias (see Fig. 3.5).

**Example 3.5** If the equilibrium hole density at the edge of a forward-bias *p-n* junction is $1 \times 10^{10}/m^3$, estimate the increase in the value of the hole density if there is a bias of 0.3 V. $T = 300$ K ($kT/q = 0.0259$ V).

***Solution:*** Based on Eq. (10), $p' = p_{n0}[\exp(qV_a/kT) - 1] = 1 \times 10^{10} \times \exp(0.3/0.0259)/m^3 = 1.5 \times 10^{15}/m^3$. §

**Example 3.6** Compute the thermal generation current density in a semiconductor if $n_i = 1.05 \times 10^{16}/m^3$, $W_j = 0.1$ μm, and $\tau_0 = 1 \times 10^{-10}$ s.

***Solution:*** Based on Eq. (13), $I_G/A_{cs} = qn_iW_j/(2\tau_0) = 1.6 \times 10^{-19}$ C $\times$ $1.05 \times 10^{16}/\text{m}^3 \times 10^{-7}$ m$/(2 \times 10^{-10}$ s$) = 0.8$ A/m$^2$. §

**Example 3.7**  Compute the value of the saturation current density in Si at $T = 300$ K. Assume $N_D = 10^{22}/\text{m}^3$, $D_p = 1.22 \times 10^3$ m$^2$/s, $n_i = 1.45 \times 10^{16}/\text{m}^3$, and $L_p = 1.76 \times 10^{-3}$ m.

***Solution:***  Based on Eq. (11), $I_0/A_{cs} = qD_pn_i^2/(N_DL_p) = 1.6 \times 10^{-19}$ C $\times$ $1.22 \times 10^3$ m$^2$/s $\times (1.45 \times 10^{16}/\text{m}^3)^2/(10^{22}/\text{m}^3 \times 1.76 \times 10^{-3}$ m$) = 1.9 \times 10^{-9}$ A/m$^2 = 1.9$ nA/m$^2$. §

## Space-Charge and Charge Storage Effects

For a time-varying signal, it is important to consider the space-charge effect in a *p-n* junction. As shown in Fig. 3.10, the presence of oppositely charged donors and acceptors in the junction region is quite similar to a parallel-plate capacitor. The resulting capacitance is called the **junction capacitance,** $C_{jun}$(F), and is given by

$$C_{jun} = \epsilon_s A_{cs}/W_j \tag{15}$$

where  $\epsilon_s$ is the semiconductor permittivity in F/m,

$A_{cs}$ is the cross-sectional area of the junction in m$^2$, and

$W_j$ is the width of the junction region in m.

The width of the junction region $W_j$ changes with the bias voltage, which also changes the junction capacitance. For an abrupt junction—i.e., a *p-n* junction made

**Fig. 3.10**
A *p-n* junction showing the effect of junction capacitance

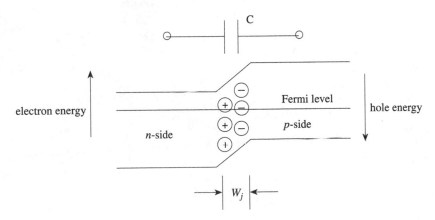

with uniform dopant densities, as shown in Fig. 3.11—the width of the junction region under reverse bias or even slightly forward bias is given by

$$W_j = \sqrt{2\epsilon_s(V_{bi} - V_a)/(qN_{\text{eff}})} \qquad \text{(16)}$$

where    $N_{\text{eff}} = N_A N_D /(N_A + N_D)$,

$V_{bi} (= \Phi_{bi}/q)$ is the built-in voltage in V, and

$V_a$ is the bias voltage in V.

Often, p-n junctions are not abrupt, especially when they are prepared by surface diffusion. In this case, the dopant densities near the junction region vary linearly with position, and the width of the junction region is given by

$$W_j = [12\epsilon_s(V_{bi} - V_a)/(qa)]^{1/3} \qquad \text{(17)}$$

where    $a$ is the gradient of the dopant density in /m$^4$.

Both Eqs. (16) and (17) suggest that the reverse-bias p-n junction acts as a voltage-controlled variable capacitor, or a **varactor** (variable-reactance device). The majority of varactors operate in reverse bias since the dynamic voltage range of a forward-bias p-n junction is quite small. Fig. 3.12 shows how the capacitance of a $p^+$-n junction varies with the substrate dopant density $N_D$. Varactors are frequently used

Fig. 3.11
A uniformly doped p-n
junction

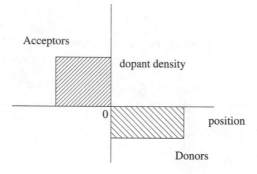

Fig. 3.12
Capacitance of a $p^+$-n
junction versus the donor
density

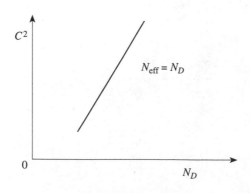

in the design of voltage-controlled oscillators (VCOs), where the oscillator frequency is set to vary with the junction capacitance.

In addition to the voltage dependence of the junction capacitance, carriers stored in a *p-n* junction under forward bias will give rise to another capacitance effect called the **diffusion capacitance.** Diffusion capacitance is due to majority carriers moving into and out of the junction region under forward bias. At a reasonably low frequency, the diffusion capacitance (F) is given by

$$C_{\text{diff}} = [qA_{cs}(L_p p_{n0} + L_n n_{p0})/2V_t]\exp(V_0/V_t) \tag{18}$$

where $L_p$ and $L_n$ are the respective diffusion lengths of the holes and the electrons in m,

$p_{n0} = n_i^2/N_D$; $n_{p0} = n_i^2/N_A$,

$V_t = kT/q$, and

$V_0$ is the forward-bias voltage in V.

The diffusion capacitance therefore increases exponentially with the forward-bias voltage. Physically, the junction region becomes narrower as the forward-bias voltage increases, which also increases the diffusion capacitance. Combining Eqs. (12) and (18), it can be shown that for a $p^+$-$n$ junction,

$$C_{\text{diff}} \approx I_0 \tau_p/(2V_t) \tag{19}$$

where $I_0$ is the saturation current in A, and

$\tau_p$ is the lifetime of the holes in s.

The **rise time** of a forward-bias *p-n* junction often dictates the frequency of operation. Since the rise time of a forward-bias *p-n* junction is given by the product of the **junction resistance** and the diffusion capacitance, Eq. (19) suggests that the rise time (and also the **switching time**) will be directly proportional to $\tau_p$. This is to be expected since in a $p^+$-$n$ junction, the dominant charge storage is due to holes, and any time delay due to the addition or removal of the holes in the junction region has to be limited by the hole **lifetime.** The total capacitance of the *p-n* junction is the sum of $C_{\text{jun}}$ and $C_{\text{diff}}$. Fig. 3.13 shows how the diode capacitance varies with

Fig. 3.13
C-V characteristics of a
*p-n* junction

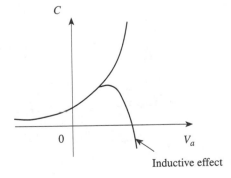

the bias voltage. At a large forward bias, inductance due to the parasitic dominates (such as the mutual inductance in the lead wires). As the junction resistance becomes small, the current is no longer capacitive. During a capacitance-voltage (*C-V*) measurement, the inductive effect actually appears as a negative capacitance. Fig. 3.14 shows the **equivalent-circuit model** for a *p-n* junction.

The stored charges, whether they are carriers or stationary charges, affect the time response of the *p-n* junction during switching. The total **delay time** of a *p-n* junction is frequently expressed as the sum of a **turn-on time** and a **turn-off time**. Short turn-on and turn-off times in a *p-n* junction are highly desirable as far as signal bandwidth and the data-transmission rate (number of switching operations per unit time) are concerned. The turn-on time of a *p-n* junction is usually quite short. The turn-off time, however, can be long in the presence of charge storage. For high-speed operation, *p-n* junctions are frequently prevented from operating in saturation simply to avoid this charge-storage effect. Heavy doping in the junction region is an effective means for reducing the carrier lifetimes and shortening the turn-off time. However, heavy doping can sometimes intensify nonideal effects, such as leakage.

**Example 3.8**   Compute the junction capacitance per unit area of a *p-n⁺* junction with the following characteristics: $\epsilon_s = 1.05 \times 10^{-10}$ F/m, $N_{\text{eff}} \approx N_A = 1 \times 10^{22}/\text{m}^3$, $V_{bi} = 0.7$ V, $V_a = -1$ V.

***Solution:***   Based on Eq. (16),

$$W_j = \sqrt{2\epsilon_s(V_{bi} - V_a)/(qN_{\text{eff}})}$$
$$= \sqrt{2 \times 1.05 \times 10^{-10} \text{ F/m} \times (0.7 \text{ V} + 1 \text{ V})/(1.6 \times 10^{-19} \text{ C} \times 1 \times 10^{22}/\text{m}^3)}$$
$$= 0.47 \times 10^{-6} \text{ m} = 0.47 \text{ μm}$$

Based on Eq. (15), $C_{\text{jun}}/A_{cs} = \epsilon_s/W_j = 1.05 \times 10^{-10}$ F/m/0.47 $\times 10^{-6}$ m $= 2.23 \times 10^{-4}$ F/m². **§**

**Fig. 3.14**
Equivalent circuit model
of a *p-n* junction

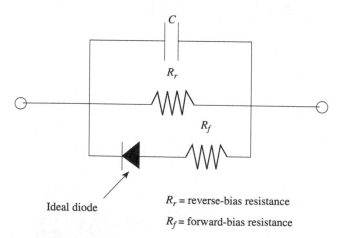

$R_r$ = reverse-bias resistance

$R_f$ = forward-bias resistance

**Example 3.9**   Compute the diffusion capacitance per unit area in the $p$-$n^+$ junction of Example 3.8. Assume $I_0/A_{cs}$ = 1.9 nA and $\tau_n$ = 1 ms. $V_t$ = 0.026 V.

**Solution:**   Based on Eq. (19), $C_{diff}/A_{cs} \approx I_0\tau_n/(2V_tA_{cs})$ = 1.9 $\times$ 10$^{-9}$ A $\times$ 1 $\times$ 10$^{-3}$ s/(2 $\times$ 0.026 V) = 3.65 $\times$ 10$^{-11}$ F/m$^2$ = 36.5 pF/m$^2$.   **§**

**Example 3.10**   Compute the rise time of a $p$-$n$ junction based on the data in Example 3.9. Assume $V_a$ = +0.6 V.

**Solution:**   Rise time $\approx R_{dyn}C_{diff}$

$$= V_tC_{diff}/I$$
$$= 0.026 \text{ V} \times 36.5 \times 10p^{-12} \text{ F/m}^2/(1.9 \times 10^{-9} \text{ A} \times \exp(0.6 \text{ V}/0.026 \text{ V}))$$
$$= 0.475 \times 10^{-12} \text{ s} = 0.475 \text{ ps.}   \textbf{§}$$

## Tunneling *p-n* Junctions

**Tunneling** occurs in a $p$-$n$ junction or a tunnel diode under both forward bias and reverse bias if the junction region is very thin (5–10 nm). The process of tunneling allows electrons to move through the potential barrier in the junction region without the electrons having sufficient energy to surmount the barrier. The current so created in principle will not dissipate any energy, and the time to pass through the barrier is almost instantaneous. The requirements for tunneling to occur are conservation of energy and momentum and the availability of empty energy states to receive the tunneling electrons, as in the case of a highly doped $p$-$n$ junction. Fig. 3.15 shows the energy band diagram under forward bias and reverse bias for a tunneling diode where the filled conduction band states in the $n$-side are directly opposite the empty valence band states in the $p$-side. Because of the high doping, the potential barrier is very thin, and tunneling in either direction is usually possible.

Quantitatively, tunneling is described by a **transmission coefficient,** $D_{12}$, which determines the probability that the tunneling event will take place. For an electron with a potential energy $E$ (eV) and a potential barrier given by $\Phi_B$ (eV), the transmission coefficient, $D_{12}$, is given by

$$D_{12} = D_b\exp(-2d_b\sqrt{2m_e^*(\Phi_B - E)}/h') \qquad (20)$$

where   $D_b$ is a constant,
   $m_e^*$ is the effective mass of the electrons in kg,
   $d_b$ is the barrier width in m, and
   $h'$ ($= 1.05 \times 10^{-34}$ J·s) is Planck's constant divided by $2\pi$.

**Fig. 3.15**
Forward and reverse
biasing of a tunneling
diode

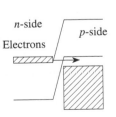

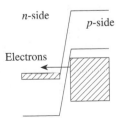

Forward bias                    Reverse bias

When a voltage $V_a$ is applied to the *p-n* junction, $\Phi_B = q(V_{bi} - V_a)$ and $d_b$ is given by the width of the depletion layer, i.e., $W_j$. The **tunneling current** (A) then becomes

$$I_{tun} = qA_{cs} \int_{E_c}^{\acute{E}} [f_t S_1 D_{12} S_2] \, dE \tag{21}$$

where  $S_1$ and $S_2$ are the respective densities of states of the energy bands in the two sides of the *p-n* junction in /m$^3$,

$f_t$ is a tunneling frequency parameter in m$^4$/s·eV,

$E_c$ is the energy at the conduction band edge in eV, and

$E'$ is the upper energy limit for tunneling in eV.

Fig. 3.16 shows the tunneling *I-V* characteristics of a heavily doped *p-n* junction. The tunneling current under forward bias is usually quite small, and it vanishes at a large forward bias when the filled conduction band states in the *n*-side are no longer facing the empty states in the *p*-side. The trailing edge of the tunneling current gives rise to a negative (**dynamic**) **resistance** effect. **Negative resistance in tunneling diodes** is frequently utilized in the design of oscillator circuits. The negative (dynamic) resistance can give rise to voltage amplification. Tunneling also occurs in a heavily doped *p-n* junction under reverse bias, as shown in Fig. 3.17. When operating in reverse bias, the device is known as a *backward diode*. A backward diode has highly nonlinear reverse-bias characteristics and is frequently used in communication circuits for signal detection and mixing. Tunneling diodes

**Fig. 3.16**
*I-V* characteristics of a
tunneling diode

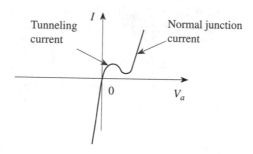

**Fig. 3.17**
*I-V* characteristics of a
backward diode

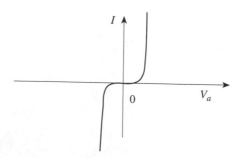

are used because of their fast response, their low noise, and the absence of charge storage. Equation (21) also suggests that the tunneling current is relatively insensitive to temperature change.

**Example 3.11**    Compute the base width of a *p-n* junction at zero bias if the dopant densities on both sides are $1 \times 10^{24}/m^3$, $\tau_s = 1.05 \times 10^{-10}$ F/m, $N_{eff} \approx N_A/2 = 0.5 \times 10^{24}/m^3$, and $V_{bi} = 0.7$ V.

*Solution:*    Based on Eq. (16),

$$W_j = \sqrt{2\epsilon_s V_{bi}/(qN_{eff})}$$
$$= \sqrt{2 \times 1.05 \times 10^{-10} \text{ F/m}(1.6 \times 10^{-19} \text{ C} \times 0.5 \times 10^{24}/m^3)} \times (0.7 \text{ V})^{1/2}$$
$$= 4.28 \times 10^{-8} \text{ m} = 42.8 \text{ nm. } §$$

**Example 3.12**    Compute the electron transmission coefficient for the tunnel diode in Example 3.11. Assume $D_b = 1$, $m_e^* = 0.91 \times 10^{-30}$ kg, and the energy of the electron $E = E_{th} = 0.026$ eV ($T = 300$ K).

*Solution:*    Based on Eq. (20),

$$D_{12} = D_b \exp(-2d_b\sqrt{2m_e^*(\Phi_B - E)}/h')$$
$$= 1 \times \exp(-2 \times 4.28 \times 10^{-8} \text{ m}$$
$$\times \sqrt{2 \times 0.91 \times 10^{-30} \text{ kg} \times (0.7 \text{ eV} - 0.026 \text{ eV}) \times 1.6 \times 10^{-19} \text{ J/eV}}/(1.06 \times 10^{-34} \text{ J·s})$$
$$= 3.82 \times 10^{-9} \text{ } §$$

## Junction Breakdown

A *p-n* junction will break down when the reverse-bias voltage is sufficiently large. This effectively limits the dynamic range of the reverse-bias *p-n* junction. Normally, breakdown is the result of carrier multiplication, or the **avalanche effect,** which occurs when the kinetic energy of the carriers is sufficient to break the energy bonds of the host atoms and release additional electrons. In a semiconductor diode, junction breakdown takes place when a **critical electric field** occurs—i.e., the electric field exceeds a critical value (of the order of $10^8$ V/m). As a result of breakdown, additional electrons and holes are produced in the depletion region, and carrier multiplication continues until the current or the power dissipation causes permanent damage to the *p-n* junction. The multiplication effect is called **impact ionization.** The commencement of impact ionization is initiated at the critical field, $\acute{E}_{cr}$ (in V/m). Table 3.2 lists the critical fields for the common semiconductors at 300 K.

Quantitatively, the **breakdown voltage,** $V_{br}$ (V), is related to the critical field $\acute{E}_{cr}$ by

$$V_{br} = \epsilon_s \acute{E}_{cr}^2/(2qN_{eff}) \tag{22}$$

| Semiconductor | $\acute{E}_{cr}$ ($\times$ 10⁸ V/m) |
|---|---|
| Si | 0.4 |
| Ge | 0.1 |
| GaAs | 0.4 |
| SiC | 2.3 |

Table 3.2
Critical fields $\acute{E}_{cr}$ for common semiconductors at 300 K

where    $\epsilon_s$ is the semiconductor permittivity in F/m, and

$$N_{\text{eff}} = N_A N_D / (N_A + N_D)$$

For a given $\acute{E}_{cr}$, the breakdown voltage decreases with increasing doping. Junction breakdown can be reversible or irreversible, depending on the type of damage. If the input voltage appears in the form of a pulse train (with a low duty cycle), breakdown is usually reversible, as long as the heating effect does not cause permanent damage to the *p-n* junction. Permanent damage may appear in the form of microchannels formed in the junction region, as shown in Fig. 3.18. These microchannels are highly conducting, and they can form shorting paths, where current tends to concentrate. The associated heating effect will expand the size of the microchannels until *p-n* junction is perforated.

Microscopic consideration of breakdown requires an analysis of the ionization rates of the carriers and is measured in terms of a parameter known as the multiplication factor, *M*. The multiplication factor, *M*, is given by the ratio of the current density at the end of the breakdown divided by the initial current density. Fig. 3.19 shows a schematic of the ionization process for electrons.

Based on Fig. 3.19, we can show that

$$d(j_n)/dx - (\alpha_i - \beta_i)j_n = -(\alpha_i - \beta_i)j_T + \alpha_i j_T \tag{23}$$

where    $j_n$ is the electron current density in A/m²,

$j_T$ ($= j_n + j_p$) is the total current density in A/m²,

$\alpha_i$ is the electron **ionization rate** in /m, and

$\beta_i$ is the hole ionization rate in /m.

Fig. 3.18
A microchannel in a *p-n* junction

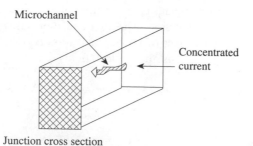

Junction cross section

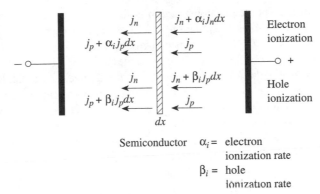

Fig. 3.19
Electron and hole
ionization in a depletion
region

For the simple case when $\alpha_i = \beta_i$, the electron multiplication factor, $M_n$, is given by

$$M_n = j_n(L)/j_n(0) = 1/[1 - \int_0^L (\alpha_i)\,dx] \tag{24}$$

where    $L$ is the length of the ionization region in m.

Equation (24) suggests that breakdown occurs when the term $\int_0^L (\alpha_i)\,dx = 1$ (note that the ionization rate $\alpha_i$ itself can be less than 1). Therefore, even a relatively small value of $\alpha_i$ can cause breakdown when it is integrated over a long distance.

Fig. 3.20 shows the electron and hole ionization rates for a GaAs *p-n* junction as a function of the electric field. As observed, the electron and hole ionization rates are quite comparable in GaAs.

Breakdown is normally undesirable because it leads to the destruction of a *p-n* junction. In specific applications, as in the case when a *p-n* junction is used as photodetector, it is common to operate the device in the breakdown mode in order to increase the detectivity of the device. Such a device is called an **avalanche photodiode**, or APD. The details of the device operation are described in Chapter 6. A *p-n* junction operating in the breakdown mode is sometimes used as a (microwave) oscillator and is called an **IMPATT (impact ionization avalanche transit time)**

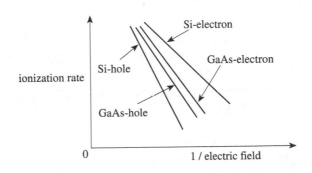

Fig. 3.20
Ionization rates in
semiconductors

**diode.** The device construction is shown in Fig. 3.21. It consists of an input region (a reverse-bias *p-n* junction), where the excess carriers are generated and breakdown occurs, and a lightly doped region that allows phase shift to occur as the excess carriers pass through. During operation, an applied ac voltage (superimposed on a dc bias voltage) temporarily drives the device into breakdown, and a current pulse (due to the excess carriers) is initiated near the input terminal. The excess carriers are then transported through the lightly doped region at saturation velocity until the excess carriers exit at the opposite terminal. Depending on the length of the lightly doped region and the saturation velocity, the IMPATT diode can operate at a very high frequency (up to tens of gigahertz). As an oscillator, the current and the applied ac voltage must be 180° out of phase to give a negative resistance effect, as shown in Fig. 3.22. Both Si and GaAs are used in IMPATT diodes.

In a highly doped *p-n* junction, $V_{br}$ can also be due to tunneling. The built-in potential $\phi_{bi}$ of a *p-n* junction presents a barrier to the carriers moving across the *p-n* junction. When the barrier width is sufficiently thin, as in the case of a highly doped *p-n* junction, electrons will tunnel through the barrier. This results in a tunneling current, and the breakdown voltage is approximately given by

$$V_{br} \approx 4E_g/q \tag{25}$$

where    $E_g$ is the energy gap of the semiconductor in eV.

**Fig. 3.21**
Structure of an IMPATT diode

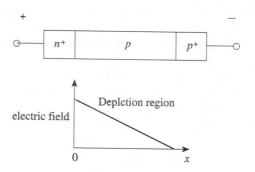

**Fig. 3.22**
Hole motion in an IMPATT diode

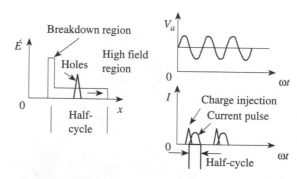

Breakdown due to tunneling normally occurs at a lower voltage when compared with avalanche breakdown, and it has a value between 5 and 6 V (avalanche breakdown voltage has a much higher value). Fig. 3.23 shows the *I-V* characteristics of a *p-n* junction during tunneling breakdown and avalanche breakdown. Avalanche breakdown is usually accompanied by extensive heating, and the temperature rise will further increase the reverse current until the device is permanently damaged.

**Example 3.13**  Compute the breakdown voltage in Si if $N_A = N_D = 1 \times 10^{22}/m^3$, $\epsilon_s = 1.05 \times 10^{-10}$ F/m, and $\acute{E} = 0.4 \times 10^8$ V/m.

*Solution:*  Based on Eq. (22), $V_{br} = \epsilon_s \acute{E}_{cr}^2/(2qN_{eff}) = 1.05 \times 10^{-10}$ F/m $\times (0.4 \times 10^8$ V/m$)^2/(2 \times 1.6 \times 10^{-19}$ C $\times 1 \times 10^{22}/m^2/2)$ V $= 10.5$ V.  **§**

**Example 3.14**  Compute the breakdown voltage in GaAs due to tunneling.

*Solution:*  In GaAs, $E_g = 1.42$ eV. Based on Eq. (25), $V_{br} \approx 4E_g/q = 4 \times 1.42$ V $= 5.7$ V.  **§**

**Example 3.15**  In Si, if the electron ionization rate is $10^6/m$, compute the minimum length of the depletion region for avalanche breakdown to be possible. Assume $\alpha_i$ is a constant. What is the maximum frequency of operation for the IMPATT diode? The saturation velocity of electrons in Si is $1 \times 10^6$ m/s.

*Solution:*  Based on Eq. (24), $M_n = \infty$ requires $\int_0^L (\alpha_i)\, dx = 1$. Since $\alpha_i = 10^6/m$, $L$ must be at least 1 µm. During avalanche breakdown, the peak current is 90° out of phase with the applied voltage. Since the total phase shift is 180°, the balance in the phase shift (90°) is achieved when the carriers transit through the depletion region. The maximum frequency of operation is $\omega = \pi v_s/L = 3.14$ rad $\times 10^5$ m/s/($1 \times 10^{-6}$ m) $= 3.14 \times 10^{11}$ rad/s. Since $\omega = 2\pi f$, $f = \omega/2\pi = 2.14 \times 10^{11}$ rad/s/($2 \times 3.14$ rad) $= 5 \times 10^{10}$ Hz $= 50$ GHz.  **§**

## Noise in a *p-n* Junction

**Shot noise** is generated whenever carriers cross a barrier, and it is the dominant noise source in a *p-n* junction. Shot noise is of a diffusion nature due to the diffusion

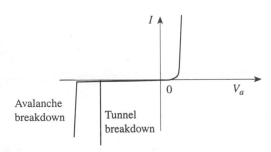

of the carriers. Shot noise is often modeled as a current source with the square of the rms value ($A^2$). It is given by

$$\tilde{i}^2 = 2q(I_d + 2I_0)\delta f \tag{26}$$

where     $I_d$ is the diode current in A,

$I_0$ is the saturation current in A, and

$\delta f$ is the frequency range in Hz.

**Generation** and **recombination noise** are also present in a *p-n* junction; they arise from the fluctuations in the carrier densities or changes in the semiconductor resistance.

**Example 3.16**     Compute the shot noise current source if a diode operates at reverse bias with a bandwidth of 50 MHz and the saturation current is 1 μA. Assume $I_d = 0$.

*Solution:*   Based on Eq. (26),

$$
\begin{aligned}
\hat{i} &= \sqrt{2q(I_d + 2I_0)\delta f} \\
&= \sqrt{2 \times 1.6 \times 10^{-19}\,\text{C} \times 2 \times 10^{-6}\,\text{A} \times 50 \times 10^6/\text{s}} \\
&= 1.78 \times 10^{-9}\,\text{A} = 1.78\,\text{nA} \quad \S
\end{aligned}
$$

## Equivalent-Circuit Model of a *p-n* Junction

Most engineering applications of a *p-n* junction involve the use of the small-signal *I-V* characteristics, and it is convenient to represent them by an equivalent-circuit model. Such a model is shown in Fig. 3.24. In the model, the device characteristics are incorporated into parameters represented by discrete or lumped passive circuit elements, such as capacitors, resistors, inductors, a voltage source, and a current source. During simulation of the model, the *I-V* characteristics are reproduced at the terminals. There are a number of advantages in using an equivalent-circuit model. First of all, the physical device is reduced to a small number of circuit

**Fig. 3.24**
An equivalent circuit of a
*p-n* junction

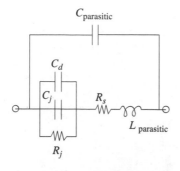

elements. For small signals, their properties are linear; i.e., superposition of signals is possible. Secondly, the device properties may easily be modified by changing the circuit parameters without changing the model. Finally, the operation of the device is transparent to the users, so they need not be concerned with the fine details of the physics of the device. Figure 3.25 shows the *I-V* characteristics of a *p-n* junction simulated using the equivalent-circuit model shown in Fig. 3.24.

**Example 3.17**    Should there be inductances in the equivalent-circuit model of a *p-n* junction?

*Solution:*    In general, the answer is yes because a magnetic field will be present whenever there is a current passing, so there can be mutual inductance (see Chapter 5). However, since the impedance of an inductor increases with frequency, only at a very high frequency will the inductance effect become important. **§**

## Heterojunctions

A heterojunction is a *p-n* junction made from two different semiconductors. In addition to a modified energy band diagram, as shown in Fig. 3.26, a heterojunction often has at its interface a potential spike and sometimes an energy trough filled with electrons. The latter will give rise to a **two-dimensional electron gas effect,** which we discuss in Section 3.7. The growth of high-quality heterostructures has been a materials problem until recently, when better-controlled MBE and MOCVD machines became available (see Section 1.7). Nevertheless, the effective growth of a heterojunction with minimum strain at the interface is still a challenge. The main difficulty has to do with the lattice matching at the interface; in a GaAlAs/GaAs heterojunction, only a lattice mismatch of about 0.04% can be achieved.

**Fig. 3.25**
*I-V* characteristics from an ideal diode model

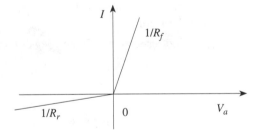

**Fig. 3.26**
Energy band diagram of a heterojunction

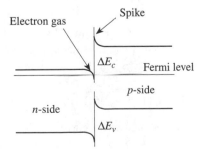

For a heterojunction made with a narrow-gap *p*-type semiconductor and a wide-gap *n*-type semiconductor, the conduction band discontinuity, $\Delta E_c$, at the interface (eV) is given by

$$\Delta E_c = \xi_1 - \xi_2 \tag{27}$$

where    $\xi_1$ and $\xi_2$ are the respective electron affinities (*electron affinity* is defined as the energy difference between the conduction band edge and the potential at infinity) in the two semiconductors in eV.

Equation (27) leads to a built-in potential, $\Phi_{bi}$, of the form

$$\Phi_{bi} = E_{g1} - \Delta E_n - \Delta E_p + \Delta E_c \tag{28}$$

where    $E_{g1}$ is the energy gap of the narrow-gap *p*-type semiconductor in eV,

$\Delta E_n$ is the energy difference between the conduction band edge and the Fermi level in the wide-gap semiconductor in eV, and

$\Delta E_p$ is the energy difference between the valence band edge and the Fermi level in the narrow-gap semiconductor in eV.

The junction width $W_j$ of a heterojunction is given by

$$W_j = W_n + W_p \tag{29}$$

where    $W_n = \sqrt{2\epsilon_1\epsilon_2 N_A V_{bi}/[qN_D(\epsilon_2 N_D + \epsilon_1 N_A)]}$,
$W_p = \sqrt{2\epsilon_1\epsilon_2 N_D V_{bi}/[qN_A(\epsilon_2 N_D + \epsilon_1 N_A)]}$,

$\epsilon_1$ and $\epsilon_2$ are the respective permittivities of the *p*-type and the *n*-type semiconductors in F/m,

$N_A$ and $N_D$ are the respective acceptor density and donor density in $/m^3$, and

$V_{bi} = \Phi_{bi}/q$.

The junction capacitance, $C_{jun}$, of the heterojunction (F) is then given by

$$C_{jun} = \sqrt{\epsilon_1\epsilon_2 N_A N_D/[2(\epsilon_2 N_D + \epsilon_1 N_A)V_{bi}]} \tag{30}$$

In the presence of an applied voltage $V_a$ in volts, $V_{bi}$ in Eq. (30) has to be replaced by $V_{bi} - V_a$. Normally, current flow in a heterojunction is considered to be of a thermionic nature. In the simplest form, the junction current $I$ (A) is given by

$$I = I_0(1 - V_a/V_{bi})[\exp(V_a/V_t) - 1] \tag{31}$$

where    $I_0$ is the same as the saturation current in A of a Schottky barrier diode (to be described in Section 3.3) with a barrier height given by $\Phi_{bi}$.

Equation (31) is somewhat different from Eq. (12), the *I-V* characteristics of a simple *p-n* junction.

**Example 3.18**   The electron affinity (eV) for the ternary compound $Ga_{1-y}Al_yAs$ varies as $4.07 - 1.06y$, where $y$ is the percent of Al. In addition, when $y < 0.45$ the energy gap varies as $1.424 + 1.247y$. Compute $\Delta E_c$ and $\Delta E_v$ when $y = 0.3$.

*Solution:*   When $y = 0.3$, $\Delta E_c = 1.06 \times 0.3$ eV $= 0.32$ eV. $\Delta E_g = 1.247 \times 0.3$ eV $= 0.374$ eV. Since $\Delta E_c + \Delta E_v = \Delta E_g$, $\Delta E_v = 0.374 - 0.32$ eV $= 0.054$ eV.   §

## 3.3   SCHOTTKY JUNCTION AND ITS ELECTRONIC PROPERTIES

A rectifying device can also be formed when a metal is in electrical contact with a semiconductor; it is called a Schottky junction, or a **Schottky diode.** The operation of the Schottky junction relies on carrier injection over the metal-semiconductor interface barrier $\Phi_B$, as shown in Fig. 3.27.

This situation is quite different from the operation of a *p-n* junction when the diffusion of carriers is mainly responsible for the current flow. Not all metal-semiconductor interfaces will become Schottky junctions. For an *n*-type semiconductor, a Schottky junction is formed if the metal work function $\Phi_m$ (the energy difference between the Fermi energy (or the Fermi level in the case of a semiconductor) and the potential at infinity) is larger than the semiconductor **work function, $\Phi_s$.** The reverse is true for a *p*-type semiconductor; i.e., the metal workfunction has to be smaller than the semiconductor work function. Schottky junctions are in many ways quite similar to the one-sided step *p-n* junction. For instance, a space-charge region exists in the semiconductor side of the Schottky junction as a result of the diffusion of the carriers. This space-charge layer is balanced by the **image charges** created at the metal surface, so the junction as a whole is neutral. Fig. 3.28 shows the energy band diagram of a Schottky junction formed on an *n*-type semiconductor. The current-voltage (*I-V*) characteristics are shown in Fig. 3.29.

**Fig. 3.27**
Electrons surmounting a
Schottky barrier

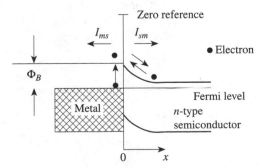

Fig. 3.28

Energy band diagram of a
Schottky junction

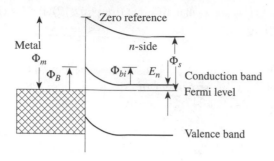

Fig. 3.29

*I-V* characteristics of a
Schottky junction

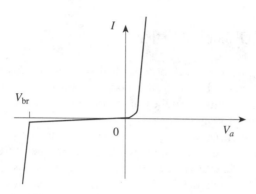

## Schottky Junction at Equilibrium

Current flow in an ideal Schottky junction is due to **thermionic emission**—i.e., thermal emission of the carriers over a potential barrier. For convenience, we consider the metal/*n*-type semiconductor Schottky junction as shown in Fig. 3.27. At equilibrium—i.e., without any bias voltage—we expect the device current to be zero. This, of course, implies that the net carrier flow across any cross section of the Schottky junction must also be zero. Carrier emission across the junction depends on the ability of the electrons to surmount the potential barrier (i.e., the energy difference between the electrons and the barrier). To compute the number of electrons crossing a barrier, we need first to determine the kinetic energy of the electrons, which is related to their velocity distribution in the direction of the current flow. We shall assume this to be the $+x$ direction. The velocity distribution $f_{vx}$ of the electrons is then given by

$$f_{vx} = \sqrt{[m_e^*/(2\pi\Phi_t)]\exp[-m_e^* v_x^2/(2\Phi_t)]} \qquad (32)$$

where    $v_x$ is the velocity of the electrons in the $+x$ direction in m/s,

   $m_e^*$ is the effective mass of the electrons in kg, and

   $\Phi_t = kT.$

The average velocity of electrons, $\langle v_x \rangle$, is the product of $v_x$ and $f_{vx}$ integrated over all velocities and is given by

$$\langle v_x \rangle = \int_0^\infty [v_x f_{vx}] \, dv_x = \sqrt{\Phi_t/(2\pi m_e^*)} \tag{33}$$

The current $I$ (A), due to these electrons can be written as

$$I = -qA_{cs} \int_0^\infty [v_x \partial n/\partial E] \, dE \tag{34}$$

where    $q$ is the electron charge in C,

$\partial n/\partial E$ is the rate at which the electron density changes with energy in the energy band in /m³·eV, and $A_{cs}$ is the cross-sectional area of the device in m².

For electrons moving from the semiconductor into the metal, the current $I_{sm}$ (see Fig. 3.27) in amperes is given by

$$I_{sm} = A^* T^2 A_{cs} \exp[(-\Phi_{bi} - E_n)/\Phi_t] \tag{35}$$

where    where $A^*$ ($= 4\pi q m_e^* k^2/h^3 = 1.2 \times 10^6$ A/m²·K²) is called the Richardson constant,

$\Phi_{bi}$ is the built-in potential in eV in the semiconductor side of the junction, and

$E_n$ is the energy difference between the semiconductor Fermi level and the conduction band edge in eV.

$I_{sm}$ is positive since the electrons are moving in the $-x$ direction—i.e., from the semiconductor into the metal (see Figure 3.27). The reverse current, $I_{ms}$, flows in the $-x$ direction and is given by

$$I_{ms} = -A^* T^2 A_{cs} \exp(-\Phi_B/\Phi_t) \tag{36}$$

where    $\Phi_B$ is the barrier height in eV.

From Fig. 3.28 we observe that $\Phi_B$ is the energy difference between the metal work function $\Phi_m$ and the semiconductor workfunction $\Phi_s$. At equilibrium, there is no net current flowing (i.e., $I_{sm} = -I_{ms}$) and Eqs. (35) and (36) give

$$\Phi_B = \Phi_{bi} + E_n \tag{37}$$

Physically $\Phi_{bi}$ is the energy barrier electrons on either side have to surmount to cross the junction interface. The band bending in the junction region is responsible for the built-in potential $\Phi_{bi}$; as in the p-n junction, it is indicative of the presence of a space-charge layer. Table 3.3 lists the Schottky barrier heights in Si and GaAs for different surface metals at 300 K.

**Example 3.19**   Compute $\langle v_x \rangle$ in a Schottky diode at 300 K. Assume $m_e^* = 0.91 \times 10^{-30}$ kg and $kT/q = 0.026$ V.

*Solution:*   Based on Eq. (33),

$$\langle v_x \rangle = \sqrt{(\Phi_t / 2\pi m_e^*)}$$
$$= \sqrt{0.026 \text{ V} \times 1.6 \times 10^{-19} \text{ C}/(2 \times 3.14 \times 0.91 \times 10^{-30} \text{ kg})}$$
$$= 2.69 \times 10^4 \text{ m/s} \quad \S$$

**Example 3.20**   Compute the value of the Richardson constant $A^*$. Assume that $m_e^* = 0.91 \times 10^{-30}$ kg.

*Solution:*   Based on Eq. (35), $A^* = 4\pi q m_e^* k^2 / h^3 = 4 \times 3.14 \times 1.6 \times 10^{-19} \text{ C} \times 0.91 \times 10^{-30} \text{ kg} \times (1.38 \times 10^{-23} \text{ J/K})^2/(6.62 \times 10^{-34} \text{ J·s})^3 = 1.2 \times 10^6 \text{ A/m}^2 \cdot \text{K}^2. \quad \S$

## Schottky Junction under Bias

Under forward bias—i.e., when the metal side is connected to the positive voltage supply—the current $I_{sm}$ flowing from the metal into the *n*-type semiconductor will increase due to a lowering in the built-in potential, as shown in Fig. 3.30.

**Table 3.3**
Values of the Schottky barrier heights in Si and GaAs for different surface metals at 300 K

| Metal | $\Phi_B$(Si) (eV) | | $\Phi_B$(GaAs) (eV) | |
|---|---|---|---|---|
| | *n*-type | *p*-type | *n*-type | *p*-type |
| Ag | 0.78 | 0.54 | 0.88 | 0.63 |
| Al | 0.72 | 0.58 | 0.80 | |
| Au | 0.80 | 0.34 | 0.90 | 0.42 |
| Cr | 0.61 | 0.50 | | |
| Cu | 0.58 | 0.46 | 0.82 | |
| Mo | 0.68 | 0.42 | | |
| Pt | 0.90 | — | 0.84 | |

(*Source:* S. M. Sze, *Physics of Semiconductor Devices*, 2d ed., (New York: J. Wiley & Sons, Inc., 1981). Reprinted with permission of John Wiley & Sons, Inc.)

**Fig. 3.30**
Schottky junction under zero bias and forward bias

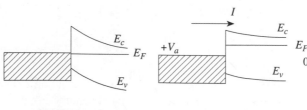

Unbiased          Forward bias

According to Eq. (37), the built-in potential (eV) becomes

$$\Phi_{bi} = \Phi_B - E_n - qV_a \tag{38}$$

where   $V_a$ is the applied voltage in V.

Combining Eqs. (35), (36), and (38), the total forward-bias current $I$ (A) is given by

$$I = I_{sm} + I_{ms} = A^*T^2 A_{cs}\exp(-\Phi_B/\Phi_t)[\exp(V_a/V_t) - 1] \tag{39}$$

$$= I_0[\exp(V_a/V_t) - 1]$$

where   $I_0 \ (= A^*T^2 A_{cs}\exp(-\Phi_B/\Phi_t))$ is the saturation current in A.

Equation (39) gives the ideal current-voltage (*I-V*) characteristics of a Schottky junction and is similar to the ideal *I-V* characteristics of a *p-n* junction (Eq. (12)). In addition to the exponential voltage and temperature dependence, the saturation current $I_0$ for a Schottky junction also depends exponentially on the barrier height $\Phi_B$. This allows us to select the *I-V* characteristics using Schottky junctions with different barrier heights, a feature not available in a *p-n* junction. Fig. 3.31 shows the sensitivity of $I_0$ to changes in $\Phi_B$ and temperature.

Tunneling also exists in a Schottky junction. Electron tunneling becomes important when the barrier height is sufficiently thin (as in the case of a highly doped semiconductor). This phenomenon is sometimes called **field emission.** The effect of tunneling or field emission increases with the electron energy and the tunneling current, $I_{tun}$, which is given by

$$I_{tun} = I_f \exp(\Phi_B/E_\infty) \tag{40}$$

where   $I_f$ is a preexponential constant that depends on the field-emission process in A, and

$E_\infty \ (=qh/4\pi\sqrt{N_D/(\epsilon_s m_e^*)})$ is a parameter dependent on the dopant density in J.

Fig. 3.32 shows how the tunneling current varies with the barrier width and the effect of the presence of field emission in a Schottky junction.

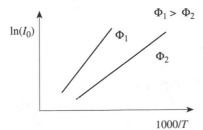

Fig. 3.31
Sensitivity of $I_0$ to barrier height and temperature

**Fig. 3.32**
Tunneling current and
field emission for different
barrier widths

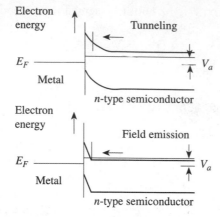

**Example 3.21**     Compute the current density in a Au/*n*-Si Schottky diode when $V_a = +0.5$ V. Assume $T = 300$ K (i.e., $\Phi_t = 0.026$ eV) and $\Phi_B = 0.80$ eV.

*Solution:*     Based on Eq. (39),

$$
\begin{aligned}
I_0/A_{cs} &= A^*T^2\exp(-\Phi_B/\Phi_t) \\
&= 1.2 \times 10^6 \text{ A/m}^2 \cdot \text{K}^2 \times (300 \text{ K})^2 \times \exp(-0.8 \text{ eV}/0.026 \text{ eV}) \\
&= 4.68 \times 10^{-3} \text{ A/m}^2 \\
J_{\text{tun}} &= I_{\text{tun}}/A_{cs} \\
&= I_0/A_{cs}[\exp(V_a/V_t) - 1] \\
&= 4.68 \times 10^{-3} \text{ A/m}^2 \times [\exp(0.5 \text{ V}/0.026 \text{ V}) - 1] \\
&= 1.05 \times 10^6 \text{ A/m}^2 \quad \S
\end{aligned}
$$

**Example 3.22**     If the tunneling current in a Au/*n*-Si Schottky diode is 10 μA, compute $I_f$ (in Eq. (40)) for $N_D = 1 \times 10^{25}/\text{m}^3$. Assume $m_e^* = 0.91 \times 10^{-30}$ kg and $\epsilon_s = 1.05 \times 10^{-10}$ F/m.

*Solution:*     Based on Eq. (40),

$$
\begin{aligned}
E_\infty &= qh/4\pi\sqrt{N_D/(\epsilon_s m_e^*)} \\
&= 1.6 \times 10^{-19} \text{ C} \times 6.62 \times 10^{-34} \text{ J·s}/(4 \times 3.14) \\
&\quad \times \sqrt{(1 \times 10^{26}/\text{m}^3)/(1.05 \times 10^{-10} \text{ F/m} \times 0.91 \times 10^{-30} \text{ kg})} \\
&= 0.86 \times 10^{-20} \text{ J}.
\end{aligned}
$$

Therefore,

$$
\begin{aligned}
I_f &= I_{\text{tun}}\exp(\Phi_B/E_\infty) \\
&= 1 \times 10^{-5} \text{ A} \times \exp(1.6 \times 10^{-19} \text{ C} \times 0.8 \text{ V}/(0.86 \times 10^{-20} \text{ J})) \\
&= 29 \text{ A}. \quad \S
\end{aligned}
$$

## Nonideal Schottky Junctions

In our previous treatment, we examined only an ideal Schottky junction. Real Schottky junctions have **interface states,** which, if present in sufficiently large quantities, can control the position of the Fermi level at the metal-semiconductor interface, as shown in Fig. 3.33. This reduces the sensitivity of the barrier height to changes in the workfunction.

In the limit, the **Schottky barrier height,** $\Phi_B$ (eV), is given by

$$\Phi_B = E_g - \Phi_0 \qquad (41)$$

where   $E_g$ is the energy gap in eV, and

     $\Phi_0$ is the location of the surface Fermi level in eV.

Note that in this case, the barrier height depends on the filling of the surface states. Hence, the equation for the ideal Schottky junction (Eq. (39)) needs some minor modifications when applied to a practical device. One modification involves the **field-induced barrier-height lowering effect,** as shown in Fig. 3.34.

This effect is caused by the image charge at the metal surface when there are excess electrons in the semiconductor. The resulting lowering in the barrier height, $\Delta\Phi_B$ (eV), is given by

$$\Delta\Phi_B = q\sqrt{q\acute{E}/(4\pi\epsilon_s)} \qquad (42)$$

where   $\acute{E}$ is the electric field in V/m, and

     $\epsilon_s$ is the semiconductor permittivity in F/m.

$\Delta\Phi_B$ is usually quite small ($\approx 0.01$ eV), except in the case when the electric field $\acute{E}$ is very large.

**Fig. 3.33**
Interface states in a
Schottky junction

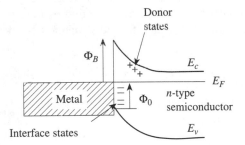

**Fig. 3.34**
Field-induced barrier-
height lowering effect

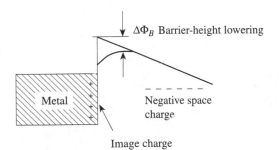

Both tunneling current (including the field-emission current) and the minority carrier injection current will modify the current-voltage (I-V) characteristics of a Schottky junction. A Schottky junction is normally a majority carrier device and is not affected by the effect of charge storage. Minority carrier current can become important at a large voltage bias when minority carrier injection takes place (for instance, when the holes generated in the depletion layer move into the n-type semiconductor in a large quantity). For an n-type semiconductor, the minority carrier current, $I_p$ (A), is given by

$$I_p = (qD_pp_{n0}A_{cs}/L_p)[\exp(V_a/V_t) - 1] \tag{43}$$

where    $D_p$ is the hole diffusivity in $m^2/s$,

   $P_{n0}$ is the equilibrium hole density of the n-type semiconductor in $/m^3$, and

   $L_p$ is the hole diffusion length in m.

Thus, Eq. (43) leads to a **minority carrier injection ratio, $\gamma^*$,** given by

$$\gamma^* = I_p/I \tag{44}$$

where    $I$ is the total current flowing across the Schottky junction in A.

The minority carrier injection ratio, $\gamma^*$, is a measure of the extent of charge storage, which gives rise to a time delay in the response of the device during current switching. Typical values of $\gamma^*$ as a function of $I$ are shown in Fig. 3.35.
   In general, current flow in a Schottky junction is given by

$$I = A^*T^2A_{cs}\exp(-\Phi_B/\Phi_t)[\exp(V_a/(\eta_IV_t)) - 1] \tag{45}$$

where    $\eta_I$ is the ideality factor (similar to the case of the p-n junction; it is usually greater than 1).

**Fig. 3.35**
Minority carrier injection
ratio versus current

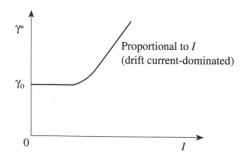

Proportional to $I$
(drift current-dominated)

**Example 3.23**   Compute the barrier-height lowering in Si if $\acute{E} = 1 \times 10^7$ V/m. Assume $\epsilon_s = 1.05 \times 10^{-10}$ F/m.

**Solution:**   Based on Eq. (42),

$$\Delta\Phi_B = q\sqrt{q\acute{E}/4\pi\epsilon_s}$$

$$= 1.6 \times 10^{-19} \text{ C}$$

$$\times \sqrt{1.6 \times 10^{-19} \text{ C} \times 1 \times 10^7 \text{ V/m}/(4 \times 3.14 \times 1.05 \times 10^{-10} \text{ F/m})}$$

$$= 0.035 \text{ eV}. \quad \S$$

**Example 3.24**   Obtain an expression for the minority carrier injection ratio, $\gamma^*$, in an ideal $n$-type Schottky diode.

**Solution:**   According to Eq. (44),

$$\gamma^* = I_p/I = (qD_p p_{n0}/L_p)/(A^* T^2 \exp(-\Phi_B/\Phi_t)) \quad \S$$

## Capacitance Effect and Equivalent-Circuit Model of a Schottky Junction

The junction capacitance, $C_{jun}$, of a Schottky diode is similar to that of a one-sided $p^+$-$n$ junction and is given by

$$C_{jun} = A_{cs}\sqrt{qN_D\epsilon_s/[2(V_{bi} - V_a)]} \tag{46}$$

where   $A_{cs}$ is the cross-sectional area of the device in m$^2$,

$V_{bi}$ is the built-in voltage in V,

$\epsilon_s$ is the semiconductor permittivity in F/m,

$N_D$ is the donor density in the $n$-type semiconductor in /m$^3$, and

$V_a$ is the applied voltage in V.

Since the ideal Schottky junction is a majority carrier device, there will not be any diffusion capacitance. Fig. 3.36 shows the equivalent-circuit model for an ideal

**Fig. 3.36**
An equivalent circuit for a Schottky junction

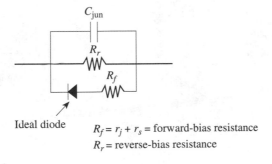

Ideal diode

$R_f = r_j + r_s$ = forward-bias resistance
$R_r$ = reverse-bias resistance

Schottky junction. The parasitic elements (capacitors and inductors) are included to account for the high-frequency effect, and the resistors $r_j$ and $r_s$ ($\Omega$) represent the dynamic junction resistance and the series resistance, respectively.

**Example 3.25**    Compare the junction capacitance of a *p-n* junction and a Schottky diode.

*Solution:*    Because of the smaller depletion-layer width, the capacitance of the Schottky junction is expected to be higher.    §

## Modification of the Barrier Height

It is possible to change the barrier height of a given Schottky diode if we change the doping near the interface. This is achieved by adding a thin layer of dopants to the semiconductor surface, as shown in Fig. 3.37.

For an *n*-type interface layer added to an *n*-type semiconductor, the effective barrier height, $\Phi_B^*$ (eV), is lowered and is given by

$$\Phi_B^* = \Phi_B - q/\epsilon_s \sqrt{n_1 a'/4\pi} \tag{47}$$

where    $\Phi_B$ is the original barrier height in eV,

$n_1$ is the interface dopant density in /m$^3$, and

$a'$ is the dopant layer thickness in m.

If a *p*-type surface layer is added to an *n*-type substrate, a potential minimum (see Fig. 3.37) exists at a distance $\Delta d$ away from the interface. The distance in meters is given by

$$\Delta d = [a'p_1 - (W - a')n_2]/p_1 \tag{48}$$

where    $p_1$ is the surface dopant density in /m$^3$,

$W$ is the depletion-layer width in m, and

$n_2$ is the substrate donor density in /m$^3$.

**Fig. 3.37**
Modification of Schottky
barrier heights

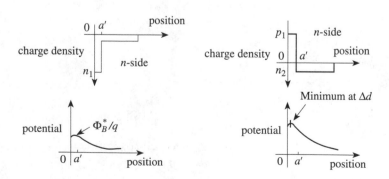

If $p_1 > n_2$ and $a'p_1 > n_2W$, Eq. (48) will give the effective barrier height

$$\Phi_B^* = \Phi_B + q^2 p_1 \Delta d^2 / 2\epsilon_s \tag{49}$$

In this limit, the change in the barrier height is proportional to the surface dopant density $p_1$.

**Example 3.26**   For an $n$-type surface dopant layer on an $n$-type Si substrate, compute the lowering in the Schottky barrier height if the surface dopant density is $1 \times 10^{24}/\text{m}^3$ and the surface-layer thickness is 10 nm. Assume $\epsilon_s = 1.05 \times 10^{-10}$ F/m.

*Solution:*   Based on Eq. (47),

$$\Delta\Phi_B = q^2/\epsilon_s \sqrt{n_1 a'/4\pi}$$
$$= q \times 1.6 \times 10^{-19} \text{ C}/(1.05 \times 10^{-10} \text{ F/m})$$
$$\times \sqrt{(1 \times 10^{24}/\text{m}^3 \times 1 \times 10^{-8} \text{ m})/(4 \times 3.14)}$$
$$= 0.043 \text{ eV} = 43 \text{ meV}. \quad \S$$

**Example 3.27**   For an $n$-type Si Schottky junction with a $p$-type surface layer, compute the location of the potential minimum $\Delta d$ if $p_1 = 1 \times 10^{24}/\text{m}^3$ and $\Delta\Phi_B = 0.02$ eV (or $\Delta\Phi_B/q = 0.02$ V). Assume $\epsilon_s = 1.05 \times 10^{-10}$ F/m.

*Solution:*   Based on Eq. (49),

$$\Delta d = \sqrt{2\epsilon_s \Delta\Phi_B/(q^2 p_1)}$$
$$= \sqrt{(2 \times 1.05 \times 10^{-10} \text{ F/m} \times 0.02 \text{ V})/(1.6 \times 10^{-19} \text{ C} \times 1 \times 10^{24}/\text{m}^3)}$$
$$= 1.62 \times 10^{-7} \text{ m} = 162 \text{ nm}. \quad \S$$

## Applications of Schottky Junctions

The Schottky junction behaves quite similarly to a *p-n* junction and is used for many related applications. Both Schottky junctions and *p-n* junctions are used for signal rectification, voltage level shifting, and as mixers and varactors. In some switching circuits, a Schottky junction is connected in parallel with the output base-collector junction of a bipolar junction transistor to prevent the latter from entering into saturation because of the smaller **turn-on voltage of a Schottky junction.** The real advantages of Schottky junctions are the simplicity in their fabrication and the absence of charge storage. Schottky junctions are used for the construction of transistors, CCDs, microwave devices, sensors, and detectors.

Not all metal-semiconductor combinations will give Schottky junctions. The ones that do provide a range of barrier heights from which different types of devices can be made. Ideally, the barrier height depends on the difference between the metal work function and the semiconductor work function. The search for a Schottky junction with a large barrier height in order to reduce the reverse current and improve the rectification properties has been an active research area for some time.

## 3.4    METAL-SEMICONDUCTOR CONTACT

The detailed properties of a metal-semiconductor interface depend on the properties of the metal and the semiconductor. When two solids are in good electrical contact, the following will occur. First of all, there must be a single energy reference, usually a reference to the potential energy at some faraway point. Secondly, the Fermi levels, which represent the average energies of the electrons in the solids, must be aligned. For most metal-semiconductor pairs, this implies some form of band bending near the metal-semiconductor interface, as illustrated in Fig. 3.38. Band bending dictates the properties at the metal-semiconductor interface and determines whether a rectifying junction or an ohmic contact is formed. For an *n*-type semiconductor, an **ohmic contact** is formed if the metal work function is smaller than the semiconductor work function (otherwise, the junction is rectifying). Table 3.4 shows a list of work functions for the common metals and semiconductors.

**Fig. 3.38**
Energy band diagram for a metal-semiconductor contact

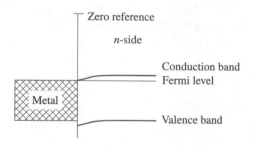

**Table 3.4**
A list of work functions for the common metals and semiconductors

| Metal/Semiconductor | Work function (eV) |
|---|---|
| Ag | 4.5 |
| Au | 4.9 |
| Pt | 5.4 |
| Mo | 4.2 |
| W | 4.5 |
| Al | 4.1 |
| Si | 4.0 |
| Ge | 4.5 |
| Ti | 4.2 |
| Ni | 5.0 |

(*Source:* Wang Shyh, *Fundamentals of Semiconductor Theory and Device Physics,* © 1989, p.131. Reprinted by permission of Prentice-Hall, Inc., Upper Saddle River, N.J.)

An ohmic contact is a resistive contact, and the current-voltage (*I-V*) characteristics are linear, as shown in Fig. 3.39. Ohmic contacts are needed when we want to minimize the effects of the contacts on a device and retain signal fidelity. Any resistance along the current path of a device will generate additional voltage drop that degrades the signal strength. Ideally, interface resistance should be zero. The actual resistance observed in a contact includes the **contact resistance,** the **shunt resistance** associated with the semiconductor substrate, and the **spreading resistance,** which exists when there is nonuniform current flow near the contact. Fig. 3.40 shows the distribution of these resistances near a metal electrode.

The contact resistance, $R_c$ ($\Omega$), is given by

$$R_c = \sqrt{R_{sh}r_c}/d_c \coth[L\sqrt{(R_{sh}/r_c)}] \qquad (50)$$

where    $R_{sh}$ is the shunt resistance in $\Omega$,

$r_c$ is the **specific contact resistance** in $\Omega \cdot m^2$,

$d_c$ is the width of the contact in m, and

$L$ is the contact length in m.

The shunt resistance, $R_{sh}$ ($\Omega$), can be expressed as

$$R_{sh} = r_{sheet}L_{sh}/d_c \qquad (51)$$

**Fig. 3.39**
*I-V* characteristics of an
ohmic contact

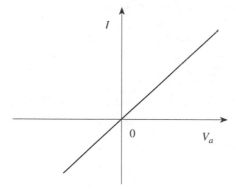

**Fig. 3.40**
Distribution of contact
resistances

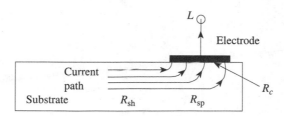

where   $r_{\text{sheet}}$ is the semiconductor **sheet resistance** in $\Omega$/square, and
$L_{\text{sh}}$ is the length of the shunt path in m.

The total resistance $R$ of a contact ($\Omega$) becomes

$$R = R_c + R_{\text{sh}} + R_{\text{sp}} \tag{52}$$

where   $R_{\text{sp}}$ is the spreading resistance in $\Omega$.

Of the three types of resistance, the spreading resistance is most difficult to determine since it also depends on the contact geometry. For planar current flow, the spreading resistance is usually negligible. In the case of a metal/$n^+$ contact, the specific contact resistance $r_c$ (contact resistance times area) is approximately given by

$$r_c \approx \exp[2V_{bn}\sqrt{\epsilon_s m_e^*/N_D}/(h')] \tag{53}$$

where   $m_e^*$ is the effective mass of the electrons in kg,
$V_{bn}\ (=\Phi_B/q)$ is in V,
$h'$ is Planck's constant divided by $2\pi$, and
$N_D$ is the donor density in /m$^3$.

For a low-resistance contact, the specific contact resistance should be less than $10^{-6}\ \Omega\cdot$cm$^2$—or, effectively, a resistance of 1 $\Omega$ for an area of 10 $\mu$m$^2$ (i.e., 10$\mu$m $\times$ 10 $\mu$m). An ohmic contact can be formed even when an interface barrier is present. If the barrier is sufficiently thin, as in the case of a highly doped interface layer, tunneling of electrons will occur and the contact will still have little resistance. An example is the Al/$n^+$-Si contact, where current flow is primarily due to tunneling. Most ohmic contacts are formed on highly doped semiconductors to minimize substrate resistance and parasitic loss. Table 3.5 shows the specific contact resistances, $r_c$, in Si and GaAs contacts.

**Example 3.28**   Compute the specific contact resistance of an Al/Si contact. Assume $\Phi_{ms} = \Phi_B = 0.1$ eV, $N_D = 1 \times 10^{22}$/m$^3$, $\epsilon_s = 1.05 \times 10^{-10}$ F/m, and $m_e^* = 0.91 \times 10^{-30}$ kg.

*Solution:*   Based on Eq. (53),

$$r_c \approx \exp[2V_{bn}\sqrt{\epsilon_s m_e^*/N_D}/(h')]$$

$$= \exp(2 \times 0.1\text{ V})$$

$$\times \sqrt{(1.05 \times 10^{-10}\text{ F/m} \times 0.91 \times 10^{-30}\text{ kg}/1 \times 10^{22}/\text{m}^3)}/(1.05 \times 10^{-34}\text{ J}\cdot\text{s})$$

$$= 1.85 \times 10^2\ \Omega\cdot\text{m}^2. \quad \S$$

| Table 3.5<br>Specific contact<br>resistances in Si and<br>GaAs contacts | W/Si | Al/Si | Au/Ge/Ni/GaAs |
|---|---|---|---|
| $r_c$ ($\mu\Omega\cdot$m$^2$) | 0.25 | 1.8 | 1.0 |

**Example 3.29** Compute the contact resistance between Al and Si. Assume $R_{sh} = 1 \times 10^{-3}$ Ω, $L = 2000$ μm, and $W = 100$ μm using the data in Example 3.25.

*Solution:* Based on Eq. (50),

$$R_c = \sqrt{R_{sh}r_c}/d_c\coth(L\sqrt{R_{sh}/r_c})$$
$$= \sqrt{1 \times 10^{-3}\ \Omega \times 1.85 \times 10^2\ \Omega\cdot m^2}/(1 \times 10^{-4}\ m)\coth(2 \times 10^{-3}\ m\ \sqrt{10^{-3}\ \Omega/1.85 \times 10^2\ \Omega\cdot m^2}$$
$$= 5.9 \times 10^8\ \Omega. \quad \S$$

## 3.5 MIS JUNCTION AND FIELD-EFFECT PROPERTIES

Similar to the reverse-bias *p-n* junction, the metal-insulator-semiconductor (MIS) junction is a very important structure in many semiconductor devices. The electronic properties of the MIS junction are essentially determined by the energy band diagram at the interface. Since different MIS junctions can have different energy band structures, it is useful to start the analysis with an ideal band structure. The most convenient energy band structure is one where the energy bands are initially flat, as shown in Fig. 3.41. This is called the **flat-band condition.**

For a real MIS junction, to achieve flat-band condition a voltage called the **flat-band voltage,** $V_{FB}$, must be applied to the metal side of the MIS junction (we assume the substrate semiconductor is kept at zero potential for the time being). The flat-band voltage, $V_{FB}$ (V), can be either positive or negative and is given by

$$V_{FB} = [\Phi_m - \Phi_s - (E_c - E_F)]/q \tag{54}$$

where   $\Phi_m$ is the metal work function in eV,

$\Phi_s$ is the semiconductor work function in eV, and

$E_c - E_F$ is the energy difference between the conduction band edge and the Fermi level in the semiconductor in eV.

### Surface Inversion

Starting with the flat-band condition, we shall show that we can change the semiconductor **surface potential,** $V_s$, in the MIS junction simply by raising or lowering the

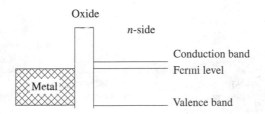

**Fig. 3.41**
Flat-band condition in a MIS junction

metal potential through an applied voltage. For convenience, we choose a p-type semiconductor substrate. Based on **Poisson's equation** ($d^2(V_a)/dx^2 = -Q_0/\epsilon_s$, where $V_a$ is the applied voltage and $Q_0$ is the (volume) charge density), it is possible to show that under flat-band condition the metal potential, $V_m$, expressed in volts is approximately given by

$$V_m = V_s + \sqrt{2\epsilon_s q N_A V_s}/C_i \tag{55}$$

where   $V_s$ is the potential at the semiconductor surface in V,

$q$ is the electron charge in C,

$\epsilon_s$ is the semiconductor permittivity in F/m,

$N_A$ is the acceptor density in /m³, and

$C_i$ is the oxide capacitance per unit area in F/m².

For a sufficiently large $V_s$, the second term, containing $\sqrt{V_s}$, can be ignored, and $V_m \approx V_s$. This suggests that the surface potential $V_s$ changes almost linearly with the applied voltage $V_a$, and we may adjust the band bending in the semiconductor to achieve charge accumulation, depletion, or even charge inversion by changing the applied voltage. This situation is illustrated in Fig. 3.42.

The respective surface charge densities for the electrons and the holes, $n_s$ and $p_s$ (/m³), are given by

$$n_s = n_{p0}\exp(V_s/V_t) \tag{56}$$

$$p_s = p_{p0}\exp(-V_s/V_t)$$

where   $n_{p0}$ is the equilibrium electron density in the p-type semiconductor in /m³,

$p_{p0}$ is the equilibrium hole density in the p-type semiconductor in /m³, and

$V_t = kT/q$.

In the case of surface inversion—i.e., when a thin sheet of carriers of the opposite charge is located at the semiconductor side of the interface—it is possible to relate

**Fig. 3.42**
MIS junction biased at accumulation depletion and inversion

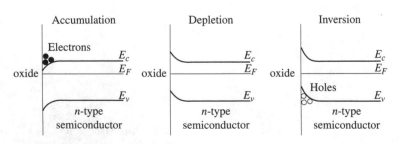

the surface charge density $Q_s$ (C/m$^2$) to the electric field $\acute{E}_s$ (V/m) at the semiconductor surface; i.e., $Q_s = \epsilon_s \acute{E}_s$. Note that $Q_s$ is per unit area. A more rigorous calculation shows that

$$\acute{E}_s = \sqrt{2}(V_t/L_{D_p})\sqrt{[\exp(-V_s/V_t) + V_s/V_t - 1]n_{p0}/p_{p0}[\exp(V_s/V_t) - V_s/V_t - 1]} \tag{57}$$

where $L_{D_p}$ is the hole-diffusion length in m.

As we shall see, our ability to form a **surface inversion** layer in a MIS structure is very important for the operation of a metal-oxide semiconductor field-effect transistor, or MOSFET. Defining **strong inversion** as the condition when $V_s = 2\phi_b$, where $\phi_b$ is the potential difference between the intrinsic Fermi level (see Chapter 2) and the equilibrium Fermi level (V), as shown in Fig. 3.43, the applied voltage $V_a$ (V) necessary for strong inversion to occur (according to Eq. (55)) is given by

$$V_a = 2\phi_b - \sqrt{4qN_A\phi_b\epsilon_s}/C_i \tag{58}$$

This is also called the **threshold voltage**, $V_T$. To achieve surface inversion, the applied voltage must overcome the flat-band voltage, $V_{FB}$, and also any interface potential drop in the insulator layer due to the oxide charges, which we have neglected so far. In addition, if the substrate has a voltage bias $V_{sub}$, then the applied voltage $V_a$, which is the same as $V_m$ (see Eq. (55)), must be replaced by $V_a - V_{sub}$ (V). Effectively, the modified threshold voltage, $V_T^*$, that includes all those effects is given by

$$V_T^* = V_T + V_{FB} + V_{sc} + V_{sub} \tag{59}$$

where $V_{sc}$ is potential drop due to oxide charges in the insulator in V.

**Example 3.30** Compute the flat-band voltage in an Al/SiO$_2$/$p$-Si structure. Assume $N_A = 1 \times 10^{22}$/m$^3$, $kT = 0.026$ eV, and $N_c = 1.04 \times 10^{25}$/m$^3$.

**Fig. 3.43**
Strong inversion in a MIS structure

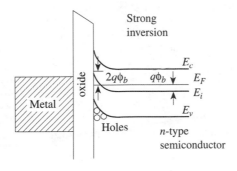

*Solution:*   From Chapter 2, Eq. (42), $E_v - E_F = kT \ln(N_v/N_A) = 0.026$ eV $\times$ $\ln(1.04 \times 10^{25}/\text{m}^3/1 \times 10^{22}/\text{m}^3) = 0.181$ eV. Therefore, $E_c - E_F = 1.12$ eV $-$ $0.181$ eV $= 0.94$ eV. Based on Eq. (54), $V_{FB} = [\Phi_m - \Phi_s - (E_c - E_F)]/q =$ $[4.1 - 4.0 - 0.94]$ V $= -0.84$ V.   §

**Example 3.31**   Compute the fractional error if we assume that the surface potential in a MIS structure is proportional to the applied voltage. It is given that $V_s = 1$ V and $C_i = 5$ mF/m².

*Solution:*   Based on Eq. (55),

$$V_m = V_s + \sqrt{2\epsilon_s q N_A V_s}/C_i$$
$$= 1 + \sqrt{2 \times 1.05 \times 10^{-10} \text{ F/m} \times 1.6 \times 10^{-19} \text{ C} \times 1 \times 10^{22}/\text{m}^3 \times 1 \text{ V}}/(5 \times 10^{-3} \text{ F/m}^2) \text{ V}$$
$$= 1.116 \text{ V}$$

The fractional error is $(1.116 - 1)/1 = 0.116$, or about 12%.   §

**Example 3.32**   Compute the voltage required to cause strong inversion in an *n*-type Si MIS structure. Assume $N_c = 2.8 \times 10^{25}/\text{m}^3$, $N_D = 1 \times 10^{22}/\text{m}^3$, $C_i = 5$ mF/m², and $T = 300$ K.

*Solution:*   From Section 2.7 and the definition of $\phi_b$, $\phi_b = (E_c - E_F)/q = kT/q \ln(N_c/N_D) = 0.026$ V $\times \ln(2.8 \times 10^{25}/(1 \times 10^{22}/\text{m}^3)) = 0.206$ V. Based on Eq. (58),

$$V_s = 2\phi_b - \sqrt{4q N_A \phi_b \epsilon_s}/C_i$$
$$= 2 \times 0.206 \text{ V}$$
$$- \sqrt{(4 \times 1.6 \times 10^{-19} \text{ C} \times 1 \times 10^{22}/\text{m}^3 \times 0.206 \text{ V} \times 1.05 \times 10^{-10} \text{ F/m})}/(5 \times 10^{-3} \text{ F/m}^2) \text{ V}$$
$$= 0.132 \text{ V}   §$$

## Capacitance Effect in a MIS Junction

The capacitance, $C_{mis}$, of a MIS junction is the series combination of the insulator capacitance, $C_i$, and the depletion layer capacitance, $C_{depl}$, in the semiconductor. $C_{mis}$ (F/m²) is given by

$$C_{mis} = C_i C_{depl}/(C_i + C_{depl}) \tag{60}$$

where   $C_i$ is the insulator capacitance per unit area in F/m², and

$C_{depl}$ is the depletion layer capacitance per unit area in F/m².

The study of the capacitance-voltage (*C-V*) characteristics of the MIS junction at different frequencies is of particular interest since a MIS junction will respond only to an ac voltage. As shown in Fig. 3.44, the measured capacitance is essentially that of the insulator during accumulation—i.e., when the substrate carriers (holes in this case) are drawn toward the oxide-semiconductor interface. This situation occurs when a negative bias is applied to the metal and the *p*-type semiconductor is grounded. With decreasing negative bias (near $V_a = 0$), the interface layer is depleted of carriers, and the depletion capacitance $C_{depl}$ becomes more important. The capacitance of the MIS junction is given by Eq. (60). As the applied voltage becomes more positive, charge inversion occurs, and a thin sheet of electrons is formed near the oxide-semiconductor interface. This happens only at low frequency, when there is enough time for carrier generation (electrons and holes) to occur. The electrons will move to the interface (under the influence of the bias voltage), whereas the holes will migrate to the substrate and neutralize the negatively charged acceptors. The overall effect is an increase in the amount of surface charge near the oxide-semiconductor interface, and the capacitance will be that of the oxide capacitance. At high frequency, however, there will be insufficient time for the carriers to be generated (as the bias voltage will change sign every half-cycle) and redistribute. The capacitance in this case will be dominated by the depletion capacitance. These features are illustrated in the capacitance-voltage (*C-V*) characteristics shown in Fig. 3.44. Any oxide charge or surface-state charge present in the MIS junction will show up in the *C-V* characteristics as a shift along the voltage axis.

**Example 3.33**  For a MIS junction formed on a *p*-type substrate, suggest the direction that the *C-V* curve will shift if there are positive charges located at the oxide-semiconductor interface.

*Solution:*  The positive interface charges will raise the surface potential. To restore the condition corresponding to zero surface charge, a negative bias has to be applied to the metal. This implies a shift of the *C-V* curve to the right (in the direction of negative bias).  §

## 3.6  MATERIALS CONSIDERATIONS

The most important materials used in the making of electronic devices are semiconductors and their oxides. In Chapters 1 and 2, we discussed the different methods

**Fig. 3.44**
*C-V* characteristics of a MIS junction formed on a *p*-type semiconductor

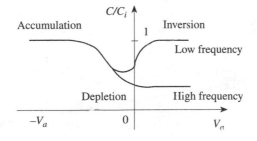

of their preparation and some of the structural and electronic properties. In this section, we look at the physical origin of defects and how these defects affect the performance of electronic devices. Si is the most widely used semiconductor material; it can be prepared as a single crystal relatively free of defects, except for point defects, which are always present. In particular, self-interstitials of Si can exist in quantities as large as $10^{22}/m^3$. O and C are also present in comparable quantities. O has the advantage of increasing the mechanical strength of Si by pinning dislocations and retarding slips. Excessive O, however, is precipitated in the form of oxides and assists in the formation of stacking faults. Most Si wafers used for IC fabrication therefore require a denuded zone about 20 $\mu$m thick that is defect free. This result is achieved through outgassing of $O_2$ at 1000 to 1200 °C using a process called gettering. Gettering involves the formation of a region full of defects deep into the wafer so that unwanted impurities can migrate there and be trapped. Defects are also known to form nucleation sites for O clusters, interstitial Si, and $SiO_2$ precipitates.

Different from Si, $SiO_2$ is mostly used for the formation of MIS junctions and for isolation and passivation. $SiO_2$ can be grown using a number of techniques, as discussed in Chapter 1; thermal oxidation is most reliable and gives the best results. Oxides possess oxide charges, which directly affect the potential distribution in a device. In MOSFETs, oxide charges are known to be responsible for changes in the threshold voltage. Oxide breakdown is another concern. Breakdown is related to pinholes in the oxide, and the density of these pinholes increases with decreasing oxide thickness. Defects also generate weak spots for tunneling current in the oxide layer. It has been found that, in general, oxide quality improves with a preoxidation annealing step and some form of backside gettering to remove the heavy metal ions, such as $Cu^{2+}$, $Ni^{2+}$, and $Na^+$.

Another semiconductor that is used quite frequently in high-speed devices and optoelectronic devices is GaAs. GaAs has a high electron mobility, but the defect density is usually quite high. The latter can result in significant variations ($> 30$ mV) in the threshold voltage in metal-semiconductor field-effect transistor (MESFET) devices (GaAs does not have a native oxide and is rarely used to form MOSFETs). GaAs, being a compound semiconductor, sometimes has problems related to crystal stoichiometry and low partial pressures for the constituents. GaAs wafers, however, can be made semi-insulating (SI) through the incorporation of deep energy levels, which—when present in a sufficiently large quantity—will pin the Fermi level to near midgap. This often has the effect of fixing the threshold voltage in MESFETs and creating a surface inversion layer. The deep energy levels are due to the presence of O, Cr, and a native defect called an **EL2 center** (an EL2 center is formed by an As atom occupying a Ga site). Device-quality GaAs epitaxial layers can be prepared using MOCVD and MBE. As mentioned in Chapter 1, the quality of the native oxides of Ga and As is quite inferior, and for device isolation $Si_3N_4$ is often used. $Si_3N_4$ can be deposited by vapor-phase deposition.

## 3.7   STRUCTURE AND OPERATION OF TRANSISTORS

**Transistors** are probably the most important devices in modern electronics. The word *transistor* results from the combination of the two words *transfer* and *resistor.*

Thus, as the name implies, a transistor has different input and output resistances. Transistors have many useful properties, the most important being the ability to amplify an electrical signal. In general, transistors are divided into two main types, depending on the principles of operation: those that utilize resistance modulation in a semiconductor channel and those that rely on physical charge transfer from the input to the output. The former includes all the **field-effect transistors** (FETs), and the latter are mainly the **bipolar junction transistors** (BJTs). Mixed BJTs and FETs are also fabricated, and they carry names such as BIMOS and BATMOS.

## Field-Effect Transistors (FETs)

*Metal-Oxide Semiconductor Field-Effect Transistor*   The metal-oxide semiconductor field-effect transistor (MOSFET) is the most widely used FET in IC applications. The basic structure of a MOSFET is shown in Fig. 3.45. It consists of a MIS input gate connected to a lateral $p^+$-$n$-$p^+$ or $n^+$-$p$-$n^+$ structure with the $n^+$ or $p^+$ regions forming what are known as the *drain* and the *source*. These are the output terminals of the MOSFET. Usually, the source is grounded. A possible fourth terminal exists in the substrate. Because of the presence of the insulator underneath the gate, the latter does not conduct any dc current. For the structure shown in Fig. 3.45, the MOSFET does not conduct a drain-to-source current unless an inverted channel is formed in the semiconductor. Such a MOSFET is *normally off* and is called an **enhancement-mode field-effect transistor** (E-MOSFET). If an inverted channel is present in the as-prepared device, the MOSFET is then *normally on* and is called a **depletion-mode field-effect transistor** (D-MOSFET). In this case, the conduction between the drain and the source is essentially ohmic.

Both $n$-channel and $p$-channel MOSFETs can be made. To determine the current flow in the channel, we usually invoke the **gradual-channel approximation.** Basically, this approximation assumes that the lateral electric field due to the drain-to-source voltage is much smaller than the vertical electric field due to the gate voltage that gives rise to surface inversion. Thus, for a MOSFET with an inverted $n$-type channel, the surface charge density, $Q_s(x)$ (C/m²), induced in the semiconductor due to a gate voltage, $V_g$ (V), is given by

$$Q_s = -C_i[V_g - 2\phi_b - V_{FB} - V_c(x)] \tag{61}$$

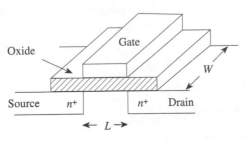

**Fig. 3.45**
Cross section of an
$n$-channel MOSFET

where    $C_i$ is the oxide capacitance per unit area in F/m$^2$,

2$\phi_b$ is the change in the surface potential in V required to generate strong inversion (see Fig. 3.43),

$V_{FB}$ is flat-band voltage in V (see Eq. (54)), and

$V_c(x)$ in V is the channel potential, which varies along the length of the channel in the $x$ direction.

The effective electron density along the channel, $n_{ss}(x)$ (/m$^2$), is then given by

$$n_{ss}(x) = Q_s - \sqrt{2\epsilon_s q N_A(V_c(x) + 2\phi_b)} \tag{62}$$

where    the second term on the right-hand side is due to the acceptor charges.

Assuming that the source is grounded, the drain-to-source current, $I_{ds}$ (A), can be written as

$$I_{ds} = q n_{ss} W_c \mu_n dV_c(x)/dx. \tag{63}$$

where    $\mu_n$ is the channel mobility in m$^2$/V·s,

$dV_c(x)/dx$ is the incremental change in the channel potential divided by the incremental change in position in V/m, and

$W_c$ is the channel width in m.

In arriving at Eq. (63), we make use of Ohm's law (Section 2.5). We can now combine Eqs. (62) and (63) and integrate $dx$ over the channel length and $dV_c$ from the source (= 0 V) to the drain (= $V_{ds}$), which leads to

$$I_{ds} = (\mu_n W_c/L)C_i \left\{ (V_g - V_{FB} - 2\phi_b - V_{ds}/2) - \left(\frac{2}{3}\right)\left[\sqrt{2\epsilon_s q N_A}/C_i\right]\left[(V_{ds} + 2\phi_b)^{3/2} - (2\phi_b)^{3/2}\right] \right\} \tag{64}$$

where    $L$ is the channel length in m, and

$V_{ds}$ is the drain-to-source voltage in V.

Equation (64) gives the current-voltage (*I-V*) characteristics of a MOSFET up to the point of **saturation.** Further increase in the drain-to-source voltage, $V_{ds}$, beyond this point will result in **channel pinch-off** since the drain-substrate *p-n* junction is at reverse bias. Pinch-off is the condition when a portion of the channel is totally depleted. It occurs when

$$V_{ds} = V_g - V_T \tag{65}$$

where    $V_g - V_{ds}$ is the voltage drop across the drain-substrate *p-n* junction in V, and

$V_T$ is the threshold voltage in V (see Eq. (59)).

The saturation current $I_{Dsat}$ can now be obtained by substituting $V_{ds}$ (Eq. (65)) into Eq. (64). The complete *I-V* characteristics of the MOSFET are shown in Fig. 3.46.

We can also include the effect of a voltage applied to the substrate. Since the source is grounded, the substrate voltage will affect only the depletion layer width in the channel. The threshold voltage $V_T$ in the presence of a substrate voltage, $V_{sub}$ (V), is therefore given by

$$V_T = V_{FB} + 2\phi_b + V_{sc} + \sqrt{2\epsilon_s q N_A (2\phi_b - V_{sub})}/C_i \qquad (66)$$

According to these equations, the MOSFET will have two modes of operation. For a small drain-to-source voltage, $V_{ds}$, the current-voltage (*I-V*) characteristics of the MOSFET are given approximately by (see Eq. (64))

$$I_{ds} \approx \beta_0 (V_g - V_T) V_{ds} \qquad (67)$$

where    $\beta_0 = \mu_n C_i W_c / L$

This is the linear mode of operation, in which the device behaves just like a voltage-controlled resistor. The output conductance $g_{ds}$ (defined as $\partial I_{ds}/\partial V_{ds}|_{V_g = 0}$) can be computed from Eq. (67) for a small $I_{ds}$. It is given by $\beta_0 (V_g - V_T)$ ($/\Omega$ or S (siemens)). For a large drain-to-source voltage, the channel is in pinch-off at the drain-substrate junction, and the saturation current, $(I_{Dsat})$ (A), is approximately given by

$$I_{Dsat} \approx \beta_0 (V_g - V_T)^2/2 \qquad (68)$$

Equation (68) represents the saturation mode of operation, since $I_{ds}$ is no longer dependent on $V_{ds}$. The **transconductance,** $g_m$, of the MOSFET ($/\Omega$ or S) gives the figure of merit of the device and is defined as $\partial I_{ds}/\partial V_g|_{V_{ds} = 0}$. In saturation, it is given approximately by

$$g_m = \beta_0 (V_g - V_T) \qquad (69)$$

To encompass both the linear mode and the saturation mode of operation for the MOSFET, the following empirical *I-V* equation is often used:

$$I_{ds} = I_{Dsat} \tanh(g_{ds} V_{ds}/I_{Dsat})[1 + \gamma' V_{ds}] \qquad (70)$$

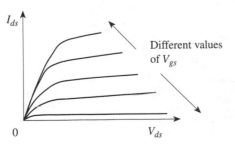

**Fig. 3.46**
*I-V* characteristics of a MOSFET

Different values of $V_{gs}$

where    $\gamma'$ is an empirical parameter in /V, and

$I_{Dsat}$ is the saturation current in A.

Most of the time, when used for signal amplification the MOSFET operates in the saturation mode. This mode can be set up using the following procedure. Initially, the MOSFET is biased at a given point in the saturation region on the dc $I$ $V$ characteristics, as shown in Fig. 3.47.

When a small ac signal, $\Delta V_g$, applied to the gate is superimposed on the dc characteristics, the incremental change in $I_{ds}$, or $\Delta I_{ds}$, is given by

$$\Delta I_{ds} = g_m \Delta V_g = \beta_0 (V_g - V_T) \Delta V_g \tag{71}$$

where we have used Eq. (69) for the value of $g_m$. If a resistive load $R_L$ is also present at the output of the MOSFET, as shown in Fig. 3.48, the voltage gain $G_v$ ($= \Delta I_{ds} R_L / \Delta V_g$) becomes

$$G_v = \beta_0 (V_g - V_T) R_L \tag{72}$$

Thus, for a large $\beta_0$, a significant voltage gain can be achieved. Most MOSFETs have a value of $g_m / W_c$ ($W_c$ is the width of the MOSFET in meters) between 10 and 20 S/m. This figure is about 10 times lower than the corresponding value in similar BJTs.

**Example 3.34**    Compute the surface-charge density near the drain region of a MOSFET if $V_g = 1$ V, $\phi_b = 0.1$ V, $V_{FB} = 0.3$ V, and $C_i = 5$ mF/m$^2$.

*Solution:*    Near the drain region, $V_c(x) = 0$. Based on Eq. (61),

$$\begin{aligned} Q_s &= -C_i[V_g - 2\phi_b - V_{FB} - V_c(x)] \\ &= -5 \times 10^{-3} \text{ F/m}^2 \times (1 - 2 \times 0.1 - 0.3) \text{ V} \\ &= 0.0025 \text{ C/m}^2, \text{ or } 4 \times 10^{16} \text{ electrons/m}^2. \quad \S \end{aligned}$$

**Example 3.35**    Compute the saturation current in a MOSFET if $V_g - V_T = 1$ V, $\mu_n = 0.02$ m$^2$/V·s, $C_i = 5$ mF/m$^2$, and $W_c /L = 100$.

**Fig. 3.47**
Bias point on the *I-V* characteristics of a MOSFET

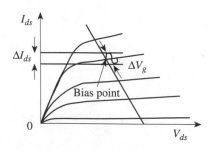

Fig. 3.48
MOSFET biasing circuit
with a load

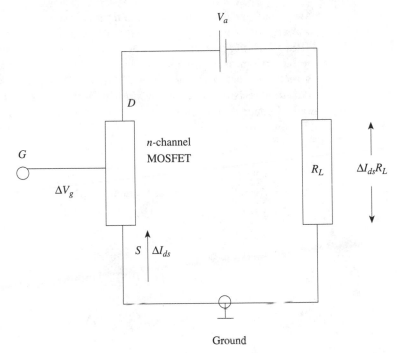

Ground

**Solution:** Based on Eq. (67), $\beta_0 = \mu_n C_i W_c/L = 0.02$ m$^2$/V·s $\times 5 \times 10^{-3}$ F/m$^2 \times 100 = 0.01$ F/V·s. Based on Eq. (68), $I_{Dsat} = \beta_0(V_g - V_T)^2/2 = 0.01$ F/V·s $\times (1$ V$)^2/2 = 0.05$ A $= 50$ mA. §

**Example 3.36** Reduce Eq. (70) to its limits **(a)** when $V_{ds}$ is small, and **(b)** when $V_{ds}$ is large.

**Solution:** Based on Eq. (70), $I_{ds} = I_{Dsat}\tanh(g_{ds}V_{ds}/I_{Dsat})[1 + \gamma'V_{ds}]$.

   **(a)** When $V_{ds}$ is small, $I_{ds} = I_{Dsat}(g_{ds}V_{ds}/I_{Dsat})[1 + \gamma'V_{ds}] = g_{ds}V_{ds}$, which is the same as Eq. (67).

   **(b)** When $V_{ds}$ is large, $I_{ds} = I_{Dsat}[1 + \gamma'V_{ds}]$. The term in the square brackets represents the fractional increase in the channel resistance as $V_{ds}$ increases. §

***Metal-Semiconductor Field-Effect Transistor*** The metal-semiconductor field-effect transistor (MESFET) shown in Fig. 3.49 is quite similar to the MOSFET,

Fig. 3.49
Cross section of an
$n$-channel MESFET

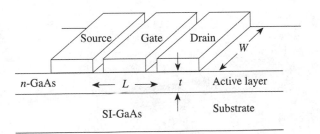

except that gate isolation is achieved through a reverse-bias Schottky junction. This, in a way, reduces the dynamic range of the input voltage. Like the MOSFET, the MESFET can operate in the enhancement mode or the depletion mode. Because of the lack of an oxide isolation layer, a MESFET has a much higher input leakage current when compared to the MOSFET. For the same reason, MESFETs are more **radiation resistant** since oxide **punch-through** is no longer possible. Using the gradual-channel approximation, as in the MOSFET, an incremental voltage drop, $\Delta V$, along the *n*-type channel of the MESFET can be written as

$$\Delta V = I_{ds}\Delta x/[q\mu_n N_D W_c(t_c - a_d(x))] \tag{73}$$

where    $\Delta x$ is the incremental channel length in m,

$q$ is the electron charge in C,

$\mu_n$ is the electron mobility in $m^2/V{\cdot}s$,

$N_D$ is the donor density in $/m^3$,

$W_c$ is the width of the MESFET in m,

$t_c$ is the thickness of the active layer in m, and

$a_d(x) = 2\epsilon_s(V(x) + V_{bi} - V_g)/(qN_D)$.

As before, we integrate $\Delta x$ over the entire channel length $L$ and $\Delta V$ from 0 to $V_{ds}$, so the drain-to-source current, $I_{ds}$, is given by

$$I_{ds} = g_0[V_{ds} - 2((V_{ds} + V_{bi} - V_g)^{3/2} - (V_{bi} - V_g)^{3/2})/(3\sqrt{V_{po}})] \tag{74}$$

where    $g_0$ is the channel conductance in $/\Omega$ or S,

$V_{po}$ $(= qN_D t_c^2/(2\epsilon_s))$ is the pinch-off voltage in V, and

$V_{bi}$ is the built-in voltage in V.

The transconductance $(= \partial I_{ds}/\partial V_g|_{V_{ds}=0})$ of a MESFET is then given by

$$g_m = g_0(\sqrt{V_{ds} + V_{bi} - V_g} - \sqrt{V_{bi} - V_g})/\sqrt{V_{po}} \tag{75}$$

In saturation, we can write

$$V_{Dsat} = V_{po} - V_{bi} + V_g$$
$$I_{Dsat} = g_0[V_{po}/3 + 2(V_{bi} - V_g)^{3/2}/(3\sqrt{V_{po}}) - V_{bi} + V_g] \tag{76}$$
$$g_m = g_0[1 - \sqrt{(V_{bi} - V_g)/V_{po}}]$$

For device modeling, the drain-to-source current can also be expressed as

$$I_{ds} = \beta'(V_g - V_T)^2(1 + \gamma'V_{ds})\tanh(g_0V_{ds}/I_{Dsat})/[1 + b(V_g - V_T)] \tag{77}$$

where   $\beta' = \mu_n \epsilon_s W/(t_c L)$,

   $V_T = V_{bi} - V_{po}$, and

   $\gamma'$ and $b$ are empirical constants in /V.

Equation (77) is, in fact, quite similar to Eq. (69), and so are the gain characteristics between MESFETs and MOSFETs. The figures of merit ($g_m/W_c$) for the two devices are also very similar.

   The gate current, $I_{\text{gate}}$ (A), represents the loss in the MESFET and is given by

$$I_{\text{gate}} = I_{gs} + I_{gd} \tag{78}$$

where   $I_{gs} = I_{go}\exp\{[V_{gs} - I_g R_g - (I_{ds} + I_{gs})R_s]/(\eta_g V_{\text{th}})\}$

   $I_{gd} = I_{go}\exp\{[V_{gd} - I_g R_g - (I_{ds} + I_{gd})R_s]/(\eta_g V_{\text{th}})\}$

   $I_{go}$ is the gate saturation current in A,

   $R_g$ is the gate resistance in $\Omega$, and

   $\eta_g$ is the ideality factor for the gate junction.

**Example 3.37**   Compare the input leakage currents per unit gate area in a Si MESFET and a Si MOSFET. Assume for the MESFET that the depletion-layer (junction) width is 1 $\mu$m and $\tau_0 = 0.1$ $\mu$s, and for the MOSFET, oxide resistivity $\approx 1 \times 10^{13}$ $\Omega \cdot$m at 1 V for a thickness of 50 nm.

*Solution:*   For a MESFET, the input leakage current will be roughly the thermal generation current in the *p-n* junction. Based on Eq. (13),

$$\begin{aligned}
I_G/A_{cs} &= qn_i W_j/(2\tau_0) \\
&= 1.6 \times 10^{-19} \text{ C} \times 1.45 \times 10^{16}/\text{m}^3 \times 1 \times 10^{-6} \text{ m}/(2 \times 10^{-7} \text{ s}) \\
&= 0.012 \text{ A/m}^2
\end{aligned}$$

The leakage current in the MOSFET is

$$\begin{aligned}
I_{\text{gate}}/A_{cs} &= V/(\rho L) \\
&= 1 \text{ V}/(1 \times 10^{13} \ \Omega \cdot \text{m} \times 5 \times 10^{-8} \text{ m}) \\
&= 2 \ \mu\text{A/m}^2 \quad \S
\end{aligned}$$

   Note that the leakage current will be more significant in the MESFET (by a factor of 6000 in this case).

**Example 3.38**   Compute the threshold voltage of an *n*-channel GaAs MESFET if $N_D = 1 \times 10^{22}/\text{m}^3$ and the undepleted channel thickness is 0.5 $\mu$m. Assume $V_{bi} = 0.8$ V.

*Solution:*   Based on Eq. (76), $V_T = |V_{po} - V_{bi}| = qN_D t_c^2/(2\epsilon_s) - V_{bi} =$ $1.6 \times 10^{-19}$ C $\times 1 \times 10^{22}/\text{m}^3 \times (0.5 \times 10^{-6} \text{ m})^2/(2 \times 1.05 \times 10^{-10} \text{ F/m}) -$ $0.8$ V $= 1.1$ V.   $\S$

## Bipolar Junction Transistor BJT

The operation of a bipolar junction transistor (BJT) is quite different from that of the MOSFET. First of all, the essential feature in a BJT is not a gate-controlled resistor, as in the MOSFET or the MESFET, but rather a voltage-controlled current source. Structurally, a BJT consists of two *p-n* junctions connected back-to-back, and it can have either a *p-n-p* structure or an *n-p-n* structure, as shown in Fig. 3.50. The three terminals of the BJT are called the *emitter,* the *base,* and the *collector.* These terminals are so named because the emitter acts as a charge injector, the base is common to the two *p-n* junctions, and the collector is the receiver of the injected charges. The fact that a BJT has three terminals rather than two (as in the case of a *p-n* junction) allows the input terminal of the device to be physically isolated from the output terminal, just as in the FETs.

To initiate the so-called trans[fer-res]istor action, current is generated at a low-resistance forward-bias *p-n* junction (the input) and then transported to a high-resistance reverse-bias *p-n* junction (the output). Since current is exponentially dependent on the voltage in a forward-bias *p-n* junction, a substantial current can flow into the output for a given input voltage bias. If a load resistor also appears across the output (which behaves as a current source), a significant voltage gain can be developed. As a current injector, the emitter is small and heavily doped. It is very different from the collector, which is much larger and often physically the same as the substrate. The base is the region where current is transported from the emitter to the collector and is normally very thin ($\sim$ 50 nm) to minimize loss. In principle, a BJT can operate in three different modes, namely, the **common-emitter mode,** the **common-base mode,** and the **common-collector mode.** For most circuit applications, the common-emitter mode is used for current amplification and the common-base and common-collector modes are more suitable for impedance matching (which gives optimal power transfer from the input to the output). The basic transistor actions in the three different modes of operation are quite similar, except for the differences in the choice of the input and the output and the bias connections.

The detailed current flow in a BJT is determined by diffusion and recombination of the minority carriers (see Section 2.7). The simplest way to describe current flow in a BJT is to focus on the minority-carrier distributions. We shall consider a *p-n-p* BJT with a forward-bias emitter-base junction and a reverse-bias collector-base junction. This is the common-base mode of operation. For low injection, if we assume $x = 0$ to be at the edge of the emitter-base junction, as shown in Fig. 3.51, the hole density in the base, $p_b(x)$ ($/\text{m}^3$), along the $+x$ direction (toward the collector) is governed by the continuity equation, which may be expressed as

$$D_{pb}d^2(p_b(x) - p_{n0})/dx^2 - (p_b(x) - p_{n0})/\tau_p = 0 \tag{79}$$

Fig. 3.50
*p-n-p* and *n-p-n* BJTs

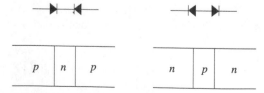

|   |   |   |   |   |   |
|---|---|---|---|---|---|
| *p* | *n* | *p* |   | *n* | *p* | *n* |

Fig. 3.51
Schematic of a BJT

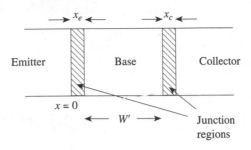

where   $q$ is the electron charge in C,

   $D_{pb}$ is the hole diffusivity in the base in m²/s,

   $p_{n0}$ is the equilibrium hole density in the base in /m³, and

   $\tau_p$ is the hole lifetime in s.

The solution to Eq. (79) depends on the boundary conditions (i.e., the values of $p_b(x)$ when $x = 0$ and when $x = W'$, the base width in m) and it has the form

$$p_b(x) = p_{n0} + A_{b1}\exp(x/L_{pb}) + A_{b2}\exp(-x/L_{pb}) \tag{80}$$

where   $A_{b1}$ and $A_{b2}$ are constants in /m³, and

   $L_{ph}\ (= \sqrt{D_{pb}/\tau_p})$ is the hole diffusion length in the base in m.

At $x = 0$, $p_b(0) = p_{n0}\exp(V_a/V_t)$ (see Eq. (10)), and if we assume at $x = W'$ that $p_b(W') \approx 0$ for a reverse-bias $p$-$n$ junction, Eq. (80) then becomes

$$p_b(x) = p_{n0}[\exp(V_a/V_t - 1)]\sinh[(W' - x)/L_{pb}]/\sinh[(W')/L_{pb}] \\ + p_{n0}[1 - \sinh(x/L_{pb})/\sinh(W'/L_{pb})] \tag{81}$$

Equation (81) can be simplified if we assume $W' \ll L_{pb}$ (a feature often found in modern BJTs). This gives

$$p_b(x) \approx p_{n0}[\exp(V_a/V_t - 1)][1 - x/W'] \tag{82}$$

Thus, in this limit the hole density drops off linearly in the base, and the base diffusion current, $I_{pb}$ (A) (see Chapter 2, Eq. (46)), is given by

$$I_{pb} = -qD_{pb}d(p_b(x))/dxA_{cs} = qD_{pb}p_{n0}[\exp(V_a/V_t - 1)]A_{cs}/W' \tag{83}$$

Note that the hole-diffusion current, $I_{pb}$, is exponentially dependent on the applied voltage, $V_a$. The corresponding electron density distributions, $n_e(x)$ and $n_c(x)$ (/m³), in the emitter and the collector are given by

$$n_e(x) = n_{p0} + n_{p0}[\exp(V_a/V_t) - 1]\exp[(x + x_e)/L_{ne}] \tag{84}$$

$$n_c(x) = n_{p0} + n_{p0}[\exp(-(x - x_c)/L_{nc})$$

where    $L_{ne}$ and $L_{nc}$ in m are the electron diffusion lengths in the emitter and in the collector, respectively,

and

$x_e$ and $x_c$ in m are the widths of the base-emitter junction and the base-collector junction, respectively.

Again, we can determine the minority-carrier diffusion currents in the emitter and in the collector; they are given by

$$I_{ne} = qD_{ne}n_{p0}A_{cs}[\exp(V_a/V_t - 1]/L_{ne} \tag{85}$$

$$I_{nc} = qD_{nc}n_{p0}A_{cs}/L_{nc}$$

where    $I_{ne}$ and $I_{nc}$ in A are the electron-diffusion currents in the emitter and in the collector, respectively,

$D_{ne}$ and $D_{nc}$ in m$^2$/s are the diffusivities of the electrons in the emitter and in the collector, respectively, and,

$L_{ne}$ and $L_{nc}$ in m are the diffusion lengths of the electrons in the emitter and in the collector, respectively.

During device operation, the emitter current, $I_e$, the base current, $I_b$, and the collector current, $I_c$, are given approximately by

$$I_e \approx I_{ne}$$

$$I_c = I_{nc} \tag{86}$$

$$I_b = I_{pb}$$

Thus far, we have considered only the diffusion currents. In a real device, recombination and generation currents (see Section 2.8) are important within the junction region and in the base. These currents have to be added to Eq. (86) if a more accurate treatment is required.

One of the figures of merit of a BJT is the **common-base current gain,** $\alpha_g$, which is defined as $\Delta I_c / \Delta I_e$. It can be expressed in terms of the product of two parameters, namely, $\gamma_e$, the emitter efficiency, and $\alpha_t$, the transport factor. The **emitter efficiency** measures the effectiveness of the emitter as a charge injector, and the **transport factor** determines how much of the injected charges goes to the output. For most modern BJTs, both $\gamma_e$ and $\alpha_t$ are about 1. The **common-emitter current gain,** $\beta_g$, is defined as $\Delta I_c / \Delta I_b$, and it measures directly the current gain (the output current divided by the input current) in the common-emitter mode of operation. In the absence of charge recombination, it is given approximately by (see Eqs. (83) and (85))

$$\beta_g \approx D_{pb}N_{Ae}x_e/(D_{ne}N_{Db}W') \tag{87}$$

where  $N_{Ae}$ is the acceptor density in the emitter in /m³, and

  $N_{Db}$ is the donor density in the base in /m³.

For a lightly doped base, $\beta_g$ can have a value between 100 and 1000, which can mean a substantial current gain even for a reasonable load. It is sometimes useful to express $\beta_g$ in terms of a **Gummel number,** $Q_G$, which is defined as the base doping per unit area:

$$Q_G = \int_0^{W'} [N_{Db}(x)] \, dx \approx W'N_{Db} \qquad (88)$$

Equations (87) and (88) suggest that $Q_G$ is inversely proportional to $\beta_g$, the current gain. $Q_G$ is, therefore, a convenient measure of the performance of a BJT.

So far, we have considered only the low injection limit when majority carriers are unimportant. At high current, the electron and hole densities in the base can become comparable near the emitter-base junction (use Eq. (10) to show that $n_b p_b \approx n_i^2 \exp(V_a/V_t)$), and

$$n_b(0) = p_b(0) \approx n_i \exp(V_a/2V_t) \qquad (89)$$

Based on Eqs. (85) and (86),

$$I_e \approx I_c \approx (qn_i D_{pb} A_{cs}/W') \exp(V_a/2V_t) \qquad (90)$$

$$I_b \approx qD_{ne}n_i^2 A_{cs}/(x_e N_{Ae}) \exp(V_a/V_t)$$

A calculation of $\beta_g$ at high injection suggests it has a $1/I_c$ dependence, which is due primarily to an increase in carrier recombination in the base. Fig. 3.52 is a plot showing how $\beta_g$ decreases with increasing $I_c$.

**Example 3.39**   Compute $\beta_g$ in a BJT with the following parameters: $N_{Ae} = 1 \times 10^{24}$/m³, $x_e = 100$ nm, $N_{Db} = 10^{20}$ /m³, $W' = 1000$ nm, $\mu_n = 0.15$ m²/V·s, and $\mu_p = 0.1$ m²/V·s. (Given that $\mu_n = D_n /kT$ and $\mu_p = D_p /kT$.)

**Solution:**   Based on Eq. (87), $\beta_g = D_{pb}N_{Ae}x_e/(D_{ne}N_{Db}W') = 0.1$ m²/V·s × $1 \times 10^{24}$/m³ × $1 \times 10^{-7}$ m/(0.15 m²/V·s × $1 \times 10^{20}$/m³ × $10^{-6}$ m) = 6.67 × $10^2$.  **§**

Fig. 3.52
Common-emitter current
gain versus collector
current

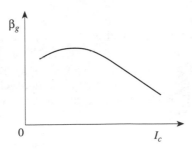

**Example 3.40**    Estimate the change in $\beta_g$ at high injection.

*Solution:*    Based on Eq. (90), $\beta_g = I_e/I_b \approx D_{pb}N_{Ae}x_e/(D_{ne}n_iW')\exp(-V_a/2V_t) = \beta_g N_{Db}/n_i\exp(-V_a/2V_t)$. Since the exponential term is negative, $\beta_g$ is inversely proportional to $I_c$.    §

## 3.8    NONIDEAL EFFECTS AND OTHER PERFORMANCE PARAMETERS

### MOSFETs

*Short-Channel Effects*    Short-channel MOSFETs suffer from a number of nonideal effects, the foremost being velocity saturation of the carriers. At a low drain-to-source bias voltage, the velocities of the electrons and the holes are proportional to the electric field, and constant mobilities are observed. This does not happen in a short-channel device at high electric field when the carrier velocities are no longer proportional to the electric field. After taking velocity saturation into account, the electron mobility, $\mu^*$, will have the following form:

$$\mu^* = \mu_0/(1 + \mu_0\acute{E}/v_s) \tag{91}$$

where    $\mu_0$ is the low-field mobility in $m^2/V\cdot s$,

   $\acute{E}$ is the electric field in V/m, and

   $v_s$ is the saturation velocity in m/s.

The drain-to-source current $I_{ds}$ of an *n*-channel MOSFET (A) can then be written as

$$I_{ds} = qn_{ss}v_n(\acute{E})W_c \tag{92}$$

where    $q$ is the electron charge in C,

   $n_{ss}$ is the electron density in the channel in $/m^2$,

   $v_n(\acute{E})$ $(= \mu^*\acute{E})$ is the field-dependent velocity in m/s, and

   $W_c$ is the channel width in m.

When $\acute{E}$ is large, $v_n(\acute{E}) \approx v_s$. Oftentimes, **velocity saturation** occurs only in a portion of the channel, as in the pinch-off region during saturation. To model this effect, the MOSFET is divided into two regions. One region has a linear velocity-field relationship and the other, a constant-saturation velocity. In the first region, the channel potential, $V_c(x)$ (V), at a distance $x$ meters from the source is given by

$$V_c(x) = V_{gt} - \sqrt{V_{gt}^2 - 2I_{ds}x/\beta_0 L} \tag{93}$$

where    $V_{gt} = V_g - V_T$,

$\beta_0 = \mu^* C_i W_c / L$, and

$L$ is the channel length in m.

In the second region, assuming that the electric field at the pinch-off point is $\acute{E}(L) = \acute{E}_p$, the saturation current $I_{Dsat}$ (A) is given by

$$I_{Dsat} = \beta_{sl} V_{sl}^2 [\sqrt{(1 + (V_{gt}/V_{sl}))^2} - \beta_{sl} V_{sl}^2 \tag{94}$$

where    $V_{sl} = \acute{E}_p L$, and

$\beta_{sl}$ is the value of $\beta_0$ at the point of saturation.

For a short-channel device, $I_{Dsat} \approx \beta_{sl} V_{gt} V_{sl}$, which suggests that $I_{Dsat}$ is $2V_{sl}/V_{gt}$ times smaller than the corresponding value in a long-channel device (Eq. (68)). Therefore, the shorter the channel, the smaller will be $I_{Dsat}$.

Short-channel MOSFETs are also affected by energetic electrons injected and trapped within the oxide layer near the drain region. This is called the **hot-electron effect**. The charges on the trapped electrons deplete the channel prematurely and lead to an increase in the saturation current (for a reduced $V_T$, see Eq. (68)).

**Example 3.41**    Compute the fractional change in the electron mobility of Si if the electric field is $1 \times 10^6$ V/m. Assume $\mu_n = 0.15$ m$^2$/V·s and $v_s = 1 \times 10^7$ m/s.

*Solution:*    Based on Eq. (91),

$$\begin{aligned}\mu^*/\mu_0 &= 1/(1 + \mu_0 \acute{E}/v_s) \\ &= 1/(1 + 0.15 \text{ m}^2/\text{V·s} \times 1 \times 10^6 \text{ V/m}/(1 \times 10^7 \text{ m/s})) \\ &= 0.985 \quad §\end{aligned}$$

**Example 3.42**    Compute the fractional reduction in the saturation current due to the short-channel effect if the saturation field is $5 \times 10^4$ V/m, the channel length is 0.8 µm, $V_g = 1.0$ V, and $V_T = 0.2$ V.

*Solution:*    The fractional reduction in the saturation current is $2V_{sl}/V_{gt} = 2 \times 5 \times 10^4$ V/m $\times 0.8 \times 10^{-6}$ m/(1 V − 0.2 V) = 0.1.    §

*Subthreshold Conduction*    For most MOSFETs, there is a small drain-to-source current even when the gate voltage is below threshold and the channel is not inverted (in depletion). This current is called the **subthreshold current**. For a long-channel device, the subthreshold current, $I_{sub}$ (A), is given by

$$\begin{aligned}I_{sub} &= -qA_{cs}D_n \Delta n/\Delta x \\ &= \epsilon_s \mu_n (W_c/L) V_t^2 (n_i/N_A)^2 \sqrt{V_t/V_s} \exp(V_s/V_t) \\ &\quad \times [1 - \exp(-V_{ds}/V_t)]/(\sqrt{2} L_{Dp})\end{aligned} \tag{95}$$

where    $\epsilon_s$ is the semiconductor permittivity in F/m,

   $\mu_n$ is the electron mobility in m²/V·s,

   $W_c/L$ is the ratio of the channel width to the channel length,

   $n_i$ is the intrinsic carrier density of the semiconductor in /m³,

   $N_A$ is the acceptor density in /m³,

   $V_s$ is the surface potential in V, and

   $L_{Dp}$ is the hole-diffusion length in m.

Since $V_s \approx V_g$ (see Eq. (55)), Eq. (95) suggests that the subthreshold current depends exponentially on the gate voltage. It is possible to operate the MOSFET in a subthreshold mode with an exponential (input) voltage dependence similar to the BJT, but this operation has to be at a very low current level. The dynamic range of the subthreshold current can still be quite appreciable (5 to 6 orders of magnitude) and is limited only by the leakage current in the MOSFET.

**Example 3.43**    Compute the range of substhreshold current density if $V_s$ varies from 0.1 V to 0.6 V. Assume $V_{ds}$ is large. $W_c/L = 100$, $V_{th} = 0.026$ V, $\epsilon_s = 1.05 \times 10^{-10}$ F/m, $\mu_n = 0.15$ m²/V·s, $n_i = 1.45 \times 10^{16}$ /m³, $V_s \approx V_g = 0.5$ V, $N_A = 1 \times 10^{22}$ /m³, and $L_{Dp} = 1 \times 10^{-7}$ m.

*Solution:*    Based on Eq. (95),

$$
\begin{aligned}
I_{\text{sub}} &\approx \epsilon_s \mu_n (W_c/L) V_t^2 (n_i/N_A)^2 \sqrt{(V_t/V_s)} \exp(V_s/V_t)/(\sqrt{2}\, L_{Dp}) \\
&= 1.05 \times 10^{-10} \text{ F/m} \times 0.15 \text{ m}^2/\text{V·s} \times 100 \times (0.026 \text{ V})^2 \\
&\quad \times \sqrt{0.026 \text{ V}/0.2 \text{ V}} \times \exp(0.2 \text{ V}/0.026 \text{ V})/(2 \times 10^{-7} \text{ m}) \\
&= 5.6 \times 10^{-12} \text{ A} \\
&= 5.6 \text{ pA}  \quad \S
\end{aligned}
$$

*Capacitance Effect*    As shown in Fig. 3.53, the capacitances of a MOSFET include (1) the gate-to-source capacitance, $C_{gs}$; (2) the gate-to-drain capacitance, $C_{gd}$; (3)

**Fig. 3.53**
An equivalent circuit of a
MOSFET

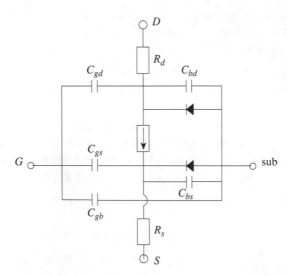

the substrate-to-source capacitance, $C_{bs}$; (4) the substrate-to-drain capacitance, $C_{bd}$, and (5) the gate-to-substrate capacitance, $C_{gb}$.

In saturation, these capacitances (F) are given by

$$C_{gs} = C_{gsf} + 2C_i W_c L/3$$

$$C_{gd} = C_{gdf}$$

$$C_{gb} = 0$$

$$C_{bs} = C_{js}(1 + \tfrac{2}{3}C_{gb}/C_{js}W_cL)/(1 + V_{\text{sub-}s}/\phi_{bi})^{mB}$$

$$C_{bd} = C_{jd}/(1 + V_{\text{sub-}d}/\phi_{bi})^{mB} \tag{96}$$

where    $C_{gsf}$ is the gate-to-source fringing capacitance in F,

$C_i$ is the gate capacitance per unit area in F/m$^2$,

$C_{gdf}$ is the gate-to-drain overlap capacitance in F,

$C_{js}$ is the substrate-to-source capacitance at zero bias in F,

$C_{gb}$ is the gate-to-substrate capacitance in F,

$C_{jd}$ is the substrate-to-drain capacitance at zero bias in F,

$V_{\text{sub-}s}$ is the substrate-to-source voltage in V,

$V_{\text{sub-}d}$ is the substrate-to-drain voltage in V,

$\phi_{bi}$ is the built-in voltage in V, and

$mB$ is a constant.

In the linear regime, the same parameters are given by

$$C_{gs} = C_{gsf} + \tfrac{2}{3}C_i W_c LV_{Dsat}(3V_{Dsat} - 2V_{ds})/(2V_{Dsat} - V_{ds})^2$$

$$C_{gs} = C_{gdf} + \tfrac{2}{3}C_i W_c L(V_{Dsat} - V_{ds})(3V_{Dsat} - V_{ds})/(2V_{Dsat} - V_{ds})^2$$

$$C_{gb} = 0 \tag{97}$$

$$C_{bs} = C_{js}[1 + \tfrac{2}{3}C_{gb}/C_{js}W_cLV_{Dsat}(3V_{Dsat} - 2V_{ds})/(2V_{Dsat} - V_{ds})^2]/(1 + V_{\text{sub-}s}/\phi_{bi})^{mB}$$

$$C_{bd} = C_{jd}[1 + \tfrac{2}{3}C_{gb}/C_{js}W_cL(V_{Dsat} - V_{ds})(3V_{Dsat} - V_{ds})/(2V_{Dsat} - V_{ds})^2]/(1 + V_{\text{sub-}d}/\phi_{bi})^{mB}$$

Although a full equivalent-circuit model of the MOSFET has to take into account all these capacitances, a simplified model need retain only the input capacitances and an output current source, as shown in Fig. 3.54.

Fig. 3.54
A simplified equivalent circuit of a MOSFET

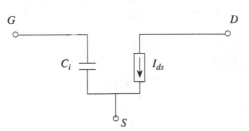

The **cutoff frequency,** $f_T$, limits the frequency response of the MOSFET and is dominated by the input capacitances. It is given by

$$f_T = g_m/[2\pi(C_g + C_p)] \tag{98}$$

where   $g_m$ is the transconductance of the MOSFET in $/\Omega$ or S,

   $C_g$ is the gate capacitance in F, and

   $C_p$ is the input parasitic capacitances in F.

Equation (98) also determines the rise time of the MOSFET.

***Control of the Threshold Voltage***   The threshold voltage of the MOSFET can be controlled by changing the dopant density near the surface of the channel. The changes in the threshold voltage $\Delta V_T$ (V) due to surface doping for enhancement-mode and depletion-mode *n*-channel MOSFETs are given, respectively, by

$$\Delta V_{TE} = \Delta V_{bi} + q\Delta n_A/C_i \tag{99}$$
$$\Delta V_{TD} = \Delta V_{bi} - \Phi_x/q - q\Delta n_D/C_i$$

where   $\Delta V_{bi}$ is the change in the built-in voltage in V,

   $C_i$ is the gate capacitance per unit area in F/m$^2$,

   $\Delta n_A$ and $\Delta n_D$ are the surface dopant densities in /m$^2$, and

   $\Phi_x$ is any dip in the surface potential in eV due to the surface dopant atoms.

The dependence of the threshold voltage on the surface dopings $\Delta n_A$ and $\Delta n_D$ in the channel provides a convenient way to adjust the threshold voltage of MOSFETs.

**Example 3.44**   Compute the change in surface dopant density if the threshold voltage of a MOSFET changes by 0.2 V. Assume the oxide thickness is 0.7 $\mu$m, $\epsilon_{\text{oxide}} = 3.9$, and there is no change in the built-in voltage.

***Solution:***   Based on Eq. (99), $\Delta n_A = \Delta V_T C_i/q = 0.2$ V $\times$ 3.9 $\times$ 8.85 $\times$ $10^{-12}$ F/m/(0.7 $\times$ 10$^{-6}$ m $\times$ 1.6 $\times$ 10$^{-19}$ C) $= 6.16 \times 10^{13}$/m$^2$.   §

***Complementary MOSFET (CMOS)***   A **complementary MOSFET,** or CMOS, is formed by stacking an *n*-channel MOSFET and a *p*-channel MOSFET together, as shown in Fig. 3.55. The two devices have a common input; i.e., their gates are connected together. Since the two MOSFETs have opposite threshold voltages (negative voltage in the *p*-channel device and positive voltage in the *n*-channel device), neither of the transistors will be conducting at zero input, so there is no standby current. Current will flow only when one of the two threshold voltages is exceeded, and beyond this point, the CMOS acts as a voltage inverter. A CMOS, therefore, has zero power dissipation except during switching, which is very useful for large-scale circuits. One of the most common problems with a CMOS is the **latch-up** effect illustrated in Fig. 3.56. The lateral parasitic *p-n-p* BJT present in the CMOS substrate is structurally connected to the vertical *n-p-n* BJT in the *p*-well. This situation induces positive feedback on the vertical *n-p-n* BJT and can produce a significant current through a CMOS. Often, breakdown will occur between the power rails even for a small substrate current. One way to reduce this problem is to increase the distance between the *p*-well and the external *p*⁺-contact.

**Example 3.45**   Explain how latch-up gives rise to positive feedback in a CMOS.

*Solution:*   As shown in Fig. 3.56, an increase in the current flowing through (the collector of) the lateral BJT also increases the (base) current through the

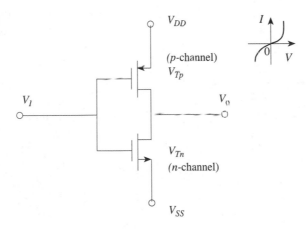

**Fig. 3.55**
Complementary MOSFET

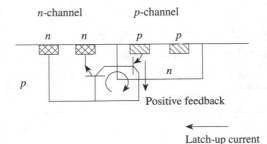

**Fig. 3.56**
Latch-up effect in CMOS

vertical BJT since the collector of the lateral BJT is connected to the base of the vertical BJT.   §

**Example 3.46**    Propose a way to remove latch-up effect in a CMOS structure.

*Solution:*   One method is to use two different wells (a *p*-well and a *n*-well) to form the complementary MOSFETs so that isolation between the two transistors is ensured.   §

*Noise in a MOSFET*   Thermal noise is one of the performance limitations in MOSFETs. The equivalent thermal noise source in a MOSFET, $\hat{i}_{th}^2$ (A$^2$), is given by

$$\hat{i}_{th}^2 = 4kTg_{do} \tag{100}$$

where   $g_{do}$ is the conductance per unit length in /$\Omega$ or S,

$k$ is the Boltzmann constant, and

$T$ is the absolute temperature in K.

According to Eq. (100), noise in a MOSFET increases with temperature; low-noise MOSFETs frequently operate at a reduced temperature in order to improve the signal-to-noise ratio.

*Scaling in MOSFETs*   **Scaling** is much more important in MOSFETs because of the need for large-scale integration. Assuming that all the device dimensions are reduced by a factor $k*$, where $k*$ is a constant, then the modified device parameters are given by

$$C_i' = C_i/k*$$

$$I_{Dsat}' = I_{Dsat}/k* \tag{101}$$

$$f_T' = k*f_T$$

$$P_{ac}' = P_{ac}/(k*)^2$$

where   $C_i$ is the oxide capacitance in F,

$I_{Dsat}$ is the saturation current in A,

$f_T$ is the cutoff frequency in Hz, and

$P_{ac}$ is the ac power dissipation in W.

The prime indicates the adjusted parameters after scaling. Downscaling in MOSFET, therefore, improves all aspects of the device performance except for the current density (which scales up as $k*$).

**Example 3.47**   Show that with scaling, (a) $P'_{ac} = P_{ac}/(k^*)^2$, and (b) $J' = J_{Dsat}k^*$, where $k^*$ is the scaling factor and $J_o$ is the *current density*.

**Solution:**
(a) Based on Eq. (101), $I'_{Dsat} = I_{Dsat}/k^*$. Since $P_{ac} = I^2_{Dsat}R_L$, where $R_L$ is the load resistor (which is not scaled), $P'_{ac} = P_{ac}/(k^*)^2$.

(b) The current density is $J_{Dsat} = I_{Dsat}/A_{cs}$. If the area is scaled down by $(k^*)^2$ and $I_{Dsat}$ is scaled down by $k^*$, the net effect is that $J'$ will scale up by $k^*$.   **§**

## MESFETs

*Velocity Saturation*   Velocity saturation limits the performance of the MESFETs. When velocity saturation becomes important, the drain-to-source saturation current $I_{Dsat}$ in a MESFET will become

$$I_{Dsat} = \beta''(V_G - V_T)^2 \tag{102}$$

where   $\beta'' = 2\epsilon_s\mu_n v_s W_c /[t_c(\mu_n V_{po} + 3v_s L)]$,

$V_T = V_{bi} - V_{po}$, and

$v_s$ is the saturation velocity in m/s.

Note that $\beta''$ is now a function of $v_s$, the saturation velocity (compare with Eq. (77)). This relationship limits the value of the transconductance of the MESFET and hence the gain in the device.

*Capacitance Effect*   The gate-to-source and gate-to-drain capacitances of the MESFET are given by

$$C_{gs} = C_{go}/\sqrt{1 - V_{gs}/V_{bi}} + \pi\epsilon_s W_c/2$$
$$C_{gd} = C_{go}/\sqrt{1 - V_{gd}/V_{bi}} + \pi\epsilon_s W_c/2 \tag{103}$$

where   $C_{go} = W_c L/2\sqrt{q\epsilon_s N_D /2V_{bi}}$,

$V_{gs}$ is the gate-to-source voltage in V,

$V_{gd}$ is the gate-to-drain voltage in V,

$V_{bi}$ is the built-in voltage in V,

$\epsilon_s$ is the semiconductor permittivity in F/m,

$W_c$ is the width of the MESFET in m, and

$L$ is the channel length in m.

The second terms in these expressions are due to the parasitic sidewall capacitances.

*Scaling of MESFETs*   The scaling of MESFETs involves parameters such as the channel length $L$, the channel doping $N_D$, and the epitaxial-layer (channel) thickness

$t_c$. Empirically, it has been found that to avoid short-channel effect, the following restrictions apply:

$$L/t_c < 3 \quad \text{and} \quad N_D L \approx 1.6 \times 10^{23} \ \mu\text{m/m}^3 \qquad \textbf{(104)}$$

In addition, gate-to-source tunneling and drain-to-source breakdown further limit an increase in channel doping (channel doping increases the transconductance (see Eq. (76) and hence the device performance), whereas the dynamic range of the input voltage (directly linked to the pinch-off voltage) limits the thickness of the epilayer.

**Example 3.48**   An *n*-channel GaAs MESFET has a channel length of 0.25 μm. What should be the value of the pinch-off voltage to avoid the short-channel effect? $\epsilon_s = 1.16 \times 10^{-10}$ F/m.

*Solution:*   Based on Eq. (104), $t_c > L/3 = 0.25 \ \mu\text{m}/3 = 0.083 \ \mu\text{m}$, and $N_D = 1.6 \times 10^{23} \ \mu\text{m/m}^3/0.25 \ \mu\text{m} = 6.4 \times 10^{23}/\text{m}^3$. Based on Eq. (74),

$$
\begin{aligned}
V_{po} &= qN_D t_c^2/(2\epsilon_s) \\
&= 1.6 \times 10^{-19} \text{ C} \times 6.4 \times 10^{23}/\text{m}^3 \times (0.083 \times 10^{-6} \text{ m})^2/(2 \times 1.16 \times 10^{-10} \text{ F/m}) \\
&= 3.04 \text{ V}.
\end{aligned}
$$

This is the minimum value of $V_{po}$.   **§**

# BJTs

***Base Resistance, Emitter Current Crowding, and Base Push-out (Kirk) Effect***   The finite base resistance in the BJT will generate a voltage drop across the base if a base current is flowing. This voltage drop changes the bias voltage at different points in the base and gives rise to a nonuniform current flow across the base-emitter junction. Because of this, the effective resistance of the base is called the **base-spreading resistance.** It is also the cause for **emitter current crowding** in parts of the base where the potential drop is highest. When the emitter current is large, carrier injection may extend beyond the base into the collector, where the doping is usually low. This results in a widening of the base (**base push-out effect**) in a lowly doped collector and is the cause for a reduction in the current gain due to a reduced transport factor, $\alpha_T$.

**Example 3.49**   Estimate the maximum base resistance if the current drop along the base of a BJT is to be less than a factor of 2. The saturation current of the base is $I_{bo} = 0.1$ pA, $V_{be} = 0.6$ V, and $V_t = 0.026$ V.

*Solution:*   The base current is $I_b \approx I_{bo}(\exp(V_a/V_t) - 1) = 1 \times 10^{-13}$ A $\times$ $(\exp(0.6 \text{ V}/0.026 \text{ V}) - 1) = 1.05 \times 10^{-3}$ A $= 1.05$ mA. For $I_b = 0.5 \times$

1.05 mA = 0.525 mA, $V_b$ = 0.581 V. The maximum base resistance along the base is approximately $\Delta V/I_b$ = (0.60 V − 0.581 V)/(1.05 × $10^3$ A) = 19 Ω.

§

**Emitter and Base Doping** A highly doped semiconductor suffers from energy-gap shrinkage, which occurs in the emitter side of the emitter-base junction of a BJT. Energy gap shrinkage is responsible for a decrease in the emitter efficiency, and the amount of shrinkage, $\Delta E_g$ (J), is given by

$$\Delta E_g = 3q^2 \sqrt{qN_{De}/\epsilon_s \cdot V_t} / 16\pi\epsilon_s \tag{105}$$

where $N_{De}$ is the donor density in the emitter in /m$^3$.

**Energy gap reduction** increases the intrinsic carrier density $n_i$ and, hence, the minority carrier density. In the emitter, this raises the reverse carrier flow from the base and decreases the emitter efficiency.

Base doping also affects current flow in the base. If the base of an *n-p-n* BJT is doped with a high acceptor density near the emitter-base junction, the internal electric field generated in the base will aid in the forward flow of the electrons through the base. Normally, we can assume a base dopant profile, $N_{Ab}$ (in /m$^3$) to be given by

$$N_{Ab} \approx N_{Ao}\exp[-(x - x_e)/\lambda_b] \tag{106}$$

where $N_{Ao}$ is the acceptor density in /m$^3$ at $x = x_e$ (see Fig. 3.51), and

$\lambda_b$ is the effective length of the dopant density gradient in m.

The modified current gain $\beta^*$ (see Eq. (87)) due to a nonuniform base doping is given by

$$\beta^* \approx \beta_g W'/2\lambda_b \tag{107}$$

where $W'$ is the base width in m.

**Example 3.50** Compute the energy gap shrinkage in Si if $N_D$ = 1 × $10^{25}$/m$^3$. $V_t$ = 0.026 V.

**Solution:** Based on Eq. (105),

$$\Delta E_g = 3q^2 \sqrt{qN_{De}/\epsilon_s V_t}/16\pi\epsilon_s$$
$$= 3 \times (1.6 \times 10^{-19} \text{ C})^2$$
$$\times \sqrt{1.6 \times 10^{-19} \text{ C} \times 10^{25}/\text{m}^3/(1.05 \times 10^{-10} \text{ F/m} \times 0.026 \text{ V})}/(16 \times 3.14 \times 1.05 \times 10^{-10} \text{ F/m})$$
$$= 3.47 \times 10^{-20} \text{ J} = 0.22 \text{ eV. } §$$

**Charge Storage and Frequency Response of a BJT** The normal operation of a BJT depends on the establishment of a minority carrier density gradient in the base.

This generates a charge storage effect, which is responsible for the time delay observed during switching. When the device is turned off, the decay in the base charge, $\Delta Q_{bs}$ (C/m$^3$), when the base current drops from $I_{b1}$ (A) to $I_{b2}$ (A) is given by

$$\Delta Q_{bs} = (I_{b1} - I_{b2})\tau_{sr}\exp(-t/\tau_{sr}) + (I_{b2} - I_{ba})\tau_{sr} \qquad (108)$$

where   $I_{ba}$ is the base current at the threshold of saturation in A, and

   $\tau_{sr}$ is the recombination lifetime of the base charge in s.

The storage time, $\tau_s$, in seconds (i.e., the time needed to remove the stored charges) is given as

$$\tau_s = \tau_{sr} \ln[(I_{b1} - I_{b2})/(I_{ba} - I_{b2})] \qquad (109)$$

The storage time is important in estimating the time delay during switching in digital circuits. The frequency response of a BJT is also limited by the time required for minority carriers to transport through the base. With respect to the common-base current gain, $\beta_g$, the cutoff frequency, $f_\beta$, is given by

$$f_\beta = (C_e r_e)/2\pi \qquad (110)$$

where   $C_e$ is the emitter capacitance in F, and

   $r_e$ is the emitter resistance in $\Omega$.

$f_\beta$ is basically limited by the input frequency response of the transistor.

A different cutoff frequency $f_\beta$ exists for the common-emitter current gain, $\beta_g$, and is given by

$$f_\beta = g_{b'e}/2\pi(C_{b'e} + C_{b'c}) \qquad (111)$$

where   $g_{b'e}$ is the base-emitter conductance in /$\Omega$ or S,

   $C_{b'e}$ is the base-emitter capacitance in F, and

   $C_{b'c}$ is the base-collector capacitance in F.

Note the presence of the base-collector capacitance in Eq. (111).

It is sometimes convenient to define a unity-gain cutoff frequency, $f_T$, when the current gain goes to zero. Effectively, $f_T \approx f_\alpha$. Fig. 3.57 illustrates the different types of cutoff frequencies in a BJT.

**Fig. 3.57**
Cutoff frequencies in a
BJT

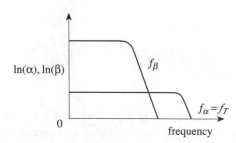

The **transit time**, $\tau_{tr}$ (s), of a transistor is the average time necessary for a carrier to transit through the base. In the physical limit, the transistor cutoff frequency, $f_{tr}$, is given by

$$f_{tr} = \langle v \rangle / W' \tag{112}$$

where   $\langle v \rangle$ is the average carrier velocity through the base in m/s, and

   $W'$ is the base width in m.

$f_{tr}$ is also called the *transit-time cutoff frequency*. Depending on the types of device configurations and their applications, different cutoff frequencies are used.

**Example 3.51**   Compute the storage time of the base charge if the base current changes from 1 mA to 1 μA. The saturation current is 10 μA, and the recombination lifetime of carriers in the base is 0.1 μs.

*Solution:*   Based on Eq. (109), $\tau_s = \tau_{sr} \ln[(I_{b1} - I_{b2})/(I_{ba} - I_{b2})] = 1 \times 10^{-7}$ s $\times \ln[(1000 - 1) \; \mu A/(10 - 1) \; \mu A] = 4.7 \times 10^{-6}$ s $= 4.7$ μs.   §

**Example 3.52**   Compute the transit-time frequency of a bipolar transistor if $\langle v \rangle = 1 \times 10^4$ m/s and the base width is 300 nm.

*Solution:*   Based on Eq. (112), $f_{tr} = \langle v \rangle / W' = 1 \times 10^4$ m/s/$(300 \times 10^{-9}$ m $= 3.3 \times 10^{10}$ Hz $= 33$ GHz.   §

***Breakdown***   **Breakdown** in a bipolar transistor normally occurs at the base-collector junction. The breakdown voltage, $BV_{cb}$ (V) in a BJT is estimated to be

$$BV_{cb} = \epsilon_s \acute{E}_{cr}^2 / (2qN_{Dc}) \tag{113}$$

where   $\acute{E}_{cr}$ is the critical field, which lies between $3 \times 10^7$ and $4 \times 10^7$ V/m, and

   $N_{Dc}$ is the donor density in the collector in /m³ for an *n-p-n* transistor.

Emitter-collector breakdown occurs at a lower voltage due to *transistor action* in a forward-bias emitter-base junction (when emitter-collector breakdown is measured, the bias voltage is applied across the emitter and the collector, and the base is floating). The emitter-collector breakdown voltage, $BV_{ce}$ (V), is given by

$$BV_{ce} = BV_{cb}(1 - \alpha_g)^{1/mb} \tag{114}$$

where   $\alpha_g$ is the common-base current gain, and

   $mb$ is a constant that varies between 2 and 5.

**Punch-through** occurs in the case of a thin base when the base is fully depleted. The punch-through voltage, $V_{pt}$, of a BJT (V) is given by

$$V_{pt} = qN_{Ab}W^2/(2\epsilon_s N_{Dc}) \tag{115}$$

where   $N_{Ab}$ is the acceptor density in the base in /m³.

In addition to avalanche breakdown and base punch-through, thermal effect due to power dissipation in the BJT produces what is known as **secondary breakdown,** which occurs when the collector current is large and takes place at a voltage somewhat smaller than the primary breakdown voltage (i.e., $BV_{cb}$).

**Example 3.53**     Estimate the breakdown voltage across the base-collector junction of a Si BJT. $\acute{E}_{br} = 1 \times 10^7$ V/m and the carrier density in the collector is $1 \times 10^{22}/m^3$.

***Solution:***     Based on Eq. (113), $BV_{cb} = \epsilon_s \acute{E}_{br}^2/(2qN_{Dc}) = 1.05 \times 10^{-10}$ F/m $\times (1 \times 10^7$ V/m$)^2/(2 \times 1.6 \times 10^{-19}$ C $\times 10^{22}/m^3) = 3.28$ V.     §

***Noise***     Noise in a BJT is similar to that in a *p-n* junction. The input and output noise current sources, $\hat{i}_1^2$ and $\hat{i}_2^2$ (A$^2$), are given by

$$\hat{i}_1^2 = 2q(I_e + 2I_b)\delta f \tag{116}$$

$$\hat{i}_2^2 = 2qI_c\delta f$$

where     $I_e$, $I_b$, and $I_c$ are the emitter, base, and collector currents in A, respectively, and

$\delta f$ is the bandwidth in /s.

***Scaling of BJTs***     Because of the more complex structure in a BJT, scaling is not directly linked to the device parameters, as in the case of the MOSFET. To achieve a higher packing density in a circuit, BJTs are scaled based on their emitter areas. Scaling is also limited by the width of the isolation regions between the neighboring BJTs. As discussed in Section 3.7, the gain characteristics of a BJT are inversely proportional to the base width.

## 3.9     NEW TRANSISTOR STRUCTURES

### Heterojunction Bipolar Transistor (HBT)

**Heterojunction bipolar transistors** make use of the energy gap differences in the emitter, the base, and the collector to enhance transistor action. For instance, a HBT uses a wide-energy-gap semiconductor in the emitter to reduce the reverse carrier flow and increase the emitter efficiency. For an *n-p-n* HBT with an energy gap difference of $\Delta E_g$ ($= E_{g1} - E_{g2}$) between the emitter and the base, as shown in Fig. 3.58, the maximum current gain, $\beta_{max}$, is

$$\beta_{max} = v_{nb}N_{De}x_e/(v_{pe}N_{Ab}W') \exp(\Delta E_v/(qV_t)) \tag{117}$$

**Fig. 3.58**
Heterojunction *n-p-n* BJT

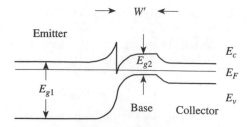

where    $v_{nb}$ is the electron velocity in the base in m/s,

$v_{pe}$ is the hole velocity in the emitter in m/s,

$N_{De}$ is the donor density in the emitter in /m³,

$N_{Ab}$ is the acceptor density in the base in /m³, and

$\Delta E_v$ is the energy difference in the valence band in eV (see Fig. 3.58).

In a practical device, only a part of the difference in the energy gap is dropped across the valence band. In some HBTs, a conduction band energy barrier is present at the emitter side of the base-emitter junction, as shown in Fig. 3.59. Such a barrier allows only the more energetic carriers to pass into the base. For those carriers that can surmount the barrier, their energies are often sufficient to allow them to pass through the base without any significant energy loss. This creates a situation called the **ballistic effect;** i.e., carriers pass through the base without suffering any collision. Ballistic BJTs offer much higher gains than conventional BJTs. The reduction in the reverse carrier flow in the emitter of the HBT also allows for a much higher doping in the base and a lower doping in the emitter. This simultaneously lowers the input capacitance (related to the frequency response) and the base resistance (related to the gain).

The propagation delay of a HBT logic gate, $\tau_d$ (s), can be approximated by

$$\tau_d = 2.5 R_B C_{b'c} + R_B \tau_B / R_L + (3 C_{b'c} + C_L) R_L \qquad \text{(118)}$$

where    $R_B$ is the base resistance in $\Omega$,

$C_{b'c}$ is the base-collector junction capacitance in F,

$\tau_B$ is the carrier lifetime in the base in s,

$R_L$ is the load resistance in $\Omega$, and

$C_L$ is the load capacitance in F.

From Eq. (118), it is obvious that the speed of the HBT is optimized if there is a thin and highly conducting base and a reduced-output capacitance for a given resistive load $R_L$. HBTs are frequently made with a AlGaAs/GaAs heterostructure, although an InGaAs/InP heterostructure is also used because of its higher electron mobility.

**Example 3.54**    Suggest the reasons why a HBT can simultaneously lower the input capacitance and the base resistance, whereas a BJT cannot do so.

Fig. 3.59
HBT structure showing
electrons surmounting the
energy barrier

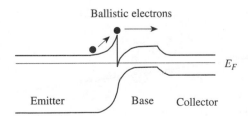

*Solution:*   Increasing the base resistance in the BJT will increase the reverse carrier flow in the base and lower the emitter efficiency. Because of the wide energy gap in the HBT, this effect is less severe, so the base resistance in the BJT can be reduced without significantly affecting its emitter efficiency.   §

## Heterojunction Field-Effect Transistor (HFET)

**Heterojunction field-effect transistors** (HFETs) are becoming more important because of their superior performance. Most HFETs are *n*-channel devices formed on GaAs substrates, as shown in Fig. 3.60. When an undoped surface layer (such as AlGaAs) is added, the device is called a HIFET. On the other hand, if a doped plane is present, as shown in Fig. 3.61, it is called a **planar-doped HFET.** In all HFETs, conduction takes place in a two-dimensional electron gas layer a little distance below the gate electrode. The region is undoped and has a high carrier mobility. A HFET usually has a high cutoff frequency, $f_T$ (Hz); the latter is given by

$$f_T = 1/[2\pi\tau_{\text{tr}}(1 + C_L/C_g)] \tag{119}$$

where    $\tau_{\text{tr}}$ is the transit time of the HBT in s,

$C_L$ is the load capacitance in F, and

$C_g$ is the gate capacitance in F.

The speed of the HFET is ultimately limited by the transit time of the device ($f_T \approx 100\text{–}200$ GHz), although the threshold voltage of the HFET is normally higher

**Fig. 3.60**
Cross section of a HFET

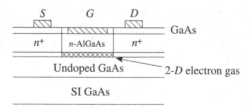

**Fig. 3.61**
A cross section of a planar-doped HFET

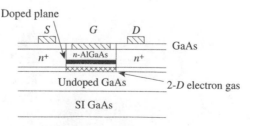

than that of the MESFET due to the presence of the undoped surface layer. The threshold voltage, $V_T$, of a uniformly doped HFET (V) is given by

$$V_T = \frac{(\Phi_B - \Delta E_c)}{q} - qN_D d_{dd}^2/(2\epsilon_1) \tag{120}$$

where    $\Phi_B$ is the interface barrier height in eV,

$\Delta E_c$ is the discontinuity in the conduction band in eV,

$N_D$ is the surface dopant density in the doped plane in /m$^3$,

$d_{dd}$ is the thickness of the AlGaAs layer in m, and

$\epsilon_1$ is the permittivity of the undoped interface layer in F/m.

For a planar-doped HFET, the threshold voltage, $V_T$ (V), is given by

$$V_T = \frac{(\Phi_B - \Delta E_c)}{q} - \frac{qn_s d_d}{\epsilon_1} \tag{121}$$

where    $n_s$ is the electron density in the two-dimensional gas in /m$^2$, and

$d_d$ is the distance between the gate and the doped plane in m.

The *I-V* characteristics of the HFET are given by

$$I_{ds} = q\mu_n n_{xs} d(V_a)/dx \tag{122}$$

where    $\mu_n$ is the electron mobility in the undoped layer in m$^2$/V·s,

$n_{xs} = n_s - \epsilon_1 V_a(x)/(qd_{eff})$,

$d(V_a)/dx$ is the gradient of the channel potential in the $x$ direction in V/m, and

$d_{eff}$ is the effective value of $d_d$ or $d_{dd}$.

It can be shown that the drain-to-source current is

$$I_{ds} = \mu\epsilon_1(W_c/L)[(V_g - V_T)V_{ds}/d_{eff} - V_{ds}^2/(2d_{eff})] \tag{123}$$

In saturation, the saturation voltage (V) and the saturation current (A) are given by

$$V_{Dsat} = (qn_s/\epsilon_1)(1 + a' - \sqrt{1 + a'^2}) \tag{124}$$

$$I_{Dsat} = qn_s\mu_n \acute{E}_s W_c(\sqrt{1 + a'^2} - a')$$

where    $\acute{E}_s = v_s/\mu_n$, and

$a' = \epsilon_1 v_s L/(qn_s\mu_n d_{eff})$.

The saturation velocity, $v_s$, of the carriers in a HFET can be considerably higher than that of the MESFETs since scattering is essentially absent in the two-dimensional electron gas. At a large gate voltage, however, the drain-to-source current will decrease considerably due to hot electrons transferred over the barrier into the AlGaAs layer.

**Example 3.55**   Compute the cutoff frequency for a HFET if the channel length is 0.8 μm, $C_L = C_g$, and $v_s = 2 \times 10^5$ m/s.

**Solution:**   Based on Eq. (119), $f_T = 1/[2\pi\tau_{tr}(1 + C_L/C_g)] = 1/(2\pi L/v_s) = 1/(2 \times 3.14 \times 0.8 \times 10^{-6}$ m/(2 $\times 10^5$ m/s)) $= 10 \times 10^9$ Hz $= 10$ GHz.   **§**

## Amorphous Silicon Thin-Film Transistor (TFT)

Amorphous Si, or α-Si, can be prepared inexpensively and is an alternative to single-crystal Si in many applications. Commercial α-Si is prepared with a high concentration of $H_2$ in order to saturate the dangling bonds and is often denoted by α-Si:H. α-Si:H can have a direct energy gap as high as 1.7 eV, and the carrier mobilities are of the order of $1 \times 10^{-3}$ m$^2$/V·s. α-Si:H has a continuum of localized donorlike states and acceptorlike states in the energy gap. The peak donorlike state density is located near the valence band edge and tails off exponentially toward the conduction band, as shown in Fig. 3.62 (simplified from Fig. 2.31).

The reverse is also true for the acceptorlike states. At equilibrium, the Fermi level is about 0.6 eV below the conduction band edge. Using a simple model, the donor charge density, $Q_D$ (C/m$^3$), and the acceptor charge density, $Q_A$ (C/m$^3$), are given, respectively, by

$$Q_D = qg_{vd}E_{donor}\exp[(E_v - E_F)/E_{donor}] \qquad (125)$$

$$Q_A = qg_{cd}E_{acceptor}\exp[(E_F - E_c)/E_{acceptor}]$$

**Fig. 3.62**
Donorlike and acceptorlike states in amorphous silicon

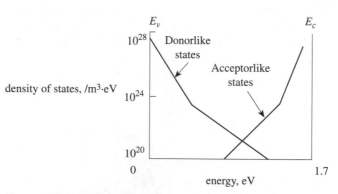

(*Source:* W.B. Jackson, R.J. Nemanich, and N.M. Amer, "Energy dependence of the carrier mobility—lifetime product in hydrogenated amorphous silicon," *Phy. Rev.,* B27, no. 8, p. 486. Reprinted by permission of the American Physical Society.)

where  $g_{vd}$ and $g_{cd}$ are the respective densities of states in the conduction band and in the valence band in $/eV \cdot m^3$,

$E_c$ and $E_v$ are the respective energies at the conduction band edge and the valence band edge in eV,

$E_F$ is the Fermi level in eV, and

$E_{donor}$ and $E_{acceptor}$ are the respective donor and acceptor ionization energies in eV.

The location of the Fermi level is given as

$$E_c - E_{F0} = E_{acceptor}/(E_{acceptor} + E_{donor}) \qquad (126)$$
$$\times \ [E_g - E_{donor}\ln(g_{vd}E_{donor})/(g_{cd}E_{acceptor})]$$

where  $E_{F0}$ is the energy of the equilibrium Fermi level in eV.

The construction of an α-Si TFT is shown in Fig. 3.63. The device looks like an inverted MOSFET, with the gate directly deposited on the glass substrate. The insulator layer is usually $Si_3N_4$ and is laid down prior to the deposition of the α-Si:H. The rest of the construction is quite similar to the MOSFET. Initially, the device is off, and as the (positive) gate voltage increases, the localized states within the energy gap are gradually filled up as the Fermi level moves toward the conduction band edge. At a sufficiently large voltage, most of the localized states are filled, and the density of the mobile charge will increase to form an inverted channel above the gate. Additional increase in the gate voltage will further increase the carrier density. Based on the energy band diagram of Fig. 3.62, we can show that the gate voltage will affect the channel in the following ways.

1.  A *subthreshold regime* occurs when the Fermi level is at the equilibrium value (0.6 eV below the conduction band edge) and the gate voltage results primarily in the filling up of the acceptorlike states.
2.  A *transition regime* takes place with the progressive filling of the donorlike tail states in addition to the filling of the acceptorlike states and the surface states. The upward movement of the Fermi level slows down, and there is a small increase in the electron density.
3.  In the *crystallinelike regime*, the Fermi level moves up close to or above the conduction band edge, and most of the impurity states are filled. The electron density is large, and the device is similar to one formed on single-crystal Si.

**Fig. 3.63**
An α-Si thin-film
transistor

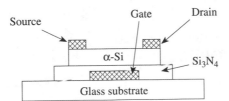

This phase occurs only at a bias voltage between 50 to 100 V for an insulator thickness of around 100 nm.

As in the case of a MOSFET, the drain-to-source current, $I_{ds}$ (A), is given by

$$I_{ds} = q\mu n_s d(V_a)/dx W_c \tag{127}$$

where    $n_s$ is the surface sheet charge density in /m$^2$, and

       $d(V_a)/dx$ is the potential gradient in the channel in the $x$ direction in V/m.

It is possible to express $n_s$ as $(\mu^{**}/\mu)n_{ind}$, where $\mu^{**}$ is a field-dependent mobility (m$^2$/V·s). This leads to

$$n_{ind} = n_{inds} - \epsilon_i V_a/(qd_i) \tag{128}$$

where    $n_{inds}$ is the surface charge density near the source in /m$^3$,

       $\epsilon_i$ is the insulator permittivity in F/m, and

       $d_i$ is the insulator thickness in m.

For a small drain-to-source voltage, $V_{ds}$ (V), the drain-to-source current (A) is given by

$$I_{ds} = q\mu^{**}W_c/Ln_{inds}V_{ds} \tag{129}$$

At saturation, the saturation voltage, $V_{Dsat}$ (V), is given by

$$V_{Dsat} = qn_{inds}d_i/\epsilon_i \tag{130}$$

The total induced charges in the channel include the localized donorlike charges and the acceptorlike charges. The surface electron density can be estimated numerically. Both the electron density and the trapped charge densities increase as the Fermi level moves toward the conduction band; the former is always about one order of magnitude larger than the latter. The surface electron density increases rapidly with the applied gate voltage up to about 20 V before it saturates at a value of the order of $10^{15}$/m$^2$.

A variation of the α-Si:H TFT is called the high-voltage thin-film transistor (HVTFT). It can be formed by adding an extended buffer region between the gate region and the drain region to enable a larger voltage drop across the output of the device. The basic operation of the HVTFT is quite similar to the TFT. Such a device (with a 40-μm buffer region) can withstand an output voltage of hundreds of volts without any drain-to-source breakdown.

## Double-Injection Field-Effect Transistor (DIFET)

The **double-injection field-effect transistor** (DIFET) is basically a hybrid between a FET and a BJT. It has a FET structure with an additional *p*-type anode added to

Fig. 3.64
A double-injection FET

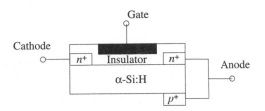

Fig. 3.64
A double-injection FET

the $n$-type substrate, as shown in Fig. 3.64. The anode and the source (the cathode) together form a forward-bias $p$-$n$ junction. The charge injection from the $p$-$n$ junction increases the electron and hole densities in the channel and results in a significant increase in the channel conductivity for a given gate voltage. The TFT version of the DIFET is particularly attractive, since the substrate $p$-type region can be formed quite readily using a thin-film technique. A considerable increase in the channel current has been reported for a given cathode-anode bias voltage (as much as 14 times when the voltage increases to 25 V).

## Integrated Bipolar Transistors

Most BJTs are designed for IC applications, and $n$-$p$-$n$ transistors are favored because of the higher electron mobility. In ICs, optimization extends beyond the individual transistors; the more efficient transistors are designed with oxide isolation to minimize the leakage current and the parasitic. The collector resistance can be reduced using a buried $n^+$ or $p^+$ layer, and ion implantation is the standard method used for doping. Polysilicon emitters are also used to improve the emitter efficiency. In addition to vertically placed transistors, lateral/substrate transistors are sometimes used. These are mostly $p$-$n$-$p$ transistors used for less critical tasks, such as load transistors and for level shifting. Faster transistors have Schottky diodes attached to the base-collector junctions to prevent the latter from entering into saturation and to minimize the charge storage effect (**Schottky transistor**). Multiple-emitter (TTL) and multiple-collector ($I^2L$) transistors have also been developed for large fan-in (multiple inputs) and fan-out (multiple outputs) in digital ICs.

## GLOSSARY

| | |
|---|---|
| **diode, or $p$-$n$ junction** | a rectifying device made of a $p$-type and an $n$-type semiconductor joined together. |
| **rectification** | the property that current flows easily in one direction but not in the opposite direction. |
| **junction** | the interface where a $p$-type semiconductor meets an $n$-type semiconductor. |
| **$n$-side and $p$-side** | the two ends of a $p$-$n$ junction. |
| **stationary charges (or space-charge region)** | the ionized donors and acceptors in a $p$-$n$ junction. |
| **built-in potential (energy)** | the energy difference corresponding to the total amount of band bending across a $p$-$n$ junction. |

| | |
|---|---|
| **I-V characteristics of a** *p-n* **junction** | the current-voltage relationship. |
| **dynamic resistance** | instantaneous resistance of a forward-bias *p-n* junction. |
| **short circuit** | a situation in which there is negligible resistance in the current path. |
| **ideal diode equation** | an equation that gives the characteristics of a *p-n* junction dominated by diffusion currents. |
| **saturation current** | the current of a *p-n* junction when the reverse bias voltage is very large. |
| **one-sided step junction** | a *p-n* junction in which one side of the *p-n* junction is more heavily doped than the other side. |
| **diffusion current** | the current in a semiconductor due to the presence of a carrier density gradient. |
| **diffusion length** | the distance over which the carrier density is dropped by a factor of $1/e$ (= 0.368). |
| **reverse-bias current** | the current in a reverse-bias *p-n* junction. |
| **thermal generation current** | the current in a *p-n* junction due to thermally generated carriers at the junction region. |
| **recombination current** | the current due to carrier recombination in the junction region under forward bias. |
| **leakage current** | in a *p-n* junction, current primarily due to leakage along the semiconductor surface. |
| **ideality factor** | a parameter that distinguishes whether diffusion current, recombination current, or leakage current is dominant in a *p-n* junction. |
| **homojunction** | a *p-n* junction formed by a single semiconductor. |
| **heterojunction** | a *p-n* junction formed by two different semiconductors. |
| **junction capacitance** | the capacitance due to the space-charge region in a *p-n* junction. |
| **varactor** | a variable-reactance device made of a *p-n* junction. |
| **diffusion capacitance** | the capacitance due to the carriers stored in a *p-n* junction under forward bias. |
| **rise time** | the time taken for a signal to come close to its full value (often considered to be 90% of the steady-state value). |
| **junction resistance** | the resistance associated with a forward-bias *p-n* junction. |
| **switching time** | the time required for a signal to go from its (peak) positive value to its (peak) negative value. |
| **lifetime** | the average time required for a minority carrier to recombine. |
| **equivalent-circuit model** | a physical model developed using resistors, capacitors, inductors, current sources, and voltage sources. |
| **delay time** | the sum of the turn-on time and the turn-off time of a *p-n* junction. |
| **turn-on time** | the time required for a *p-n* junction to become forward biased when it is initially reverse biased. |
| **turn-off time** | the time required for a *p-n* junction to become reverse biased when it is initially forward biased. |
| **tunneling** | the process in a *p-n* junction, whereby electrons move through a potential barrier in the junction without having sufficient energy to surmount the barrier. |
| **transmission coefficient** | a parameter that determines the probability that tunneling will take place. |
| **tunneling current** | the current resulting from electrons tunneling through a barrier. |
| **dynamic resistance** | the incremental resistance of a nonlinear device. |
| **negative resistance** | the dynamic resistance of a device when the incremental current becomes negative for a positive change in the applied voltage. |
| **tunneling diode** | a *p-n* junction that exhibits a tunneling effect. |

| | |
|---|---|
| **avalanche effect** | carrier multiplication in a *p-n* junction when the reverse-bias voltage exceeds a critical value. |
| **critical electric field** | the minimum electric field needed to induce junction breakdown in a *p-n* junction. |
| **impact ionization** | the situation in which energetic electrons moving at a high velocity collide with atoms and create additional electron-hole pairs. |
| **breakdown voltage** | the voltage needed to initiate avalanche effect. |
| **microchannels** | narrow conducting channels in a *p-n* junction that give rise to leakage. |
| **ionization rate** | the number of electron-hole pairs created per unit distance a carrier has traveled. |
| **avalanche photodiode** | a photodiode in which the avalanche effect takes place in the junction region and generates additional gain in the device. |
| **impact ionization avalanche transit time (IMPATT) diode** | a power device in which carrier ionization and transit time delay in the device result in an output current and an output voltage that are out of phase and give rise to a negative resistance effect. |
| **shot noise** | noise of a diffusion nature due to carriers crossing a potential barrier. |
| **generation and recombination noise** | noise in a *p-n* junction due to fluctuations in the carrier densities or changes in the semiconductor resistance. |
| **two-dimensional electron gas effect** | the existence of a layer of electrons confined to a potential minimum in a semiconductor device. |
| **Schottky diode** | a rectifying metal-semiconductor junction. |
| **work function** | the energy required for an electron in a solid to overcome the surface potential. |
| **image charge** | the charge created at a metal surface to offset the electric field created by free charges located outside the metal. |
| **thermionic emission** | the emission of carriers over the surface potential of a solid by virtue of their thermal energy. |
| **field emission** | refers to tunneling of electrons in a high field region such as a *p-n* junction under reverse-bias. |
| **interface states** | defect states present at the interface of a metal-semiconductor contact. |
| **Schottky barrier height** | the interfacial barrier height in a metal-semiconductor junction. |
| **field-induced barrier-height lowering effect** | the barrier height lowering due to image charge at the metal surface when there are excess electrons in the semiconductor side of a Schottky diode. |
| **minority carrier injection ratio** | a measure of the extent of charge storage, which gives rise to a time delay in the response of the device during current switching. |
| **turn-on voltage of a Schottky junction** | the forward voltage required to give a reasonable forward current (say, $\approx 1$ μA). |
| **ohmic contact** | a metal semiconductor junction that exhibits very low resistance when current flows in either direction. |
| **contact resistance** | the resistance of the contact itself. |
| **shunt resistance** | the resistance associated with the semiconductor substrate. |
| **spreading resistance** | the resistance due to nonuniform current flow in the substrate near the contact. |
| **specific contact resistance** | the contact resistance of a unit-area cross section, which is proportional to the contact area. |
| **sheet resistance** | the resistance of a semiconductor with a unit-area cross section. |
| **flat-band condition** | the situation when the energy bands are flat throughout the device, which also implies that there is no space charge within the device. |

| | |
|---|---|
| **flat-band voltage** | the voltage needed to achieve flat-band condition. |
| **surface potential** | the potential at the surface of a semiconductor. |
| **Poisson's equation** | an equation relating the second derivative of potential to the charge density. |
| **surface inversion** | the situation when there is charge reversal at the semiconductor surface in a metal-oxide-semiconductor structure. |
| **strong inversion** | the case when the change in the surface potential is twice the potential difference between the intrinsic Fermi level and the equilibrium Fermi level in the semiconductor substrate. |
| **threshold voltage** | the voltage needed to achieve surface inversion. |
| **EL2 center** | a trap center formed by an As atom occupying a Ga site. |
| **transistor** | a three-terminal device capable of creating signal amplification when suitably biased. |
| **field-effect transistor** | a transistor whose operation depends on current modulation in a semiconductor channel due to an applied voltage. |
| **bipolar junction transistor** | a transistor whose operation depends on charge flow across a base region from a low potential region to a high potential region. |
| **enhancement-mode field-effect transistor** | a field-effect transistor that is normally off (nonconducting). |
| **depletion-mode field-effect transistor** | a field-effect transistor that is normally on (conducting). |
| **gradual channel approximation** | in a field-effect transistor, the assumption that the lateral electric field (due to the drain-to-source voltage) is much smaller than the vertical electric field (due to the gate-to-substrate voltage). |
| **saturation** | in a field-effect transistor, the situation when the channel is partly depleted and the drain-to-source current is no longer dependent on the bias voltage. |
| **channel pinch-off** | the situation when the channel in a field-effect transistor is fully depleted. |
| **transconductance gain** | in a field-effect transistor, the ratio between the output current and the input voltage. |
| **radiation resistant** | the quality of a device that makes it immune to the effects of high-energy radiation. |
| **punch-through** | the case when permanent conduction paths are created by electrons moving through an oxide layer. |
| **common-emitter mode** | the biasing of a bipolar junction transistor that gives maximum current gain. |
| **common-base mode** | the biasing of a bipolar junction transistor that gives maximum voltage gain. |
| **common-collector mode** | the biasing of a bipolar junction transistor that gives optimal impedance transfer. |
| **common-base current gain** | the ratio of the collector current to the emitter current in a bipolar junction transistor. |
| **emitter efficiency** | the fraction of the emitter current due to carrier flow from the emitter to the base. |
| **transport factor** | the fraction of current originating in the emitter that reaches the collector in its passage through the base. |
| **common-emitter current gain** | the ratio of the collector current to the base current. |
| **Gummel number** | a parameter inversely proportional to the common-emitter current gain, defined as the base doping per unit area. |

| | |
|---|---|
| **velocity saturation** | the situation when carrier velocity is no longer proportional to the applied voltage. |
| **hot electron** | energetic electrons that are injected and trapped within the oxide layer near the drain region of a field-effect transistor, resulting in premature depletion in the channel and an increase in the saturation current. |
| **subthreshold current** | in a field-effect transistor, the small drain-to-source current flow when the gate voltage is below threshold and the channel is not inverted (actually in depletion). |
| **cutoff frequency** | the limiting frequency of a field-effect transistor, beyond which the device will no longer produce a response. |
| **complementary MOSFET** | a composite device formed by stacking an $n$-channel field-effect transistor and a $p$-channel field-effect transistor together; a device with low power dissipation, ideally suited to the implementation of digital circuits. |
| **latch-up** | a phenomenon that gives rise to breakdown in a complementary field-effect transistor. |
| **scaling** | the process whereby devices are reduced in dimension in order to increase performance and packing density. |
| **base-spreading resistance** | the effective resistance in the base, which can cause nonuniform current flow in the base of a bipolar junction transistor. |
| **emitter current crowding** | the nonuniform current flow due to the base-spreading resistance. |
| **base push-out effect** | related to current injection into the base from the emitter that extends the base into the collector, where the doping is low. |
| **energy gap reduction** | a phenomenon that occurs in a highly doped semiconductor; responsible for a decrease in the emitter efficiency in a bipolar junction transistor. |
| **transit time** | the average time necessary for a carrier to transit through the base in a bipolar junction transistor. |
| **breakdown** | occurring in a junction under reverse bias when carrier multiplication or tunneling takes place and results in significant current flow. |
| **punch-through** | breakdown in the base of a bipolar junction transistor when the base is fully depleted. |
| **secondary breakdown** | breakdown due to thermal effect in a reverse-bias $p$-$n$ junction. |
| **heterojunction bipolar transistor** | a transistor that uses a wide-energy-gap semiconductor in the emitter to reduce the reverse carrier flow and increase the emitter efficiency. |
| **ballistic effect** | the situation in which carriers pass through the base of a heterojunction transistor without suffering any collision. |
| **heterojunction field-effect transistor** | a field-effect transistor with an undoped surface layer (such as AlGaAs) placed directly under the gate to produce a two-dimensional electron gas effect. |
| **planar-doped HFET** | a field-effect transistor with a doped plane placed under the gate to produce a two-dimensional electron gas effect. |
| **double-injection field-effect transistor** | a hybrid between a field-effect transistor and a bipolar junction transistor relying on charge injection from a $p$-$n$ junction to increase the electron and hole densities in the channel, which can result in a significant increase in the channel conductivity for a given gate voltage. |
| **Schottky transistor** | a bipolar junction transistor with a Schottky diode connected to the base-collector junction to prevent the latter from entering into saturation. |

**REFERENCES**   Grove, A. S. *Physics and Technology of Semiconductor Devices,* New York: John Wiley & Sons, Inc., 1967.

Neudeck, G. W. *Volume II: The p-n Junction Diode,* Modular Series on Solid State Devices. Reading, Mass.: Addison-Wesley Publishing Company, 1989

Shur, M. *Physics of Semiconductor Devices,* Upper Saddle River, N.J.: Prentice Hall Inc., 1990.

Streetman, B. G. *Solid State Electronic Devices,* Upper Saddle River, N.J.: Prentice-Hall Inc., 1980.

Sze, S. M. *Semiconductor Devices Physics and Technology.* New York: John Wiley & Sons, Inc., 1985.

**EXERCISES**   ## 3.2  Concept of a *p-n* Junction

1. A diffused *p-n* junction can be formed by adding *p*-type dopants to the surface of an *n*-type substrate. Name the important parameters needed to characterize this type of *p-n* junction.

2. What is the physical significance of the built-in voltage $V_{bi}$? Compute its value when $N_A = N_D = 10^5 \times n_i$ ($n_i$ is the intrinsic carrier density) and $T = 300$ K. Is it possible for the built-in voltage of a *p-n* junction to be larger than the energy gap of the semiconductor?

3. Are there any roles played by the *n*-side and the *p*-side as far as the electrical properties of a *p-n* junction are concerned?

4. For a Si *p-n* junction, as shown in Fig. 3.2, estimate the series resistance if the cross-sectional area of the *p-n* junction is 1 mm $\times$ 1 mm, the length of the *p*-side and the *n*-side are both 0.5 mm long, and their dopant densities are $1 \times 10^{22}/\text{m}^3$. Assume $\mu_n = 0.135$ m$^2$/V·s and $\mu_p = 0.05$ m$^2$/V·s.

5. If the saturation current of a *p-n* junction is 10 nA and the applied voltage is 0.2 V, compute the dynamic resistance, $R_{\text{dyn}}$, at 300 K. At what bias voltage will $R_{\text{dyn}}$ be doubled its value at 300 K? What happens to the equation for $R_{\text{dyn}}$ if the ideality factor is $\eta_I$?

6. Estimate the diffusion current density in a *p-n* junction if the carrier density gradient across the *p-n* junction is $2 \times 10^{29}/\text{m}^4$. The hole-diffusion constant and the electron-diffusion constant are $3 \times 10^{-3}$ m/s and $1 \times 10^{-3}$ m/s, respectively.

7. Using Table 3.1, compute the saturation current density $J_0$ for a GaAs *p-n* junction if the dopant densities on the two sides are $N_D = 1 \times 10^{21}/\text{m}^3$ and $N_A = 1 \times 10^{22}/\text{m}^3$. Assume $L_n = L_p = 10$ μm.

8. Experimentally, suggest how you would determine the ideality factor, $\eta_I$, and the saturation current, $I_O$, from the *I-V* characteristics of a *p-n* junction.

9. Based on Table 3.1, for the same dopant densities on the two sides of the *p-n* junction, suggest which of the three types of semiconductors (Si, Ge, or GaAs) will give the smallest saturation current.

10. At zero bias, compute the depletion layer width of a Si *p-n* junction if the dopant densities on the two sides are $1 \times 10^{23}/\text{m}^3$. Compute the zero-bias junction capacitance

per unit area, $C_{jun0}$. Determine the bias voltage required to increase $C_{jun}$ to five times its value at zero bias, i.e., $5C_{jun0}$. Assume an abrupt junction and $\Phi_{bi} = 0.82$ eV.

11. Compare Eqs. (16) and (17) and explain why the linearly graded $p$-$n$ junction has a larger voltage dependence on the bias voltage.

12. For a forward-bias $p$-$n$ junction, estimate the voltage required to increase the ideal diffusion capacitance to three times its value at zero bias.

13. Using the equivalent-circuit model of a $p$-$n$ junction, as shown in Fig. 3.14, show how the derived $I$-$V$ characteristics (from the model) compare with the measured characteristics.

14. Compare the forward- and reverse-tunneling $I$-$V$ characteristics of a highly doped $p$-$n$ junction.

15. Based on Table 3.2, can you explain the differences in the values of the critical electric field, $\acute{E}_{cr}$, observed in the different semiconductors?

16. Compute the breakdown voltage in a Si $p$-$n$ junction if the dopant densities on the two sides are $1 \times 10^{24}/m^3$ and $\acute{E}_{cr} = 0.4 \times 10^8$ V/m.

17. If the microchannels in a leaky $p$-$n$ junction have a conductivity that is $1 \times 10^9$ times that of the normal $p$-$n$ junction, suggest what fraction of the junction area is occupied by microchannels if the leakage current density is 4000 times the normal saturation current density.

18. In a reverse-bias $p$-$n$ junction, if $\alpha_i = \beta_i = 2 \times 10^4$/cm, compute the multiplication factor, $M_n$, assuming a depletion region width of 100 $\mu$m.

19. Explain the out-of-phase current and voltage waveforms observed in an IMPACT diode, as illustrated in Fig. 3.22.

20. Compare the differences in the breakdown voltages due to tunneling in Si and in GaAs.

21. Relate the parameters used in the equivalent-circuit model of a $p$-$n$ junction (as shown in Fig. 3.24) to the physical structure of the device.

22. Construct the energy band diagram of a Si-Ge heterojunction. Explain what may happen at the junction interface.

### 3.3 Schottky Junction and Its Electronic Properties

23. Highlight the differences between a Schottky junction and a $p$-$n$ junction.

24. Compute the thermionic saturation current density from a metal electrode at 1000 °C assuming a metal work function of 4 eV.

25. Compute the built-in voltage, $V_{bi}$, for a Au/$n$-type GaAs Schottky junction. Assume $E_n = 0.2$ eV. What will be the saturation current density $J_0$ at 300 K? Compute the current density if the Schottky junction is forward biased by 0.4 V.

26. Compute the barrier-height lowering in a Si Schottky junction if the electric field is 10 kV/m.

27. Compute the minority-carrier injection ratio $\gamma^*$ in a Schottky junction having the following parameters: $D_p = 0.01$ m$^2$/s, $N_D = 1 \times 10^{21}/m^3$, $A_{cs} = 1 \times 10^{-2}$ cm$^2$, $L_p = 1$ $\mu$m, $T = 300$ K, $n_i = 1 \times 10^{16}/m^3$, $A^* = 1.2 \times 10^6$ A/m$^2$·K$^2$, and $\Phi_B = 0.8$ V.

28. Physically, explain the changes in the minority carrier injection ratio $\gamma^*$ at different current levels, as shown in Fig. 3.35.

29. Estimate the lowering in the barrier height of a metal/$n$-type GaAs Schottky junction if an interface dopant layer of 0.1 $\mu$m (with a donor density of $1 \times 10^{23}/m^3$) is present.

30. Estimate the increase in the barrier height of a metal/$n$-type GaAs Schottky junction if the interface acceptor density is $1 \times 10^{23}/m^3$. The thickness of the dopant layer is 50

nm, and the substrate donor density is $1 \times 10^{21}/m^3$. For simplicity, assume the depletion layer width to be 1 μm.

31. List the advantages and disadvantages in using a Schottky junction instead of a *p-n* junction.

## 3.4  Metal-Semiconductor Contact

32. What are the requirements to produce an ohmic contact when a metal is deposited on a semiconductor?
33. Outline the different types of resistances that are present in an ohmic contact.
34. Define specific contact resistance and how it depends on the doping of the substrate.

## 3.5  MIS Junction and Field-Effect Properties

35. Compute the flat-band voltage in an Al/SiO$_2$/Si junction in the absence of oxide charges. The substrate is *n*-type and has a dopant density of $1 \times 10^{22}/m^3$ and $T = 300$ K.
36. Describe the phenomenon of surface inversion and compute the threshold voltage in an Al/SiO$_2$/*p*-Si junction. Assume a substrate acceptor density of $1 \times 10^{21}/m^3$, an oxide capacitance per unit area of 1000 μF/m$^2$, and $T = 300$ K.
37. Explain what happens to the threshold volatge of a MIS (*n*-type) junction if positive charges are introduced to the oxide layer.
38. Explain what is meant by *strong inversion*.
39. Explain the frequency dependency of the *C-V* curves shown in Fig. 3.44.

## 3.6  Materials Considerations

40. For a MIS junction, what are the effects if the substrate semiconductor has (a) a large energy gap, and (b) a high dopant density? Explain your answers.
41. List the advantages (if any) in replacing SiO$_2$ by Si$_3$N$_4$ as the insulator in a MIS junction.
42. In the formation of MIS junctions, why is it sometimes advantageous to use a thin semiconductor layer on top of an insulating substrate rather than a uniformly doped substrate?
43. Explain *gettering* in semiconductors.

## 3.7  Structures and Operation of Transistors

44. Compute the threshold voltage for a Si MOSFET if $V_{FB} = 0.3$ V, $\phi_b = 0.1$ V, $V_{sc} = 0.15$ V, $N_A = 1 \times 10^{21}/m^3$, $V_{sub} = -0.5$ V, and $C_i = 1$ mF/m$^2$. Assume $V_{FB} = 0.2$ V.
45. Show how Eq. (64) can be simplified to Eqs. (67) and (68).
46. Graphically, sketch $I_{ds}$ versus $V_{ds}$ for a MOSFET, as given by Eq. (70).
47. Compare the threshold voltage, $V_T$, in a MESFET and a MOSFET.
48. Compare the transconductance, $g_m$, of a MESFET and a MOSFET.
49. Why is the common-emitter mode of operation in a BJT better suited for current amplification than the other modes?
50. Compare the input impedance of a BJT and a MOSFET.
51. Suggest how you would optimize the common-emitter current gain, $\beta_g$, of a BJT as given by Eq. (87).
52. Explain the behavior of $\beta_g$ at different collector current levels.

## 3.8   Nonideal Effects and Other Performance Parameters

**53.** What is meant by velocity saturation in semiconductors? Suggest how it comes about physically.

**54.** Using Eq. (95), show that $I_{sub}$ is approximately exponentially dependent on $V_g$.

**55.** Based on Eq. (99), suggest how the change in the built-in voltage, $\Delta V_{bi}$, is affected by the surface dopant density of a MOSFET.

**56.** Explain the *I-V* characteristics of the CMOS circuit shown in the inset of Fig. 3.55.

**57.** Using Fig. 3.56, discuss how latch-up effect arises in CMOS circuits.

**58.** Suggest how the transconductance of a MOSFET would vary with scaling.

**59.** Based on Eq. (104), attempt to justify the two empirical criteria used to avoid the short-channel effect in MESFETs.

**60.** How does the nonuniform base resistance effect affect the current gain in a BJT?

**61.** Compare $BV_{ce}$ and $BV_{cb}$ and explain why $BV_{ce}$ is usually smaller. If $BV_{ce}/BV_{cb} = 0.5$ and $\alpha_g = 0.9$, compute the value of *mb* in a given BJT.

**62.** Using Eq. (105), compute the energy gap shrinkage in a GaAs BJT if $N_{De} = 1 \times 10^{25}$ /m$^3$ and $T = 300$ K.

**63.** Using Eq. (109), estimate the storage time in a BJT if $I_{b1} = 1$ mA, $I_{b2} = 0.1$ μA, $I_{ba} = 0.2$ mA, and $\tau_{sr} = 0.5$ μs.

**64.** Compare the common-base cutoff frequency, $f_\alpha$, the common-emitter cutoff frequency, $f_\beta$, the unity-gain cutoff frequency, $f_T$, and the transit-time cutoff frequency, $f_{tr}$, as illustrated in Fig. 3.57.

## 3.9   New Transistor Structures

**65.** What are the advantages of HBTs over BJTs? Show that Eq. (117) gives the maximum current gain in a HBT.

**66.** Verify Eqs. (120) and (121) for the threshold voltages of the different HFET structures.

**67.** Using Eq. (124), plot the saturation current and voltage as a function of $n_s$, the electron density in the two-dimensional electron gas. Assume $L = 1$ μm, $\mu_n = 0.8$ m$^2$/V·s, $d_{eff} = 20$ nm, $W_c = 10$ μm, $\epsilon_1 = 4$, and $v_s = 1 \times 10^5$ m/s.

**68.** What are the advantages and disadvantages in using thin-film transistors?

**69.** Based on Fig. 3.62, explain the existence of the subthreshold region, the transition region, and the crystallinelike region in an amorphous Si thin-film transistor.

**70.** Explain the operation of the double-injection field-effect transistor.

# CHAPTER 4

# Optical Properties of Solids

## 4.1  INTRODUCTION

The study of the interactions of light with solids is becoming more important in view of the recent growth in information technology. This study is also prompted by the need in other areas such as communications and television broadcasting to increase the signal and data bandwidths and to reduce the cost of transmission using lightwave technology (a collective term used for data transmission using light as the carrier). The study of light-solid interactions is not new; it has been investigated for hundreds of years (as early as the time of Galileo). It became more important near the turn of the last century, when the development of physics turned from classical physics (which is primarily mechanics and electromagnetics) to modern physics and quantum mechanics. The landmark experiment related to the study of light was perhaps the study on the photoelectric effect, in which Einstein explained the principle of duality in light using the phenomenon of photoemission. In addition, around this time quantum mechanics was developed and was rigorously applied to the study of solid-state physics, including the area of optoelectronics. Ever since, a large number of important optical effects in solids have been observed, and many have found their ways in device applications. A good example of an important optoelectronic effect observed in solids is stimulated light emission, which has been used to develop solid-state lasers. At another front, the rapid progress in materials technology, especially in crystal growth and purification, has also contributed significantly to the advances in device fabrication. These include the growth of high-quality single crystals, epitaxial layers, and complex (multiple) heterostructures, all of which have been studied extensively.

In this chapter, we first outline the fundamental properties of light and described how they contribute to light-solid interactions. To do so, we review the optical constants and present a model that relates the optical constants to the

optical properties of solids. In connection with these treatments, we also discuss such phenomena as light reflection, light absorption, and light emission. Since the optical properties of a metal, a semiconductor, and an insulator are quite different, we consider them separately. The second half of this chapter deals with optical devices, since they are important extensions to the topics that we have studied. We look at device structures, principles of operation, and important applications. These devices include light detectors, solid-state light sources, light polarizers, display and imaging devices, and energy-conversion devices. The last part of this chapter focuses on the materials aspects of guided lightwave technology and provides an introduction to optical holography, a subject of significant importance for three-dimensional imaging and defect diagnostics.

## 4.2  REVIEW OF THE PROPERTIES OF LIGHT

Light is certainly no stranger to us. From the day we are born, light comes in through our eyes as an essential source of communication with the outside world. The human vision, in simple terms, involves specific **light receptors** lining the back of the optical cavity of the eye. These light receptors are responsible for triggering the nerve fibers, which transform the light signal (or the images) into an electrical signal. The electrical signal in the nerve fibers is then transmitted to the visual cortex region of the brain, where the images are processed. The phenomenon of light perception in humans (and also in animals) represents a highly developed and efficient optoelectronic imaging and transmission system. As we know, the light receptors in human eyes are rather specific in their response and cover only a small range of the entire electromagnetic spectrum, which we commonly called the **visible light.** Fig. 4.1 shows the wavelengths of visible light versus the corresponding frequencies. The colors of visible light are perhaps best represented by the colors in a rainbow. Because of this somewhat limited frequency range in our vision, the study of the optical properties of solids, especially in relationship to applications, is biased toward semiconductors. (Semiconductors have an energy gap that covers the visible range.) This is perhaps one reason why semiconductors are so important in optoelectronic devices. In addition to the visible light, the useful electromagnetic spectrum also includes **microwave radiation** (mw); **infrared light** (IR); **ultraviolet light** (UV), and other higher-energy radiation, such as **X rays** and **gamma rays.**

Fig. 4.1
The frequency and wavelengths of visible light

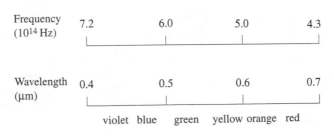

| Frequency ($10^{14}$ Hz) | 7.2 | 6.0 | 5.0 | 4.3 |

| Wavelength (μm) | 0.4 | 0.5 | 0.6 | 0.7 |

violet  blue     green    yellow orange   red

Except for the visible light, the rest of the electromagnetic spectrum is invisible to us, even though it has been widely used in engineering applications.

Light is first and foremost a form of energy, and in physics we learned that it exists as traveling electromagnetic waves. Because of the principle of duality, light sometimes behaves as particles and sometimes behaves as waves. When being considered as particles, light is referred to as **photons.** This concept of light allows us to physically count the photons and define such terms as **light intensity, $I_0$,** which is a measure of the light energy or the power delivered per unit area, and **light flux, $\phi_L$,** which is defined as the density of the photons per unit time. The energy associated with a single photon, $E_{photon}$, is given in joules, and its value depends on the oscillatory frequency of the changing electromagnetic field:

$$E_{photon} = h\nu \qquad (1)$$

where    $h\ (= 6.63 \times 10^{-34}\ \text{J·s})$ is Planck's constant, and

$\nu$ is the frequency of the electromagnetic wave in Hz.

The concept of light as particles invokes a velocity for the photons. In vacuum (free space), light travels at a velocity $c$ equal to $3 \times 10^8$ m/s. This velocity is also the group velocity of an electromagnetic wave (representing the velocity of energy transfer). According to physics, the **wavelength** of light is given by $\lambda_L = c/\mu$. Together with Eq. (1), we have

$$E_{photon} = hc/\lambda_L \qquad (2)$$

where    $\lambda_L$ is the wavelength of light in m.

For convenience, the product $hc$ is expressed as 1.24 eV/μm, and Eq. (2) allows for an easy conversion between the light energy (in eV) and its wavelength (in μm).

Now that we have described what light is, we may visualize light shining on a solid as a collection of photons hitting the solid. Of course, some of the photons will enter the solid and participate in light-solid interactions. The flow of the photons is measured by the light flux. If somehow all these photons are absorbed, then the power (energy per unit time) delivered to the solid is given by the product of the light flux and the "average" energy of the photons. Similarly, we may visualize light emission from a solid as a stream of photons emerging from the solid surface. Light emission is measured in watts (W), similar to the power rating used for a lightbulb.

**Example 4.1**    What are the energies of light with wavelengths equal to (a) 400 nm, and (b) 800 nm? What are the colors?

***Solution:***

(a) Based on Eq. (2), $E_{photon} = 1.24/\lambda_L$ eV. When $\lambda_L = 400$ nm or 0.4 μm, $E_{photon} = 3.1$ eV (this is near-ultraviolet light).

(b) When $\lambda_L = 800$ nm or 0.8 μm, $E_{photon} = 1.55$ eV (this corresponds to red light). **§**

**Example 4.2**   What is the number of photons per unit time emerging from a solid if the power output is 10 W and the wavelength is 0.65 μm?

***Solution:***   Based on Eq. (2), a photon with a wavelength of 0.65 μm corresponds to an energy $E_{photon}$ = 1.24 eV/μm/0.65 μm = 1.9 eV. The number of photons per unit time present in 10 W is 10 W/(1.9 eV × 1.6 × $10^{-19}$ V/eV) = 0.33 × $10^{20}$ (photons)/s.   **§**

## 4.3   LIGHT AND SOLID INTERACTIONS

The optical properties of a solid are determined by how light is transmitted, is reflected, or is absorbed as it passes through the solid. Since light is sometimes an electromagnetic wave, we can expect there will be interactions between the electric-field component of the incoming light and the charged particles, namely, the electrons and the ions in the solid. In simple terms, we may assume the electrons and the ions to have a number of energy states. These are primarily the electronic states of the electrons or the vibrational states of the lattice ions. Light-solid interactions therefore involve the excitation and relaxation of these energy states.

Fig. 4.2 shows the excitation of an electron from a low-energy state to a high-energy state through the absorption of a photon. Normally, the electron stays temporarily in the excited state but it will eventually relax to the initial state and emit the excess energy. During the energy transitions, both energy and momentum have to be conserved (except in some amorphous solids, where momentum conservation is not critical). For a single electron, energy conservation is achieved if the energy difference between the final energy state and the initial energy state is exactly the same as the photon energy. Momentum conservation is more complicated and often the lattice vibrations are a source of (crystal) momentum. The latter therefore can play a significant role in providing the necessary balance in momentum for the light-solid interactions.

### The Optical Constants

In addition to light absorption and emission, photons passing through a solid also polarize the valence electrons and the lattice ions (see Section 2.10). This process gives rise to induced dipoles, which change the dielectric constant and the index of refraction. Such changes are reflected in the measurement of the **permittivity.**

**Fig. 4.2**
Electronic absorption and emission of a photon

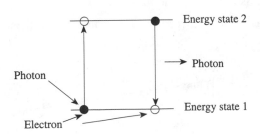

In general, how light passes through a solid is determined by optical constants. Normally, these constants are called the **index of refraction,** $\eta_s$, and the absorption coefficient, $\alpha^*$ (/m); the former affects the direction of the light path, whereas the latter is a measure of the light attenuation. To explain the physical origins of these constants, we shall examine the electric field component, $\acute{E}(x, t)$ (V/m), of a traveling electromagnetic wave. It is given by

$$\acute{E}(x, t) = \acute{E}_0 \exp(j\omega(t - N_c x/c)) \tag{3}$$

where    $\acute{E}_0$ is the peak amplitude of the electric field in V/m,

   $j\ (=\sqrt{-1})$ is the imaginary number,

   $\omega$ is the angular frequency in rad/s,

   $N_c$ is the complex index of refraction,

   $t$ is time in s,

   $x$ is the direction of light propagation in m, and

   $c$ is the velocity of light in m/s.

For most solids, $N_c$ is a complex number and is given by $\eta_s - jk$, where $\eta_s$ (the real part of $N_c$) is the index of refraction and $k$ (the imaginary part) is the extinction coefficient. In terms of $\eta_s$ and $k$, Eq. (3) can be written as

$$\acute{E}(x, t) = \acute{E}_0 \exp(j\omega(t - \eta_s x/c))\exp(-\alpha^* x) \tag{4}$$

where    $\alpha^*\ (= \omega k/c)$ is the absorption coefficient in /m.

From Eq. (4), we can see how the index of refraction, $\eta_s$, and the absorption coefficient, $\alpha^*$, affect a traveling electromagnetic wave. Basically, $\eta_s$ alters the phase (or the velocity) of the traveling wave, whereas $\alpha^*$ gives a measure of the attenuation (or the damping). Both $\eta_s$ and $\alpha^*$ can also be expressed in terms of the permittivity, $\epsilon_s$, and the conductance, $\sigma$. The Maxwell equation, when applied to a traveling wave, has the form

$$d^2\acute{E}/dx^2 - \sigma\, d\acute{E}/dt - \epsilon_s\mu_s\, d^2\acute{E}/dt^2 = 0 \tag{5}$$

where    $\epsilon_s$ is the permittivity of the solid in F/m,

   $\mu_s$ is the permeability of the solid in H/m (henries/meter), and

   $\sigma$ is the conductivity in S/m.

Combining Eqs. (4) and (5) gives

$$N_c^2 = (\eta_s - jk)^2 = c^2\mu_s\eta_s - jc^2\sigma\mu_s/\omega \tag{6}$$

The real and the imaginary parts of $N_c$ are then given by

$$\eta_s^2 - k^2 = \epsilon_s/\epsilon_0 \tag{7}$$

$$2\eta_s k = \sigma/(\omega\eta_s)$$

where    $\epsilon_0$ (= $8.86 \times 10^{-12}$ F/m) is the permittivity of a vacuum.

Equation (7) assumed a nonmagnetic solid (i.e., $\mu_s = \mu_0$) and used the relationship $c = 1/\sqrt{\mu_0\epsilon_0}$. To see how $\eta_s$ and $\epsilon_s$ are related, we take the simple case that there is no attenuation (i.e., $k \approx 0$). Equation (7) becomes

$$\eta_s = \sqrt{\epsilon_s/\epsilon_0} \tag{8}$$

$$\sigma \approx 0$$

According to Eq. (8), $\eta_s$ is the square root of the dielectric constant. The fact that $\sigma \approx 0$ simply reflects the properties of an insulator.

## A Simple Model for Studying Light-Solid Interactions

Classically, an electron in a solid is bound to the nucleus by a Hooke's law force, and the damping force is proportional to the velocity of the oscillating electron. In one dimension, the equation of motion in the $x$ direction for an electron under an oscillating electric field $\acute{E}_0\exp(j\omega t)$ is given by

$$m_0 \, d^2x/dt^2 + 2\pi\lambda_s \, dx/dt + F_h x = -q\acute{E}_0\exp(j\omega t) \tag{9}$$

where    $m_0$ is the mass of the electron in kg,

$\lambda_s$ is the damping constant in kg/s,

$F_h$ is the Hooke's law force constant in N/m,

$x$ is the instantaneous position of the electron in m,

$\acute{E}_0$ is the peak amplitude of the electric field in V/m, and

$\omega$ is the angular frequency of the ac electric field in rad/s.

In formulating Eq. (9), we have assumed that the wavelength of the oscillating electric field is much greater than the dimension of the atom. The solution of Eq. (9) is given by

$$x(t) = q/(4\pi^2 m_0)\acute{E}_0\exp(j\omega t)[(\omega_0^2 - \omega^2 - j2\pi\omega\lambda_s)/((\omega_0^2 - \omega^2)^2 + (2\pi\lambda_s\omega)^2)] \tag{10}$$

where    $\omega_0 = \sqrt{F_h/m_0}$

To relate the oscillatory motion of the electron $x(t)$ to the optical constants, we make use of the fact that the current density is $J = -n_0 q \, dx/dt$ (see Section 2.5) and that the complex conductivity is $\sigma_c = J/(\acute{E}_0 \exp(j\omega t))$. We can then show that

$$\sigma_c = \sigma + j\omega\epsilon_s \qquad (11)$$

where  $\sigma = n_0 q^2/(4\pi^2 m_0)[2\pi\lambda_s\omega^2/((\omega_0^2 - \omega^2)^2 + (2\pi\lambda_s\omega)^2)]$, and

$\epsilon_s = n_0 q^2/(4\pi^2 m_0)[(\omega_0^2 - \omega^2)/((\omega_0^2 - \omega^2)^2 + (2\pi\lambda_s\omega)^2)]$.

Equation (11) can now be substituted into Eq. (7) to determine the optical constants. Fig. 4.3(a) shows how $\eta_s$ and $k$ vary with $\omega$. We can divide the figures into two transparent regions (I and III, when $k \approx 0$), an absorption region (II, when $\omega = \omega_0$), and a reflecting region (IV, when $\eta_s \approx 0$). As expected, the transparent region implies little or no light-solid interactions; the absorption region reflects a maximum energy transfer from light to the solid; and the reflecting region is due to the out-of-phase light-solid interactions. Fig. 4.3(b) shows the absorption characteristics of Ge. Interestingly enough, there is quite a strong resemblance between the observed characteristics and those arising from the model.

## Light Refraction and Transmission

The direction of light changes as it is transmitted from one medium into another. This is the result of a change in the light velocity and is called **light refraction.** The index of refraction $\eta_s$ is a measure of the ratio of the light velocities in the two media. For light passing from air into a solid,

$$\eta_s = c/v_i \qquad (12)$$

where  $v_i$ is the light velocity in the solid in m/s.

**Fig. 4.3**
(a) A plot of $\eta_s$ and $k$ versus frequency; (b) absorption characteristics of Ge

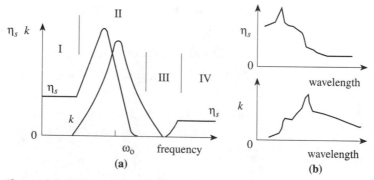

**(a)**  **(b)**

(Source: J.C. Phillips, "The fundamental optical spectra of solids," in *Solid State Physics—Advances in Research and Applications,* F. Seitz and D. Turnbull, eds., 1966. Reprinted by permission of the Academic Press, N.Y.)

Since at a moderate light frequency, $\eta_s > 1$, Eq. (12) would suggest that light always travels at a lower velocity inside a solid. Typical values of $\eta_s$ for most solids lie between 1.5 and 2.5 and are shown in Table 4.1.

As the light velocity changes, the light path also changes. Based on **Fermat's principle** of minimum *optical path* (see Example 4.3 for verification), the change in the light path is measured by the angle of refraction, $\theta_r$. $\theta_r$ is related to $\eta_s$ through **Snell's law,** which is given by

$$\sin(\theta_r) = \sin(\theta_i)/\eta_s \tag{13}$$

where    $\theta_i$ is the angle of incidence.

Both $\theta_i$ and $\theta_r$ are measured with respect to the normal—i.e., the line perpendicular to the air-solid interface. Since $\eta_s > 1$, $\theta_i > \theta_r$ implies that **light transmission** from air into a solid will bend inward toward the normal. Fig. 4.4 shows how the light path bends as it moves into a denser medium. Such an optical effect is best illustrated by the *apparent depth* of an object at the bottom of a pool of water. Because of Snell's law, light from the bottom of the pool will bend outward away

| Solid | $\eta_s$ | $\epsilon_s/\epsilon_o$ |
|---|---|---|
| Water | 1.33 | 80.3 |
| Fused silica | 1.46 | 3.7 |
| Diamond | 2.42 | 16.5 |
| NaCl | 1.5 | 5.6 |
| Si | 3.42 | 11.7 |
| Ge | 4.0 | 16 |
| GaAs | 3.3 | 13.2 |
| InSb | 3.96 | 17 |
| CdTe | — | 11 |
| CdS | 2.4 | 11.6 |

Table 4.1
Values of the index of refraction and the dielectric constant, $\epsilon_s/\epsilon_o$, for selected solids

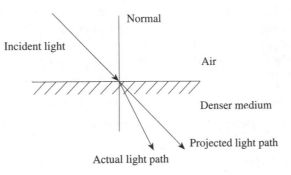

Fig. 4.4
Light path from air into a denser medium

from the normal, making the apparent depth of the pool much shallower than it actually is.

**Example 4.3**    Defining the *optical path* as the product of the distance $d$ traveled by light times the index of refraction, $\eta_s$, and assuming that light always follows the minimum optical path, verify Snell's law.

*Solution:*    The sum of the optical path in two different media, medium 1 and medium 2, can be written as $[d]_s = (\eta_s d)_1 + (\eta_s d)_2$. As shown in Fig. E4.1, $[d]_s = \eta'_s \sqrt{h'^2 + (p - x)^2} + \eta_s \sqrt{h''^2 + x^2}$. When $d[d]_s/dx = 0$ (for minimum optical path), $\eta'_s\sin(\phi) = \eta_s\sin(\phi')$. This is the same as Eq. (4) if we set $\eta'_s = \eta_{\text{air}} = 1$ and $\phi' = \theta_r$ and $\phi = \theta_i$.   §

**Fig. E4.1**

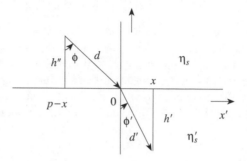

Before light from air enters into a solid, the discontinuity at the air-solid interface will reject a portion of the incoming light. This is called **reflection.** Light reflection at the air-solid interface depends on $N_c$, the complex index of refraction, and is measured in terms of a parameter called the reflectance, $R^*$. In general, **reflectance** from air into a solid is given by

$$R^* = ((\eta_s - 1)^2 + k^2)/((\eta_s + 1)^2 + k^2) \qquad (14)$$

where    $k = c\alpha^*/\omega$.

Equation (14) is called **Fresnel's formula.** Since $\eta_s > 1$, $R^* < 1$, and there is always some finite reflection from the surface of a solid. In the case when $\eta_s \gg 1$, $R^* \approx 1$, and there is total reflection (reflecting region). The quality of light reflected from a solid surface is often of interest to engineers, and this property is closely related to the properties at the interface. *Specular* reflection refers to light reflected from a smooth surface, whereas *diffused* reflection is caused by light reflected from a rough surface. Quantitatively, light reflection is measured by the relative light intensity, $I_\theta$ (W/m$^2$), of the reflected light. As an example, for a rough surface the reflected light intensity, $I_\theta$, is related to the angle of reflection $\theta_r$ by the cosine law:

$$I_\theta = I_0\cos(\theta_r) \qquad (15)$$

where    $I_0$ is the peak intensity in W/m$^2$ (when $\theta_r = 0$).

Thus, light reflection is zero when $\theta_r = \pi/2$, i.e., when the light is glancing through the surface. Another term commonly used for light reflection is **brightness,** which is defined as the intensity of the reflected light per unit area projected normal to the light path. It can be shown that brightness, as defined by Eq. (15), is independent of the viewing angle.

**Example 4.4**     When light passes from a medium with a high index of refraction to one that has a lower value, total reflection occurs at the critical angle, $\theta_{cr}$. Compute the critical angle for light transmission from glass to air. Assume $\eta_s$(glass) $= 1.46$.

*Solution:*    Based on Eq. (13), $\sin\{\theta(\text{glass})\}\eta_s(\text{glass}) = \sin\{\theta(\text{air})\} \times 1$. For total reflection, $\theta(\text{air}) = 90°$ and the critical angle is $\theta_{cr}(\text{glass}) = \sin^{-1}(1 \times 1/1.46) = 43.3°$.   §

**Example 4.5**     Compare $R^*$, the reflectance, in silica glass ($\eta_s = 1.46$) to that in pure PbO ($\eta_s = 2.60$). Explain briefly why ornamental glassware has a small percentage of PbO present.

*Solution:*    Based on Eq. (14), $R^*(\text{glass}) = (1.46 - 1)^2/(1.46 + 1)^2 = 0.035$. Similarly, $R^*(\text{PbO}) = (2.60 - 1)^2/(2.60 + 1)^2 = 0.198$. PbO is therefore much more reflective, which this gives PbO its ornamental value.   §

## 4.4   THE ABSORPTION PROCESS

**Absorption** occurs when light of an appropriate wavelength falls on a solid, and it occurs throughout the solid. If the solid is sufficiently thin, we can actually measure the amount of attenuation after light has passed through. As we mentioned earlier, light absorption in a solid usually involves either the excitation of the electronic states or the **vibrational states.** We look first at the excitation of the electronic states. Consider in general the case when there are two energy states, $E_1$ and $E_2$, in a solid, as shown in Fig. 4.2. The upward transition of an electron from $E_1$ to $E_2$ will require the absorption of energy: $\Delta E = E_2 - E_1$. This is energy conservation, as we mentioned earlier. If we assume that this energy is supplied by a photon, then the number of photons absorbed per unit time is the absorption rate, $R_{12}$, given by

$$R_{12} = B_{12}I_0(v)N_{E_1} \tag{16}$$

where     $B_{12}$ is the transition probability from $E_1$ to $E_2$ in m²/J·s,

$I_0(v)$ is the photon intensity in the incoming radiation in J/m², and

$N_{E_1}$ is the number of electrons with energy $E_1$.

For light absorption to occur, $B_{12}$ in Eq. (16) must be finite (nonzero). Physically, $B_{12}$ is the coupling parameter between the initial energy state $E_1$ and the final energy state $E_2$. In other words, it describes how "compatible" the two energy states are with respect to the transition. There is, of course, the other requirement on momentum conservation that has to be met. Conservation of momentum frequently involves lattice vibrations and how easily the lattice vibrations couple to the light waves. At equilibrium, the absorption rate, $R_{12}$, is balanced by the emission rate, $R_{21}$, that is, $R_{12} = R_{21}$.

As expected, optical absorption in a solid usually does not occur uniformly, and the absorption rate, $R_{12}$, is a function of position—i.e., the distance into the solid. Absorption is measured by a parameter known as the **absorption coefficient, $\alpha*$** (Eq. (4)), and it is defined as the fractional energy absorbed by the solid per unit distance along the light path. In terms of the light flux, $\phi_L(x)$, the absorption coefficient (/m) is given by

$$\alpha* = -[\Delta\phi_L(x)/\phi_L(x)]/\Delta x \qquad (17)$$

where    $\phi_L(x)$ is the light flux at position $x$ in /m$^2$·s,

$\Delta\phi_L(x)$ is the incremental change in the light flux at position $x$, and

$\Delta x$ is the incremental change in position ($x = 0$ is assumed to be at the surface of the solid).

The negative sign in Eq. (17) indicates that the light flux, $\phi_L(x)$, attenuates exponentially inside the solid, as shown in Fig. 4.5. This may be expressed as

$$\phi_L(x) = \phi_{L0}\exp(-\alpha*x) \qquad (18)$$

where    $\phi_{L0}$ is light flux at $x = 0$ in /m$^2$·s.

Although $\alpha*$ is a measure of the light absorption within a solid, its integral over frequency, $\Gamma_{osc}$ ($= [\alpha*(f)]\delta\nu$) (/m·s), is also a very useful parameter. $\Gamma_{osc}$ is called the **oscillator strength,** and a large $\Gamma_{osc}$ implies a greater probability for light absorption to occur.

**Example 4.6**    Green light at a wavelength of 0.5 μm is allowed to shine on a piece of Si. Estimate the distance inside the Si when the light intensity is dropped by a factor of 100.

**Fig. 4.5**
Light attenuation into a solid

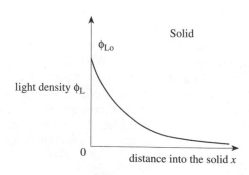

**Solution:** From Fig. E4.2. $\alpha^* = 9 \times 10^3$/cm when the wavelength is 0.5 $\mu$m. Based on Eq. (18), $x = \ln(\phi_{L0}/\phi_L(x))/\alpha^* = \ln(100)/(9 \times 10^3/\text{cm}) = 5.1 \times 10^{-3}$ cm $= 51$ $\mu$m. **§**

Fig. E4.2

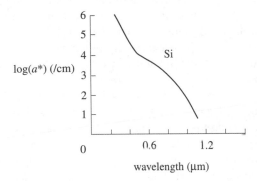

Example 4.7    Suggest how the index of refraction, $\eta_s$, of a lossless solid physically varies with the dielectric constant, $\epsilon_r \; (= \epsilon_s/\epsilon_0)$.

**Solution:** Based on Eq. (8), for a lossless solid $\alpha^* = k \approx 0$ and $\eta_s = \sqrt{\epsilon_s/\epsilon_0}$. Thus, $\eta_s$ increases with increasing $\sqrt{\epsilon_r}$. **§**

In the previous section, we pointed out that a simple model can be used to describe light-solid interactions, and we can identify the transparent regions, an absorption region, and a reflecting region. We shall now describe the more detailed absorption mechanisms in a metal, a semiconductor, and an insulator and examine whether they may explain the different regions in the model.

## Light Absorption in a Metal

A metal does not have an energy gap, so light absorption starts with the polarization of the atoms or ions. At low frequency, the electric-field component of light polarizes the valence electrons with respect to the nuclei, which increases the dielectric constant of the solid. The effect is more pronounced in a metal because of the large number of valence electrons. Any increase in the dielectric constant also increases the index of refraction and, hence, the reflectance. Thus, for a long wavelength, a metal is reflecting and has a lustrous appearance. The large reflectance in this frequency range can also be explained by the fact that electromagnetic waves are always excluded from a perfect conductor (a perfect conductor simply cannot sustain any finite electric field inside it without invoking an infinite current). A metal is not a perfect conductor, but it is usually a very good one. In addition, a metal may also absorb light through free-carrier absorption. During free-carrier absorption, the electrons in the conduction band are excited to the higher energy states within the same energy band, even though they soon return to the initial states with the release of the excess energy in the form of heat (absorption region).

Another type of direct light absorption in a metal is by lattice vibrations; it is known as **Restrahlen absorption.** During Restrahlen absorption, both the transverse and the longitudinal vibrations can be excited, and the energies involved are usually quite small (between 10 mV and 50 meV). Even with this amount of energy, a considerable change in the index of refraction is frequently observed. More energetic light, such as UV light, falling on a metal results in the emission of the electrons. This process is called **photoemission.** Fig. 4.6 shows a schematic of the photoemission process. During photoemission, the electrons have to overcome the surface barrier of the metal; the amount of energy required is called the **work function** (see Section 3.4).

Table 4.2 lists the workfunction values for common metals. As observed, the values (between 4 and 6 eV) are in the UV and the X-ray range. Photoemission in a metal can provide a source of electrons, and the principle has been used to construct photomultiplier tubes where the photoemitted electrons are allowed to produce additional electrons from nearby cathodes through impact ionization (this increases the current response in the photomultiplier tube). In addition to photoemission, there is also band-to-band absorption in the higher energy bands. X rays and gamma rays are usually able to penetrate a thin sheet of metal with minor attenuation if the thickness is less than 1 μm. This ability has been used to produce X-ray masks

**Fig. 4.6**
Photoemission in a metal

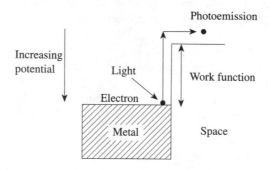

**Table 4.2**
Values of the
work function in the
common metals

| Metals | Work function (eV) |
| --- | --- |
| Cu | 4.45 |
| Au | 4.89 |
| Pt | 5.36 |
| Mo | 4.20 |
| W | 4.54 |
| Al | 4.08 |
| Ti | 4.17 |
| Ni | 5.03 |

(*Source:* Shyh Wang, *Fundamentals of Semiconductor Theory and Device Physics,* © 1989, p. 131. Reprinted by permission of Prentice-Hall, Upper Saddle River, N. J.)

using metal films of different thicknesses. Very energetic radiation (8–15 eV) will penetrate through a metal with no attenuation, since at these energies, the electrons will be too "slow" to respond to light (transparent region).

## Light Absorption in a Semiconductor and an Insulator

Light absorption in a semiconductor and that in an insulator are quite similar because both have an energy gap. In the pure material, there is no absorption when the light energy is less than the energy gap (transparent region). At higher energy, band-to-band absorption will occur, and electrons and holes are created (absorption region). This process can result in a significant increase in the conductivity, which we commonly called **photoconductivity.** Physically, light absorption in a semiconductor or an insulator may be viewed as the breakup of the covalent bonds in the solid with the formation of electrons and holes. Photoconductivity, $\Delta\sigma_{ph}$ (S/m) is given by

$$\Delta\sigma_{ph} = \Delta nq\mu_n + \Delta pq\mu_p \tag{19}$$

where    $\Delta n$ and $\Delta p$ are the respective increases in the electron density and the hole density in /m$^3$,

$q$ is the electron charge in C, and

$\mu_n$ and $\mu_p$ are the electron mobility and hole mobility, respectively, in m$^2$/V·s.

As expected, both the electron mobility and the hole mobility will change with light absorption due to the changes in the charge states of the atoms or ions. In the majority of cases, the mobility change is much smaller than the corresponding change in the carrier density. Photoconductivity occurs only when the photon energy, $E_{photon}$, exceeds the value of the energy gap. For most semiconductors, $E_{photon}$ is in the IR, or the visible range, which is why a semiconductor often has a specific color. Insulators, with a larger energy gap, will respond only to UV light; often, the photoconductivity effect is considerably less due to the presence of trap states, which can remove the carriers.

**Example 4.8**    A Si sample has a thickness of 100 μm and a conductivity of 10 S/m. If we assume light is absorbed primarily within 2 μm from the surface and gives a photoconductivity of 1000 S/m, estimate the fractional change in the sample conductivity if the conductivity measurement is made along the semiconductor surface. Ignore any change in conductivity beyond the surface layer.

*Solution:*    The lateral conductivity of the sample will be the parallel combination of the conductivity of the surface layer and that of the rest of the Si sample. The fractional change is (2 μm × 1000 S/m + 98 μm × 10 S/m)/(100 μm × 10 S/m) = 2.98.

Note that this represents close to a threefold increase in the conductivity of the sample.   §

With light energy larger than the energy gap, absorption in a semiconductor or an insulator depends on the absorption coefficient, $\alpha^*$. Fig. 4.7 shows how the absorption coefficient varies with light energy for the common semiconductors (insulators are of a lesser interest since most of the time their properties are dominated by the effects of defects and impurities).

As observed, the absorption coefficient increases by several orders of magnitude over a small energy range. Physically, this implies a rapid attenuation of the light flux near the semiconductor surface and is known as surface absorption. As a result of surface absorption, the surface carrier densities can be much higher. When $\alpha^*$ is large, very thin samples can absorb most of the incoming light, which is the reason thin-film solar cells can be very good solar-energy converters. Theoretically, the absorption coefficient, $\alpha^*$, in a semiconductor varies with the photon energy, $E_{\text{photon}}$, as

$$\alpha^* = A_{ab}(E_{\text{photon}} - E_g)^n \qquad (20)$$

where    $A_{ab}$ is a constant for a given semiconductor in $/m \cdot eV^n$,

$E_{\text{photon}}$ is the photon energy in eV, and

$n$ is a constant equal to 0.5 or 3/2 when the transition does not involve a change in the electron momentum) and equal to 2 or 3 otherwise.

Near the band edge, light absorption is also affected by the interactions of the electrons and the holes. At low temperature, **excitons**—weakly bound electron-hole pairs—are formed, and the absorption threshold will shift from the value of the energy gap to a lesser value. The energy difference corresponds to the energy of the excited state of the bound electron and hole pair. Excitons are unstable and will dissociate at high temperature. In the absorption region, $\alpha^*$ is large and so is $k$ ($= c\alpha^*/\omega$). This leads to a large reflectance (see Eq. (14)) (reflecting region). Light absorption near the band edge is also affected by the presence of a strong electric field. This effect is called the **Franz-Keldysh effect.** In the presence of a strong

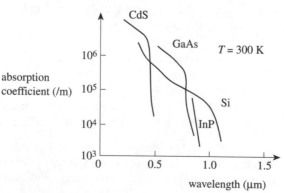

**Fig. 4.7**
Absorption characteristics
of semiconductors

absorption
coefficient (/m)

(Source: S.M. Sze, *Semiconductor Devices Physics and Technology*, Bell Telephone Laboratories, Inc., 1985. Reprinted by permission of J. Wiley & Sons, Inc., N.Y.)

electric field, $\acute{E}$, the electronic states in the energy bands will somehow extend beyond the band edge into the energy gap, as shown in Fig. 4.8.

This situation increases their overlap and can result in an increase in the absorption coefficient $\alpha^*$. For instance, in GaAs the change in $\alpha^*$ (or $\Delta\alpha^*$) can be as much as 20% for $\acute{E} = 4 \times 10^6$ V/m. The Franz-Keldysh effect will give rise to modulation in $\alpha^*$ due to the interference of the electron waves. The modulation frequency is of the order of $10^{12}$ Hz (in the IR range). A change in $\alpha^*$ also changes the index of refraction, $\eta_s$; Fig. 4.9 shows the change in the index of refraction, $\Delta\eta_s$, as a function of the photon energy for GaAs near the absorption edge. $\Delta\eta_s$ has been found to depend linearly on the electric field and the Franz-Keldysh effect is used in the making of optical modulators, devices with transmission properties that can vary with the applied voltage.

A similar effect, called the **Stark effect,** has been used to form light valves (switches) in a quantum-well structure (a quantum-well structure uses layers of different semiconductors to produce a series of submicron potential wells). The Stark effect generally involves a shift in the energy of an electronic state, such as an exciton state in a semiconductor, in the presence of an electric field. Light passage through the device can, therefore, be modulated by changing the internal electric

**Fig. 4.8**

Overlapping of electronic states at different biases

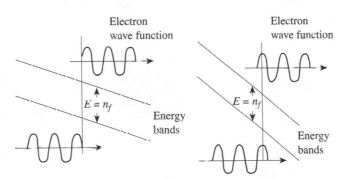

**Fig. 4.9**

Changes in the absorption coefficient in GaAs

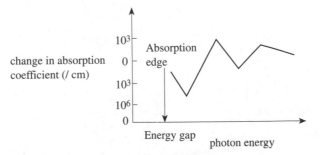

(*Source:* M. Jaros, *Physics and Applications of Semiconductor Microstructures* (New York: Oxford University Press, 1989). Reprinted by permission of Oxford University Press.)

field and hence the energy of the absorption peak. A device called a SEED (**self-electro-optic device**), shown in Fig. 4.10, relies on light absorption in the presence and the absence of the Stark effect.

The device is essentially a semiconductor waveguide transmitting light at a frequency corresponding to the zero-field energy of an exciton state in the semiconductor. With the *p-i-n* structure shown (where *i* is an intrinsic region in the center of the *p-n* junction), an applied voltage at low light intensity is dropped entirely across the intrinsic region, and very little light is absorbed since the absorption peak of the exciton state is now shifted by the strong electric field in the intrinsic region (Stark effect). At high light intensity, however, the applied voltage no longer drops exclusively across the intrinsic region due to photoconductivity in the adjacent regions, and the absorption peak of the exciton state returns to its zero-field value. This situation induces a strong absorption. The transition from high absorption to low absorption via an applied voltage results in the bistable (on-off) characteristics of the device, as shown in Fig. 4.11.

**Example 4.9**    The absorption coefficient in GaAs at $E_{\text{photon}}$ is 1.48 eV is $1 \times 10^6$/m. Estimate the constant $A_{ab}$ in Eq. (20). The energy gap for GaAs = 1.42 eV and $n = 0.5$.

***Solution:*** Based on Eq. (20), $A_{ab} = \alpha^*/\sqrt{E_{\text{photon}} - E_g} = 1 \times 10^6/\text{m}/\sqrt{1.48 \text{ eV} - 1.42 \text{ eV}} = 4.08 \times 10^6/\text{m·eV}^{1/2}$.  **§**

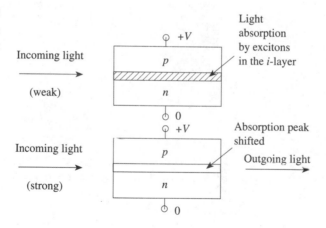

**Fig. 4.10**
Light transmission controlled by Stark effect

**Fig. 4.11**
Input and output characteristics of SEED

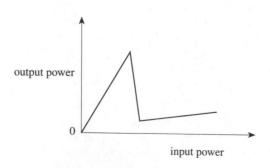

**Example 4.10**   Compute the overlap of the electronic states in the energy bands of GaAs due to the Franz-Keldysh effect if $\acute{E}$ is $4 \times 10^6$ V/m. Assume the attenuation constant of the electronic states within the energy gap is $1 \times 10^6$/m; the energy gap of GaAs is 1.42 eV, and the amplitude of the electron states is unity at the band edges.

*Solution:*   For an applied electric field equal to $4 \times 10^6$ V/m, the energy bands will be tilted, and the spatial separation between the electronic states due to the potential gradient $\acute{E}$ is equal to $E_g/(q\acute{E}) = 1.42$ eV$/(1.6 \times 10^{-19} \times 4 \times 10^6$ V/m$) = 3.5 \times 10^{-7}$ m $= 0.35$ μm. The overlap of the electronic states will occur at one-half of that value, i.e., 0.35 μm/2 $= 0.18$ μm. Thus, the entire overlap will be $\int_{0.18}^{\infty} [\exp(-x \text{ μm}/1 \text{ μm})]\, dx = 0.84$.   **§**

Light absorption due to impurities in a semiconductor results in what is called **extrinsic absorption.** Extrinsic absorption usually involves less light energy than that of the energy gap. Electrons at the impurity centers can be excited to and from one of the energy bands and result in a change in the photoconductivity. Extrinsic absorption is usually much weaker than **band-to-band absorption.** In some cases, the impurity centers are called **sensitizing centers,** and they can have a significant influence on the lifetime of the minority carriers. A carrier, when captured by a sensitizing center, will be put away temporarily until it is subsequently released. This effectively increases its lifetime. Phosphorescent materials, for instance, are frequently doped with sensitizing centers in order to increase the minority carrier lifetime and prolong the light-emission process. Most sensitizing centers are effective in capturing only one type of carrier, leaving the other type essentially free. This slows down the luminescence process and, for a display device, is particularly helpful in generating a persistent image.

The color of a semiconductor is due to the absorption of a portion of the incoming light. It can be due to light that is partially reflected or transmitted, if viewed in transmission. For instance, a semiconductor such as CdS has an energy gap of 2.45 eV and will appear reddish-yellow due to light transmission because both the green and the blue components of the light (2.5–3.5 eV) are absorbed. On the other hand, a semiconductor with a smaller energy gap, such as Si ($E_g \approx 1.1$ eV), will absorb all the visible light and has a dull metallic appearance. Light absorption by impurities can also change the color of a semiconductor. Intrinsic CdS, which normally is reddish-yellow, becomes more reddish through the addition of Cu. Cu, as an impurity, introduces trap states in CdS and reduces the value of the absorption peak to a value as low as 1.7 eV (this energy corresponds to the energy of yellow light). In some infrared light (IR) sensors, light absorption arises from shallow traps. For example, a Ni-doped Ge infrared light detector has trap states at an energy 0.23 eV above the valence band. Light absorption by these traps releases the holes in the valence band and changes the conductivity of the semiconductor.

IR detectors are sometimes made with InSb ($E_g = 0.17$ eV), PbSe ($E_g = 0.15$ eV), and PbS ($E_g = 0.19$ eV). In these semiconductors, photoconductivity is due to band-to-band absorption. Long-wavelength IR detectors are usually fairly noisy at room temperature and require external cooling to minimize thermal noise. Very long wavelength, or far infrared, light detectors require even narrower energy gaps, which exist only in the ternary compounds, such as HgCdTe and PbSnTe. Cooling is essential in these devices because of the heavy background noise.

**Example 4.11**    If we want to form a GaInP crystal with an energy gap of 1.5 eV, what percentage of GaP in GaInP has to be used? For simplicity, let us assume that the energy gap depends linearly on the composition and that the energy gaps of GaP and InP are 2.26 eV and 1.35 eV, respectively.

*Solution:*    Based on the assumption suggested, we use the following equation for the energy gap: $E_g = 1.5$ eV $= [y \times 2.26 + (1 - y) \times 1.35]$ eV, where $y$ is the percent of InP. Solving the equation gives $y = 0.17$. Therefore, the composition should be 17% GaP and 83% InP.    §

**Example 4.12**    InSb is an IR detector, and it has an energy gap of 0.17 eV. What is the probability that an electron in the valence band of InSb is thermally excited across the energy gap at 300 K?

*Solution:*    Since $E_g > kT$, we can assume that the probability of carrier excitation is $\exp(-E_g/(kT)) = \exp(-0.17$ eV$/0.026$ eV$) = 1.4 \times 10^{-3}$. This suggests that there will be 14 electrons excited from the valence band to the conduction band per 10,000 electrons in the valence band.    §

As expected, insulators are colorless (transparent to visible light) because of their large energy gaps. Intrinsically, insulators are poor current conductors, even though the addition of suitable metallic impurities can sometimes alter their optical properties considerably. For instance, ruby, a Cr-doped $Al_2O_3$ insulator, will appear red as a result of light absorption between the atomic states of the $Cr^{3+}$ ions. The absorption peak is at 694.3 nm. A similar situation exists with the Nd:YAG (Nd-doped yttrium aluminum garnet) crystal, which has an absorption peak at 1060 nm. For most insulators, light absorption depends on the transition probability between the impurity states; most of the time, the transition probability is quite small. This is the case with the ruby crystal as well as with the Nd:YAG crystal.

Just as in the case of a metal, a semiconductor or an insulator will give rise to photoemission when the photon energy exceeds the workfunction. Electrodes used in many photoemission tubes are made with semiconductor oxides, which usually have a large workfunction and are less susceptible to surface contamination and erosion. As the light energy increases further, there is less and less difference between a metal, a semiconductor, and an insulator as far as the light-solid interactions are concerned. For very high-energy radiation, we also expect a semiconductor or an insulator to be reflecting and, subsequently, transparent.

**Example 4.13**    What will be the color of a $Zn_{0.5}Cd_{0.5}S$ crystal if the energy gap varies linearly with the composition of the constituents?

*Solution:*    The energy gap of the $Zn_{0.5}Cd_{0.5}S$ crystal is $E_g = 0.5 \times (3.7 + 2.4)$ eV $= 3.05$ eV. This is in the near UV, and the solid will be essentially colorless.    §

## 4.5 THE EMISSION PROCESS

Light emission occurs in a solid in a manner opposite to the absorption process. Light emission occurs not only after electron excitation, but also naturally when a solid is at a temperature above that of its surroundings. This is called **blackbody radiation** (see Chapter 6). In general, during light emission, the emission rate, $R_{21}$ (/m³·s), can be separated into a spontaneous component and a stimulated component:

$$R_{21} = [A_{sp} + B_{21}I_0(\nu)]N_{E_2} \tag{21}$$

where    $A_{sp}$ is a constant related to the spontaneous emission in /s,

       $I_0(\nu)$ is the light (photon) intensity in J/m²,

       $B_{21}$ is the transition probability between energy states $E_2$ and $E_1$ in m²/J·s, and

       $N_{E_2}$ is the number of electrons in the upper energy state $E_2$.

The *spontaneous component,* as given by the first term of Eq. (21), is independent of the "coupling" between the two energy states. This is very different from the *stimulated component* (the second term in the same equation), which, in addition to its dependence on the energy states, also depends on the photon density ($I_0(\nu)$ can come from an external source). Most naturally occurring processes, such as fluorescence, are due to the spontaneous emission, and they occur randomly. Let us for the time being focus only on the **spontaneous emission** and consider the case of light emission in a semiconductor or an insulator when there are more electrons returning from the higher-energy states than the other way around, i.e., when $R_{21} \gg R_{12}$. This situation is illustrated schematically in Fig. 4.12. The emission rate (/m³·s) is given by

$$R_{21} = N_{E_2}/\tau_r \tag{22}$$

where we have replaced $1/A_{sp}$ by the radiation lifetime, $\tau_r$.

**Fig. 4.12**

Light absorption and emission in a semiconductor

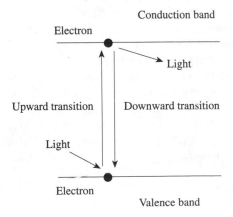

As in the case of light absorption, conservation of energy and momentum has to be observed during the spontaneous emission. Often, other processes in a semiconductor or an insulator also compete with spontaneous emission. In some of these processes, there is no light output, and the excess energy is transferred directly to the solid in the form of heat (energy transfer is most effective if it occurs in a sequence of small steps, or cascades). Thus, spontaneous emission depends on the relative effectiveness of the **radiative emission** (light-emitting) process and the **nonradiative emission** (heat-generating) process. Quantitatively, it is governed by the values of the radiative lifetime, $\tau_r$, and the nonradiative lifetime, $\tau_{nr}$. Light emission occurs only if the radiative lifetime, $\tau_r$, is much shorter than the nonradiative lifetime $\tau_{nr}$, i.e., $\tau_r < \tau_{nr}$. During light emission, if the downward transition of the electrons happens to involve no momentum change, it is called a **direct transition.** The **radiative lifetime** for direct transition is usually quite short. On the other hand, if the downward transition involves a momentum change, it is then called an **indirect transition.** In indirect transition, the radiative lifetime is much longer. As an example, the direct downward band-to-band transition in GaAs has a radiative lifetime of about 1 ns, whereas indirect downward band-to-band transition in Ge has a radiative lifetime in excess of 10 ms. Quite obviously, GaAs is a better material for light emission.

Frequently, a semiconductor is called a **direct gap semiconductor** if direct transition dominates. Otherwise, it is called an **indirect gap semiconductor.** In a direct gap semiconductor, radiative recombination of the electrons (and the holes) can also take place via **recombination centers,** which are either the ionized dopants or the impurities. This type of recombination occurs infrequently, and the radiative lifetime, $\tau_r$ (s), of a semiconductor or an insulator with $N_t$ recombination centers per cubic meter is given by

$$\tau_r = 1/(\sigma' v_{th} N_t) \tag{23}$$

where    $\sigma'$ is the **capture cross section** of the recombination centers in /$m^2$, and
           $v_{th}$ is the thermal velocity of the carriers in m/s.

Equation (23) suggests $\tau_r$ is inversely proportional to the density of the recombination centers, and any increase in the density of the recombination centers will shorten the radiative lifetime of the carriers. In fact, this happens to GaAs, and a shallow impurity, such as Te, will form a very effective radiative recombination center. The emitted light, however, will be at an energy significantly less than the energy gap. The capture cross section (see Eq. (23)) is usually determined experimentally, and it can vary over many orders of magnitude, especially when the recombination center can have different charge states. Semiconductors with an indirect energy gap, such as GaP, usually emit light in the presence of an **isoelectronic impurity.** An isoelectronic impurity has the same valence as the host semiconductor, and it forms a shallow energy state located near the band edge. When it captures a minority carrier, it becomes a radiative recombination center. Typical isoelectronic impurities found in GaP include N, Cd–O, and Zn–O complexes.

**Example 4.14**    Estimate the capture cross section of an impurity in Ge if the radiative lifetime is 10 ms and the impurity density is $1 \times 10^{20}/m^3$. Assume $v_{th} = 0.7 \times 10^7$ m/s.

***Solution:***   Based on Eq. (23): $\sigma' = 1/(\tau_r v_{th} N_t) = 1/(0.01 \text{ s} \times 0.7 \times 10^7 \text{ m/s} \times 1 \times 10^{20}/\text{m}^3) = 1.4 \times 10^{-25} \text{ m}^2.$   **§**

When nonradiative recombination occurs in a semiconductor, the dominant recombination mechanism is called **Auger recombination.** During Auger recombination, the excess energy is transferred from one carrier to another, with a small amount of heat generated in each transfer step. The probability that Auger recombination will occur increases with the square of the carrier density. In GaAs, impurities such as Cu are well-known centers for Auger recombination. Another form of nonradiative recombination occurs when there is a set of closely spaced impurity states present in the energy gap. Heat is generated as the electrons move down successively from one impurity state to another. Grain boundaries in polycrystalline semiconductors are known to possess such energy states, and they form very effective nonradiative recombination centers. A third kind of nonradiative recombination is found in heterostructures, especially those formed on ternary epitaxial layers. In these structures, the lattice mismatch between the epitaxial layer and the substrate can form nonradiative recombination centers at the interface. In addition, nonradiative recombination centers also originate from misfit dislocations and trap states.

## Luminescence

Light emission occurs naturally in some solids; the process is called **luminescence.** There are many different forms of luminescence. For instance, **photoluminescence** is light emission after a solid has been illuminated with energetic light, whereas **electroluminescence** (or **cathodoluminescence**) refers to light emission in a solid due to electron injection and recombination. Other forms of luminescence include **fluorescence,** which takes place concurrently with the absorption process; **phosphorescence,** which occurs when light emission persists for a long time after the excitation has terminated; and **bioluminescence,** which exists only in living organisms.

Photoluminescence is primarily light emission due to the presence of impurities in a solid (although there is also band-to-band photoluminescence). For instance, an ionic solid such as KCl, which has an energy gap of 9.4 eV, absorbs light with energy around 5 eV when doped with $Tl^+$ ions, and the photoluminescence peak occurs at an energy near 4 eV due to relaxation from these centers. Cu-doped ZnS is another solid that exhibits photoluminescence (it absorbs light at 3.68 eV and luminesces at 2.4 eV and 2.8 eV). The detailed photoluminescence mechanism involves radiative recombination by electrons from the donor states to the $(Cu^+)$ acceptor states located in the energy gap, as shown in Fig. 4.13.

Because of their particular role in the photoluminescence process, the $Cu^+$ ions are called **activators.** Photoluminescence output can be increased by increasing the solubility of the activators. This is sometimes achieved by adding coactivators, which help to maintain charge neutrality in the semiconductor. In the case of $Cu^+$-doped ZnS, a Group III ion, such as $Ga^{3+}$, will serve well as a **coactivator.** Cu-doped ZnS has a commercial value, since it emits visible light. Phosphorescence operates on a principle similar to that of photoluminescence, except the emission process will last a much longer time due to the presence of traps. It has been used

Fig. 4.13
Light emission from Cu⁺
centers in ZnS

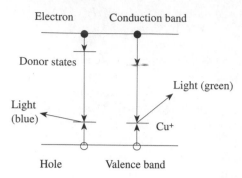

commercially in products such as display panels and various forms of low-level light sources.

Electroluminescence differs from phosphorescence because it uses an ac or a dc electric field as the source of excitation. Under a high electric field, electrons and holes created due to impact ionization near the electrodes are captured at the impurity centers and recombined radiatively, giving out the characteristic light. **Junction luminescence** is a form of electroluminescence, but in this case light emission is due to carrier injection and recombination in a *p-n* junction rather than near an electrode. Devices made from junction luminescence are called **light-emitting diodes** (LEDs), and they form an important light source in photonics. The operation of LEDs depends on charge injection into the junction region during forward bias (the positive voltage of the power supply is connected to the *p*-side of the *p-n* junction), and recombination is primarily from band to band. In some *p-n* junctions, impurity centers can also play an important role in the radiative recombination process. For instance, when GaP LEDs are doped with Cd and S atoms, green light is emitted due to direct transition between the $S^{2-}$ acceptor states and the $Cd^{2+}$ donor states. Similarly, O-complexes present in GaP LEDs shift the emission peak in the light output from green to red.

LEDs made with indirect gap semiconductors suffer from the fact that the emission efficiency is much lower than that of the direct gap semiconductors. LEDs made from ternary compounds with a mixture of direct gap semiconductors and indirect gap semiconductors can sometimes improve the emission efficiency by retaining the direct gap with an intermediate materials composition. The variation of the energy gap in $GaAs_xP_{1-x}$ is shown in Fig. 4.14.

The demarcation between direct gap and indirect gap occurs when $x \approx 0.45$. Isoelectronic impurities such as N can improve the emission efficiency of a GaP LED, as shown in Fig. 4.15. The N atom acts as an electron trap in GaP and enhances the radiative recombination process. LEDs made with $Ga_xAsP_{1-x}$ containing suitable dopants can emit green light, yellow light (with the presence of N atoms), and red light (with the presence of Zn–O and Cd–O complexes). In addition to the $Ga_xAsP_{1-x}$ LEDs, the quaternary $Ga_xIn_{1-x}As_{1-y}P_y$ LEDs have also been studied extensively. These LEDs can emit light over a broad range of frequencies between 0.8 to 1.65 μm. Usually, in LEDs some focusing mechanisms are needed to increase the directional intensity of the light output. The simplest focusing mechanism is perhaps

Fig. 4.14
Energy gap versus fraction
of GaP in GaAs

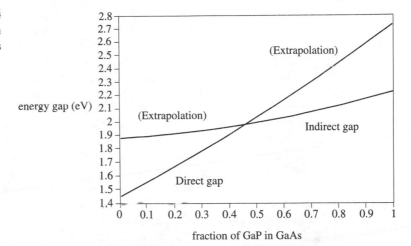

energy gap (eV)

fraction of GaP in GaAs

Fig. 4.15
Quantum efficiency in
GaP with and without N

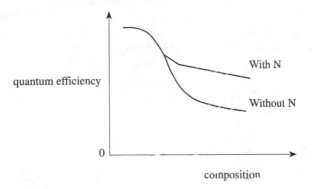

quantum efficiency

composition

a hemispherical lens, which is commonly used in the surface-emitting LEDs. It is also possible to increase the light output of an LED by increasing the dopant densities and hence the forward current. Increases in the dopant densities, however, can generate an adverse effect called **bandtailing,** or energy gap shrinkage. The electronic states in the energy bands of these highly doped semiconductors interact with the dopant states and extend the band edges into the energy gap. This bandtailing effectively reduces the size of the energy gap. In addition, the lattice potential is also screened by the high density of ionized dopants, and the modified energy states result in a shrinkage in the energy gap. As far as the LED output is concerned, a dopant density in excess of $1 \times 10^{24}/m^3$ results in light emission with an energy reduced by about 10 to 20 meV.

## Stimulated Emission

Luminescence in a solid is a random process. This works fine as a light source because in a light source we are interested in the average light output from the solid. A more intense form of light emission occurs if the electrons in a solid are somehow excited continuously and do not relax and emit light spontaneously, but rather they remain in their excited states for an extended period. This situation results in a buildup of the excited electrons and creates an unstable condition until at some stage, a massive number of excited electrons are "stimulated" to return to their ground states. In some solids, because of the long relaxation time of the electrons, the emission process can become very intense when triggered externally. Triggering is usually achieved either through an incoming photon or some form of injected microwave radiation; the process is called **stimulated emission.** Stimulated emission rarely occurs naturally and only takes place in a solid under very specific conditions. One such condition exists in a multienergy-level solid when there are more electrons in the upper energy states than in the lower energy states and is called **population inversion.** When stimulated emission occurs, the output light can be very intense and highly monochromatic (at a single frequency). Stimulated emission also requires that the spontaneous emission rate must be lower than the stimulated emission rate (see Eq. (22)). A number of mechanisms, including the use of radiation (optical pumping) and carrier injection in a *p-n* junction, have been used to achieve population inversion. A light source based on the principle of stimulated light emission is called a laser (light amplification by stimulated emission of radiation). When properly designed, the output from a laser can be very powerful and is capable of performing energy-intensive tasks such as welding.

The simplest structure that can be used to explain stimulated emission is a three-level energy system, as shown in Fig. 4.16. We assume initially that most of the electrons reside in level 1, the lowest energy level. According to the **Boltzmann statistics** some, but a small quantity of, electrons will populate level 2, the next higher energy level. Very few electrons will populate level 3, the highest energy level. For population inversion to occur, however, an energy pumping mechanism is required. This mechanism can be achieved through illumination, using a light source that excites electrons from level 1 to level 3. As a result of this excitation, the electron population in level 3 will increase, and the population in level 1 will decrease. Downward transition of electrons from level 3 occurs spontaneously, and the electrons are distributed between levels 1 and 2. If level 2 happens to have a very long electron lifetime, as in the case of a large capture cross section for the

**Fig. 4.16**
Three-level energy system
in a solid

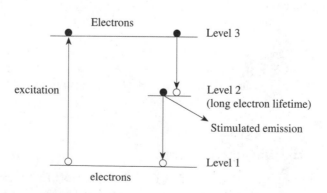

electrons, the downward transition from level 2 to level 1 will never catch up with the depletion of electrons from level 1 due to the excitation. Effectively, this implies an indirect transfer of electrons from level 1 to level 2. The mechanism is referred to as **pumping.** At some stage, level 2 will have the largest number of electrons of the three energy levels in the system. This situation, which we called population inversion, is illustrated in Fig. 4.17. It is useful to note that population inversion in a solid normally requires a three-level system (except in the case of a *p-n* junction), since any pumping of a two-level system will, at best, result in phosphorescence.

Population inversion is an unstable condition and is the prerequisite for stimulated emission to occur. Downward transition from level 2 to level 1 is triggered if a photon of the right energy (equal to the energy difference between levels 1 and 2) exists somewhere within the solid. Such a photon may enter the solid from outside or be introduced in the form of an input optical signal. In either case, the incident photon multiplies as the excited electrons make their downward transition. This transfer of energy from the electrons to the photons ends up with an amplified light signal of a very high intensity and narrow bandwidth.

Measures are often taken to further enhance the number of output photons by polishing the two ends of a laser so that multiple light reflections occur, as in a **resonant cavity.** If the round-trip light path inside the laser is adjusted to give a multiple of the photon wavelength, then the reflected light is reinforced every time it passes through the cavity. Such a reflector-type laser system, as shown in Fig. 4.18, allows a very high intensity laser beam to be built up in the cavity. Unidirectional light output from the cavity is achieved if one end of the cavity is partially reflecting.

**Fig. 4.17**
A three-level energy
system showing
population inversion

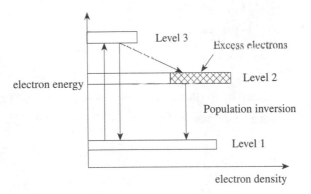

**Fig. 4.18**
A reflector-type cavity for
a ruby laser

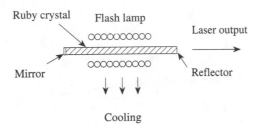

Not all solids can be made into lasers. Some, such as sapphire $Al_2O_3$, for instance, are good host materials, and the addition of $Cr^+$ will give a three-level energy system similar to the one shown in Fig. 4.17. The $Cr^+$-doped laser is called a *ruby laser* because of the color of the light output (the peak wavelength occurs at 694.3 nm). Other types of solid-state lasers include the Nd:YAG laser and semiconductor diode lasers, which are more like LEDs.

A semiconductor diode laser operates using an entirely different principle. Figure 4.19 shows the structure of a typical **laser diode.** The forward-biased *p-n* junction allows electrons and holes to be injected across the junction region and results in a form of population inversion in the energy bands. Light emission is due to carrier recombination across the energy gap, which is confined to the vicinity of the junction region. The pumping mechanism in this case is the forward-bias voltage, and the light output has an energy approximately equal to the size of the energy gap. To achieve population inversion, a **threshold current density** is normally required. This parameter is very important in the operation of a diode laser, and it determines the power efficiency.

To examine this effect, we set up the rate equations for the electron density, $n'$ ($/m^3$), and the photon density, $S_p$ ($/m^3$):

$$dn'/dt = J/(qt_c) - n'/\tau_{eff} - R_{stim} \tag{24}$$

$$dS_p/dt = R_{stim} + Yn'/\tau_r - S_p/\tau_{ph} \tag{25}$$

where   $J$ is the current density passing through the junction in $A/m^2$,

$q$ is the electron charge in C,

$t_c$ is the thickness of the active layer in m,

$\tau_{eff}$ $(= \tau_r\tau_{nr}/(\tau_r + \tau_{nr}))$ is the effective lifetime of the electrons in s,

$R_{stim}$ is the stimulated emission rate in $/m^3 \cdot s$,

$Y$ is the fraction of spontaneous emission that is coupled to the cavity, and

$\tau_{ph}$ is the photon lifetime in s.

We assume $R_{stim}$ is given by

$$R_{stim} = cd(g_c)/dn'(n' - n_0')S_p \tag{26}$$

**Fig. 4.19**
A *p-n* junction showing
stimulated emission

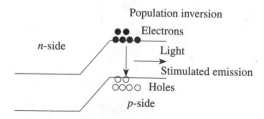

where    $c$ is the velocity of light in m/s,

$g_c$ is the gain coefficient of the cavity, and

$n_0'$ is the value of $n'$ that gives zero gain ($g_c = 0$).

To solve for $n'$, the boundary conditions can be obtained by setting $R_{stim} = 0$ in Eqs. (24) and (25), when spontaneous emission dominates:

$$n' = \tau_{eff} J/(q t_c) \tag{27}$$

$$S_p = Y \tau_{eff} \tau_p J/(\tau_r q t_c) \tag{28}$$

When stimulated emission dominates, $Y \approx 0$, and in the steady state,

$$R_{stim} = c d(g_c)/dn'(n' - n_0')S_p = S_p/\tau_p \tag{29}$$

or

$$n' = n_0' + 1/(c d(g_c)/dn' \tau_p) \tag{30}$$

Substituting Eq. (29) into Eq. (28) gives the following steady-state solution for the photon density:

$$S_p = \tau_p(J - J_T)/(q t_c) \tag{31}$$

where    $J_T = n_0' q t_c/\tau_{eff}$.

Equation (31) defines the threshold current density, $J_T$, needed to achieve stimulated emission, and Fig. 4.20 shows the $S$-$J$ plot for a typical semiconductor diode laser.

For effective laser action to occur, it is necessary to confine the area of the light output to a narrow interfacial layer near the junction region. **Light confinement** poses an additional requirement of beam confinement, which can be achieved by increasing the dopant density near the edges of the $p$-$n$ junction. Highly doped semiconductors are known to give a lower index of refraction, which increases light confinement within the junction region. A graded structure built using epitaxial

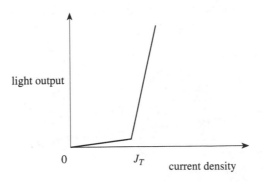

**Fig. 4.20**
Light intensity versus current density

light output

0          $J_T$          current density

layers of ternary compounds, such as $Ga_xAl_{1-x}As$ around a GaAs diode laser, can also generate a similar effect. $Ga_xAl_{1-x}As$ has a wider energy gap and hence has a lower index of refraction. Such a sandwich structure is commonly called a **double heterostructure.** Most laser diodes have polished ends, which are required for the resonant cavity effect mentioned earlier.

For a laser cavity between two reflecting planes, as shown in Fig. 4.21, it is possible to show that the transmitted light energy, $E_{trans}$, and the reflected light energy, $E_{refl}$, within the cavity (J) are given by

$$E_{trans} = tr_1 tr_2 A_{pf} E_{inc}/(1 - A_{pf}^2 R_1^2) \tag{32}$$

$$E_{refl} = (R_2 + R_1 tr_1 tr_2 A_{pf}^2/(1 - A_{pf}^2 R_1^2))E_{inc} \tag{33}$$

where    $E_{inc}$ is the incident light energy in J,

tr$_1$ and tr$_2$ are the transmission coefficients at interface 1 and interface 2, respectively,

$R_1$ and $R_2$ are the reflection coefficients at interface 1 and interface 2, respectively, and

$A_{pf}$ is the propagation factor (= exp[(gain per unit length/2 + $jk'$)L]), where $j$ is the imaginary number $\sqrt{-1}$, $k'$ is the wave number in /m, and $L$ is the length of the cavity in m.

Since spontaneous emission in a diode laser requires that there should be an energy output even for a vanishing input, this condition is equivalent to

$$A_{pf}^2 R_1^2 = 1 \tag{34}$$

Furthermore, constructive wavefronts will only be observed if

$$jkL = 2\pi \tag{35}$$

or

$$L = m\lambda_L/2\eta_s \tag{36}$$

where    $m$ is an integer,

$\lambda_L$ is the wavelength of the light output in m, and

$\eta_s$ is the index of refraction.

**Fig. 4.21**
A laser cavity showing
multiple reflections

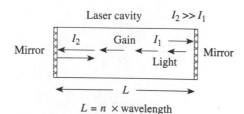

Equation (36) suggests that the output light spectrum of the diode laser will have isolated peaks separated by $\Delta\lambda_L = \lambda_L^2/(2L(\eta_s + \lambda_L\Delta\eta_s/\Delta\lambda_L))$, as shown in Fig. 4.22.

Diode lasers can operate both in the **pulse mode** when driven by a step current and in the **continuous mode** using a sinusoidal current source. Fig. 4.23 shows the output photon density in the two cases. During buildup, the diode laser will have a time delay, $\tau_d$ (s) (since initially the photon density $S_p = 0$), which is approximately given by

$$\tau_d = \tau_{bu}\ln(J/(J - J_T)) \tag{37}$$

where    $\tau_{bu}$ is the buildup time constant in s.

$\tau_d$ is the time lapse before the initiation of a response from a diode laser, as shown in Fig. 4.23.

**Example 4.15**   Compute the threshold current $J_T$ of a GaAs diode laser, if $n' = 1.5 \times 10^{18}/cm^3$, $t_c = 500$ nm, and $\tau_{\text{eff}} = 3$ ns.

**Fig. 4.22**

Laser output spectrum

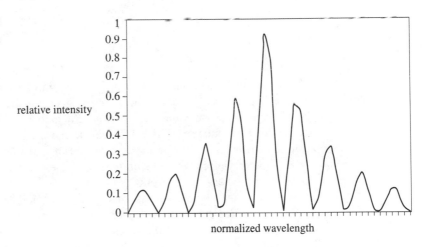

relative intensity

normalized wavelength

**Fig. 4.23**

Laser driven by a step current and a sinusoidal current

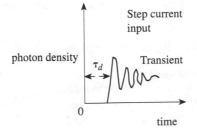

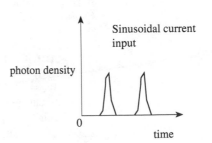

*Solution:* Based on Eq. (31), $J_T = n'qt_c/\tau_{\text{eff}} = 1.5 \times 10^{18}/\text{cm}^3 \times 1.6 \times 10^{-19}$ C $\times 5 \times 10^{-5}$ m/3 $\times 10^{-9}$ s $= 4 \times 10^7$ A/m$^2$. §

**Example 4.16**   A GaAs diode laser has a peak output wavelength $\lambda_L$ located at 868 nm. If $L = 310$ μm and $\eta_s = 4.5$, compute $\Delta\lambda_L$.

*Solution:*   Based on Eq. (36),

$$
\begin{aligned}
\Delta\lambda_L &= \lambda_L^2/(2L\eta_s) \\
&= (8.68 \times 10^{-7} \text{ m})^2/(2 \times 3.1 \times 10^{-4} \text{ m} \times 4.5) \\
&= 2.7 \times 10^{-10} \text{ m} = 0.27 \text{ nm} \quad §
\end{aligned}
$$

**Example 4.17**   Compute the delay time, $\tau_d$, of a diode laser if $J = 1.5J_T$ and $J_T = 4 \times 10^7$ A/m$^2 \cdot \tau_{bu} = 3$ ns.

*Solution:*   Based on Eq. (37),

$$
\begin{aligned}
\tau_d &= \tau_{bu}\ln[J/(J - J_T)] \\
&= 3 \times 10^{-9} \text{ s} \times \ln[1.5 \text{ A/m}^2/(1.5 - 1) \text{ A/m}^2] \\
&= 3.3 \times 10^{-9} \text{ s} \\
&= 3.3 \text{ ns} \quad §
\end{aligned}
$$

## 4.6   IMAGERS

One of the most important areas in person-machine interface is in image capture. There are numerous devices from which images can be recorded, ranging from photographic plates to video recorders. Most modern video recorders are made with **solid-state imagers.** These include the more conventional **photodiode** and **phototransistor** arrays and a newer area sensor called a **charge-coupled device (CCD) imager.** Before we move on to discuss these individual light sensors, let us first look at how an imager works.

An imager may be viewed as analogous to a pinhole camera, from which a picture is captured and projected over a reduced area in which the photosensitive elements are located. To give the necessary picture quality, an imager must maintain a high degree of spatial resolution. Thus, an imager is a device that performs the combined task of light sensing and signal conversion. The task of light sensing depends on the photosensitive elements that transform the light signals into the electrical signals. Data capture often requires some form of storage capability.

### Thin-Film Electrophotography

Thin-film **electrophotography** is widely used in photocopiers for the reproduction of documents and pictures. The key component in electrophotography is a photosensitive thin film, which must possess the appropriate electronic properties to give the required spatial resolution and image quality appearing in the document or the picture. The image is first captured on the thin film and then is transferred to a piece of paper electronically. A toner (a solution consisting of carbon black particles) is used to fix the image. In the following paragraphs, we describe the principle behind electrophotography for a Se thin-film photoconductor deposited on a metallic substrate. The sequence of steps is illustrated in Fig. 4.24.

1. The photosensitive Se thin film is first sensitized, or charged, in the dark through corona discharge so that the upper surface becomes positively charged. Simultaneously, the metallic substrate on which the thin film has been deposited is negatively charged. A field strength of at least $10^7$ V/m is required.

2. The document or picture to be copied is then placed some distance above the Se thin film, and a light source is used to transfer the image onto the Se thin film through light reflection. The reflected light creates electron and hole pairs and discharges the Se thin film in places where the images are absent. Thus, a positively charged replica of the image is reproduced on the Se thin film.

3. A negatively charged toner (consisting of carbon black particles) is deposited on the Se thin film, reproducing the image of the document or picture. A piece of slightly positively charged paper placed near the Se thin film picks up the toner. Once the toner is fixed (through a heat cycle), the paper retains the carbon particles, and the copying process is completed.

4. For reset, the Se thin film is flooded with a low-intensity light until the image is totally erased. The surface potential is reduced to about $10^6$ V/m in the standby mode.

As expected, the photosensitive Se thin film must by highly insulating in the dark and have good uniformity throughout. Furthermore, the **dark current** must be very small, and the film properties should not change appreciably with repeated cycles of charging and discharging. Commercial Se thin films have $p$-type conductivity, and carrier trapping can become a problem, since the electron release time from the traps can be as long as several minutes. The residue charge buildup in the film due to trapping is responsible for the "ghosting" effect when the document or picture is changed repeatedly.

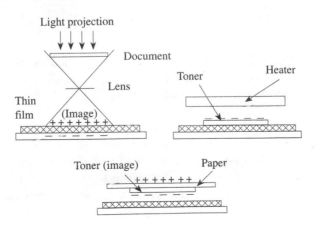

**Fig. 4.24**
Steps in
electrophotography

**Example 4.18**     Suggest why Se thin films are unsuitable for use in electrophotography of documents with a red background.

*Solution:*     Although Se thin films have an absorption threshold near 1.7 eV, the electrophotographic sensitivity or photoconductivity threshold may not coincide with the absorption threshold. Se has a photoconductivity threshold in the green range ($\approx$2.4 eV) and is not sensitive to red or yellow light. Reflected red or yellow light from the background of the document will appear as black, the same as the prints in the document or picture.     §

## Semiconductor Photodetectors

The simplest solid-state **photodetector** is a piece of photoconducting semiconductor. Photoconductivity is a result of carrier excitation due to light absorption, and the figure of merit depends on the light-absorption efficiency. The photon-induced current, $I_{pi}$ (A) for a $p$-type photoconductor is given by

$$I_{pi} = q\Delta p\mu_p\acute{E}A_{cs} \tag{38}$$

where     $\acute{E}$ is the electric field in V/m,

$\Delta p$ is the excess hole density in m$^3$,

$\mu_p$ is the hole mobility in m$^2$/V$\cdot$s, and

$A_{cs}$ is the cross-sectional area of the device in m$^2$.

By definition, $\Delta p = G\tau_p$, where $G$ is the generation rate (/m$^3\cdot$s) and $\tau_p$ is the hole lifetime (s). If the length, $L$, of the photoconductor is reasonably short, then $\mu_p = L/(\tau_t\acute{E})$, and Eq. (38) can be written as

$$I_{pi} = \tau_pqGA_{cs}L/\tau_t \tag{39}$$

where     $\tau_t$ is the **transit time** across the photoconductor in s.

The ratio $\tau_p/\tau_t$ represents the photoconductivity gain of the photoconductor. The **bandwidth** of a photoconductor can be determined from the following rate equation:

$$d\Delta p/dt = -p/\tau_p + G \tag{40}$$

For a sinusoidal signal with a time dependence of $\exp(j\omega t)$, where $\omega$ is the angular frequency of the input signal in radians per second and $j$ is the imaginary number $\sqrt{-1}$, Eq. (40) then becomes

$$\Delta p = G\tau_p/(1 + j(\omega\tau_p)) \tag{41}$$

According to Eq. (41), the bandwidth of a photoconductor is proportional to $1/\tau_p$. Thus, in the design of a photoconductor, there has to be a trade-off between the signal bandwidth, which is proportional to $1/\tau_p$, and the photon-induced current, $I_{pi}$, which is proportional to $\tau_p$ (Eq. (39)). Photoconductors are very sensitive to noise, and three types of noise sources can be identified: (1) thermal noise, whose power is dependent on the absolute temperature; (2) shot noise, whose power is dependent on the current flowing through the photoconductor; and (3) $1/f$ noise, which predominates at low frequency.

For use as discrete devices, the photodiodes and the phototransistors are the prime candidates in imaging, since they both have good light conversion efficiencies. Photodiodes and phototransistors essentially operate along the same principles as the conventional diodes and transistors (see Chapter 3). The only difference lies in the fact that the base current is replaced by the photocurrent generated in the emitter-base junction.

Photodiodes can be divided into two major types. The $p$-$i$-$n$ diode has an extended intrinsic region within the $p$-$n$ junction, and this increases the area for light absorption and hence the efficiency. One drawback of $p$-$i$-$n$ diodes has to do with the increased transit time of the carriers because of the larger separation between the $p$-side and the $n$-side of the junction. The longer transit time across the intrinsic region can significantly slow down the response of the device. Direct gap semiconductors used in $p$-$i$-$n$ diodes usually have a larger absorption coefficient and a somewhat narrower intrinsic region. A second type of photodiode is called the **avalanche photodiode** (APD). This device operates at a large reverse-bias voltage, so carrier multiplication occurs within the depletion layer. This leads to an internal gain for the device, but the response time of the APD is again slowed down by the carrier multiplication process.

Two parameters commonly used to describe the performance of a photodiode are the **noise-equivalent power** (NEP) and the **detectivity**, $D^*$. NEP is the ideal input power at which the photocurrent is exactly equal to the noise current, and it corresponds to the minimum detectable signal level. The detectivity, $D^*$, is given by

$$D^* = \sqrt{A_{cs}\delta\nu}/\text{NEP} \tag{42}$$

where    $A_{cs}$ is the cross-sectional area of the device in m$^2$, and

$\delta\nu$ is the bandwidth in Hz.

Similar to APDs, phototransistors have the advantage of providing internal gain. Both photosensitive bipolar transistors and photosensitive field-effect transistors have been developed, so an internal gain of 100 or better can be achieved in the photosensitive bipolar transistor. The main problem with this device is that a large bias is needed to achieve this level of gain, and the parasitic will increase the response time considerably.

Imagers built with discrete devices are made either in the form of a line array or an area array. In both cases, shift registers are used to output the video signals. A line imager will operate as an area imager if mechanical scanning is also provided. Most area imagers will capture an image in two dimensions, and the signals are shifted out either serially or sequentially before multiplexing and storage. Imagers built using photodiodes and phototransistors are called **vidicons.** Vidicons produce good output current linearity and are ideally suited for small-area imagers. Some problems arise when the array is too large and when the individual device has to be addressed. Even when multiplexed, the interconnections in a large array can be very complex and can generate a significant overhead both in the chip area as well as in the circuitry. In addition, the power-loading effect of a large array of active elements will make the system very noisy.

A better light-sensing device is the CCD imager, which operates in the charge domain rather than in the current domain. CCDs were initially developed for analog sampled-data circuits and were used in the implementation of analog filters. The small pixel size of the CCD, with a provision for optical input, and the dynamic charge-transfer capability make the device very suitable for high-density imaging. In addition, CCDs are developed based on the metal-oxide-semiconductor (MOS) process, a mature technology for very large scale integration (VLSI). In fact, 4M-pixel CCD area array imagers became available quite recently, and the overall trend in CCD integration follows closely that of the dynamic random access memories (DRAMs).

A charge-coupled device (CCD) imager is made up of an array of metal-oxide-semiconductor **(MOS) capacitors,** as shown in Fig. 4.25. The CCD imager relies on charge generation in the depletion region of the MOS capacitor. The potential in the MOS capacitor is, to a first order of approximation, linearly dependent on the voltage applied to the surface electrode (see Section 3.5), and a potential well can easily be formed in the semiconductor using a suitable set of voltage pulses, as illustrated schematically in Fig. 4.26.

In this mode of operation, the semiconductor under the surface electrodes must be fully depleted, and any charges generated in the depletion region will flow into the potential wells formed by the voltage pulses. The charges (for instance, the photo-excited electrons in an $n$-channel CCD) are confined laterally, since the neighboring MOS capacitors are at a low potential and will not leave the potential well. The amount of charge generated is proportional to the light intensity as well as to the period during which light is allowed to fall on the CCD. A CCD imager

**Fig. 4.25**
A CCD shown as an array of MOS capacitors

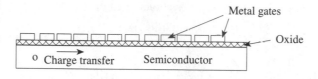

**Fig. 4.26**
Potential well in a surface ($p$-channel) CCD

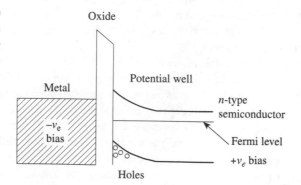

operates in the dynamic mode; i.e., the pictures are captured frame by frame before they are transferred to allow for continuous-time operation. The potential wells where the charges reside are controlled by the clock voltages, and the charges are shifted out serially from the array. During charge transfer, the most important mechanism is probably the fringing field effect caused by the voltage pulses. Other important mechanisms include the self-induced drift effect due to the mutual repulsion of the stored charges and the thermal diffusion of the charges.

To form an isolated potential well, three to four electrodes are required; this set is called a **pixel.** A pixel represents a single point in the image. Based on the concept of a pinhole camera, a CCD area imager allows for the capture of a picture over a much-reduced image area. The minimum pixel area in a CCD corresponds to three or four MOS capacitors, or a size somewhere on the order of 10 μm. This size gives a $500 \times 500$ dot-matrix array, a chip size of roughly 25 cm$^2$ (2 in. by 2 in.). However, some form of interlacing of the image (by a superposition of two or more nonoverlapping images captured at consecutive times) could easily reduce the area of the imager by a factor of 2 or more. The output charge is usually delivered to a small sensing capacitor that converts charges to a voltage. Fig. 4.27 shows a typical CCD area imager. As observed, the only additional area (other than the CCD) needed is for the on-chip clock drivers and the charge storage buffer elements if the voltages have to be downloaded sequentially to a single output.

The problems associated with the design of an imager—other than the size requirement—also involve signal-to-noise ratio and the residual charge left over in the wells. Noise in CCDs can be due to charge generation by the dark current and charge retention by the interface traps. Both can be minimized by using a relative pure semiconductor or, if possible, operating the CCDs at a reduced temperature. For an electronic still or video system, a signal-to-noise ratio of at least 40 is required. CCDs used under low light intensity or for infrared imaging are particularly prone to thermal noise. For these devices, light intensifiers, such as detector multipliers, are often used at the front end of the detector to increase the signal gain. Typically, the size of a CCD pixel is 100 μm$^2$, and it has a dynamic range (peak signal over noise) of 4000 and a charge transfer inefficiency of less than $1 \times 10^{-4}$ between successive MOS capacitors.

Another important measure for a good CCD imager is in light conversion. Normally, light comes in through a semitransparent electrode made with metal or oxide

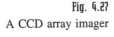

**Fig. 4.27**
A CCD array imager

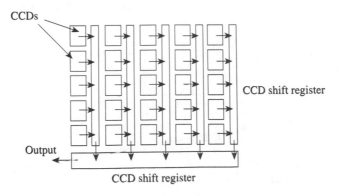

or through the back surface of the wafer after thinning. Very thin metal films of Au or Al in the thickness range of 5 to 30 nm are used, and the resistivity depends on the film thickness. A cross-sectional view of such a CCD pixel is shown in Fig. 4.28.

The optical transmittance, however, is much smaller in the thicker film, even though the loss in the underlying $SiO_2$ layer may be negligible. An alternative approach is to use a layer of thin polysilicon film as the electrode. Polysilicon is structurally compatible with $SiO_2$, but the resistivity is at least 100 times greater than a metal film. Other types of transparent electrodes include ITO (indium tin oxide) and other forms of indium oxide. ITO is quite transparent to visible light (it has a transmittance of 80% to 95%) and can have a resistivity as low as $1 \times 10^{-6}$ $\Omega \cdot m$. In ITO thin films, the addition of Sb will act as a dopant to increase the film conductivity. ITO films may be deposited by evaporation, sputtering, or other chemical techniques. In addition to the formation of conducting electrodes, ITO thin films are often used as antireflection coatings, since the index of refraction ($\eta_{ITO} \approx 2.1$) is reasonably close to the square root of the index of refraction for most popular semiconductors (for example, $\eta_{Si} \approx 3.5$).

**Example 4.19**    Compute the photoconductivity gain in a $p$-type photoconductor if the hole lifetime is $\tau_p = 50$ μs and the length of the photoconductor is 1000 μm. Assume the average velocity of the holes is $1 \times 10^5$ m/s.

*Solution:*    The transit time across the photoconductor is $\tau_t = L/v = 1 \times 10^{-3}$ m/$10^5$ m/s $= 10$ ns. By definition, the photoconductivity gain is $\tau_p/\tau_t = 50 \times 10^{-6}$ s/$10^{-8}$ s $= 5000$.    §

Note that $1 \times 10^5$ m/s is the value of the saturation velocity for most semiconductors, so the photoconductivity gain we computed is an upper limit.

**Example 4.20**    Compute the NEP for a photoconductor at 300 K over a bandwidth of 10 kHz.

*Solution:*    NEP is the ideal input power at which the photocurrent is exactly equal to the noise current. Let us assume the noise source is due to thermal noise, which is given by $\langle v_{th} \rangle^2 = 4RkT\delta v$, where $R$ is the resistance of the photoconductor and $\delta v$ is the bandwidth. Therefore, NEP $= \langle v_{th} \rangle^2/R = 4kT\delta v = 4 \times 1.38 \times 10^{-23}$ J/K $\times 300$ T $\times 10^4$/s $= 1.7 \times 10^{-16}$ W.    §

**Fig. 4.28**
A four-phase CCD

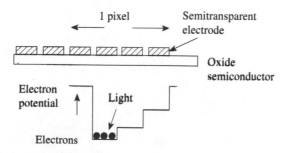

**Example 4.21**   If a CCD line imager has 500 pixels, determine the transfer inefficiency per pixel needed to achieve a charge loss of no more than 10% in the imager.

*Solution:*   The ratio of output charge over the input charge $(Q_o/Q_i)$ is 0.9. If the transfer inefficiency per pixel is $\eta_e$, then $Q_o/Q_i = (1 - \eta_e)^n$, where $n = 500$. This gives $\eta_e = 1 - 0.9^{1/500} = 2.1 \times 10^{-4}$.   §

Note that less than 0.02% per pixel of the captured charge will be lost during charge transfer.

**Example 4.22**   Compute the thickness of ITO needed to form an antireflection coating for Si, given that $\eta_{Si} = 3.5$.

*Solution:*   Since $\eta_{Si} = 3.5$ and $\eta_{air} = 1$, $\eta_{ideal} \approx \sqrt{3.5} = 1.89$. $\eta_{ITO} = 2.1$ should be close enough to 1.89.

For maximum light transfer, the thickness of the antireflection coating should be equal to $\lambda_L/4$. Assuming white light has an average wavelength of 0.6 μm, the thickness of the ITO layer ought to be around 0.15 μm.   §

## 4.7   DISPLAYS

### Cathode-Ray Tube

Earlier we studied the different ways that light may be generated from a solid. Light generation by bombardment using an energetic electron beam is known as *cathodoluminescene.* A typical example of cathodoluminescence is observed in a **cathode-ray tube** (CRT). A cross-sectional view of a CRT is shown in Fig. 4.29. The source of electrons in this case is an **electron gun,** which is a heated metal electrode at the back of the tube. Electrons are emitted when the thermal energy of the electrons exceeds the work function of the metal electrode. The emitted electrons are then attracted toward a positive electrode placed just behind the CRT screen. The electrons acquire kinetic energy (tens of thousands of electron volts) during the passage to the screen and bombard the phosphor (usually a ZnS/CdS mixture) deposited in front of the electrode. Visible light is emitted through recombinations

**Fig. 4.29**
A cross-sectional view of a CRT

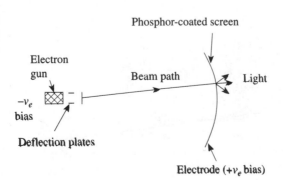

of electrons and holes in the phosphor, resulting in luminescence on the screen. The phosphor is a semiconductor thin film, and it serves as the medium for both recombination and luminescence. The lifetime of the minority carriers will be responsible for the time delay in luminescence.

Phosphors are often deliberately made to have long minority-carrier lifetimes to give a prolonged effect on the image while the signals are being refreshed. This situation is referred to as the **persistence** of the phosphor. By changing the composition of ZnS and CdS in the phosphor, the emitted light can have different colors. Most phosphors have a conversion efficiency of around 20%. In a straightforward operation of the CRT as an oscilloscope, the vertical axis of the screen (connected to a set of vertical deflection plates) displays the signal strength, whereas the horizontal axis (connected to the horizontal deflection plates, driven by a sawtooth voltage waveform) traces the time scan. A typical CRT display is shown in Fig. 4.30.

The principle behind the CRT also applies to the television tube. The electron beam of a TV tube is sequentially scanned horizontally from top to bottom (at several megahertz) to form a single frame of the picture. In some TV formats, a single picture is scanned twice to form two nonoverlapping images, thus improving the picture quality. Each picture is, therefore, composed of a matrix of bright and dark spots (up to 500,000 for a 25-in. tube). The picture is refreshed after a complete scan across the screen. The contrast of the image is produced by the density of the dots. Color pictures may be produced using three separate electron guns serving each of the RGB (red, green, and blue) colors. A mixture of these colors provides hue. Some color TVs use a *shadow-mask tube* to enhance the contrast, as illustrated in Fig. 4.31. A **shadow mask** is a metal sheet with a large number of drilled holes,

**Fig. 4.30**
A CRT display

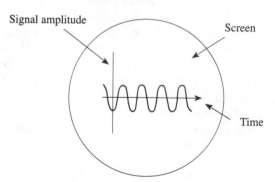

**Fig. 4.31**
Shadow mask highlighting an RGB pixel

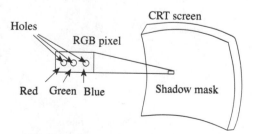

each aligned to a set of three different phosphor dots deposited on the screen. The dots are made to give the RGB colors (depending on the composition of the phosphor). These dots are selectively illuminated by each of the three electron guns. Again, a color picture is formed by a suitable combination of the three basic colors.

**Example 4.23**

An electron gun made with tungsten has a workfunction of 4.5 eV and an effective area of 0.02 m × 0.02 m. If it operates at 1000 °C, what is the electron current that may be drawn from this source?

*Solution:* From Chapter 3, Eq. (35), the thermionic current, $I_{th}$, at 1000 °C is approximately $4\pi q m_0(kT)^2 A_{cs}\exp(-\Phi_B/\Phi_t)/h^3$ = 4 × 3.14 × 1.6 × $10^{-19}$ C × 0.91 × $10^{-30}$ kg × (1.38 × $10^{-23}$ J/K × 3273 K)$^7$ × 0.02 m × 0.02 m × exp(−4.5 × 1.6 × $10^{-19}$ V/(1.38 × $10^{-19}$ J/K × 3273 K))/(6.62 × $10^{-34}$ J·s)$^3$ = 0.6 × $10^{-9}$ A = 0.6 nA. §

**Example 4.24**

If a TV set has 500 × 500 pixels and the sweep frequency is 5 MHz, how long does it take to scan the whole screen once?

*Solution:* There are 500 × 500, or 250,000, pixels on the screen. To achieve a sweep frequency of 5 MHz, the maximum time delay at each pixel has to be less than 0.1 μs (twice the sampling rate). For 0.25 million pixels, the total time delay is 0.25/0.1 = 2.5 s.

Note that this is the time needed to display the whole picture and is also the refresh time of the phosphor. §

Alphanumeric displays often use a different principle. Light generation in alphanumeric displays relies on current injection into semiconductors in what is called *electroluminescence*. A semiconductor will emit light when electrons are injected from an electrode through an interface barrier into a region where holes are present. The recombination mechanism is illustrated in Fig. 4.32. Both rectifying contacts and *p-n* junctions will give light emission. The light-emitting *p-n* junctions are the same as LEDs, and the emitted light has an energy peak roughly the same as the energy gap of the semiconductor. Most commercial LEDs are made with GaP, and they emit either red light or green light, which are the colors of more common alphanumeric displays. It is much more difficult to form stable blue LEDs because of the presence of native defects in the wide-energy-gap semiconductors. GaN, with

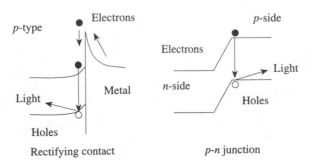

**Fig. 4.32** Light-emission mechanisms

an energy gap of 3.4 eV, is one of the few wide-energy-gap semiconductors that will give out blue light during cathodoluminescence. Attempts have also been made to use Zn–Mn–Se multiple-well structures with light emission in the ZnSe well at 2.7 eV (blue light). In addition, multiple photon absorption is found in some phosphors that will also luminesce in blue. $YF_3:Yb^{3+}-Tm^{3+}$ is an example. When illuminated with infrared light, the $Yb^{3+}$ ions in the $YF_3$ host will absorb the infrared light and, through multiple transitions, transfer the energy to the $Tm^{3+}$ ions. The return of the excited $Tm^{3+}$ ions to their ground states gives out light at around 2.6 eV.

**Example 4.25**    CdS is an $n$-type semiconductor. Suggest how you may form an LED with it. What will be the energy of the light output? Describe the shortcomings in a CdS LED.

*Solution:*    CdS will form LEDs with $p$-type CdTe and $p$-type ZnTe. Their energy gaps are 1.56 eV and 2.26 eV, respectively. Light output from these LEDs will have energy peaks near the values of the energy gaps. In the heterojunction device, lattice misfits will be a major problem that gives rise to nonradiative recombination.    §

## Light Switching

The passage of light through a solid may be controlled through anisotropic polarization of the dipoles. Ferroelectric solids affect polarized light by changing the electric-field component and, hence, the polarization angle. For instance, polarized light passing through a ferroelectric solid will experience a phase rotation if a voltage is applied across the solid. A larger bias increases the polarization and hence generates a larger phase shift. If a polarizer plate is put behind the solid away from the illumination, it is possible to control light transmission simply by altering the bias voltage across the ferroelectric solid. This process is illustrated in Fig. 4.33. An automatic light shutter for a camera may be constructed if a sensor placed in the camera measures the light intensity and uses the signal to control the bias voltage.

Light switching is also used in optical memories. Amorphous semiconductors, especially those containing Se and Te, are both light sensitive and capable of existing in a high reflectivity state and a low reflectivity state through structural

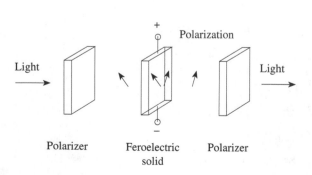

**Fig. 4.33**
Light polarization by a biased ferroelectric solid

Light

Polarization

Light

Polarizer    Feroelectric solid    Polarizer

transformation. For instance, Te compounds will recrystallize at a reasonably low temperature (below the melting temperature, $T_m$) if placed under irradiation. This allows us to produce either the amorphous phase or the crystalline phase without actually going through the melting process; the process is called **photocrystallization.** With phase transformation, the absorption coefficient of Te, for instance, will change dramatically, as shown in Fig. 4.34. The change in the transmission properties allows the process to be used for optical recording.

Fig. 4.35 shows the ideal transformation cycle. Curve A shows crystallization from the amorphous phase to the crystalline phase in the presence of a laser pulse. The smaller absorption constant in the amorphous phase and the high thermal capacity prevent the temperature from reaching the melting point, $T_m$. Crystal regrowth is therefore achieved through direct photocrystallization. In the reverse process, using the same laser pulse, the higher absorption constant and smaller heat capacity will bring the solid above its melting point, and recrystallization can be prevented if the cooling rate is sufficiently rapid. Such a crystallization and amorphization cycle allows us to change the Se or Te compound from a high reflective state to a low reflective state, thus generating the bistable properties of the recording elements.

$Sb_2Se_3$ and $Sb_2Te_3$ have dramatically different transmittivities between the amorphous phase and the crystalline phase, as shown in Fig. 4.36. To achieve phase transformation, which occurs at 160 °C in $Sb_2Se_3$, a relatively thin $Sb_2Se_3$ layer,

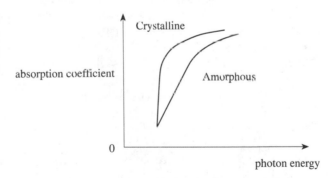

**Fig. 4.34**
Absorption coefficient for Te with different crystal structures

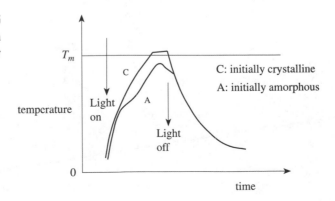

**Fig. 4.35**
An ideal transformation cycle in an optical memory

**Fig. 4.36**
Transmissivity for $Sb_2Se_3$
thin film

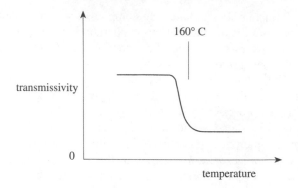

perhaps with a double-layer construction, is often used. The composite structure consists of a PMMA (polymethyl-methacrylate) layer placed on top of the $Sb_2Se_3$ thin film (less than 100 nm) deposited on a $Bi_2Te_3$ substrate. During recording, a laser light with a wavelength of 830 nm is allowed to pass through the PMMA and induces structural changes in the $Sb_2Se_3$. $Sb_2Se_3$ serves as the active layer and is thermally stable up to 160 °C. The $Bi_2Te_3$ substrate acts as a light reflector. The overall transmission efficiency of the system varies between 10% and 30%.

In general, the design considerations for an optical disc system are as follows:

1. High data-storage density of $10^8$ bits/s can be reached and is limited by the playback bit-error rate and the precision in timing and positioning control. Direct read after write operations, as in DRAMs, should offer a reduced bit-error rate (say to 1 part in $10^{12}$ bits).
2. The average laser power output is of the order of 25 mW, which requires a materials sensitivity of about 10 mW with a threshold value of a few milliwatts. The read-beam power should also be at least a factor of 3 less than the write-beam power.

Another form of light switching involving a similar principle uses α-Si:H instead of Se compounds. In α-Si:H thin films, the effusion of $H_2$ by laser heating can result in the formation of **silicon bubbles.** These bubbles have an area of less than 1 μm in diameter and a reflectivity of only a few percent, compared with 40% in the original film. α-Si:H thin films are potentially useful for optical recording through laser irradiation. The energy required for writing a single bit is quite high, but a storage capability in excess of $10^{10}$ bits/disk may be achieved.

**Example 4.26**    A diode laser with an output power, $P$, of 15 mW is used to heat up a semiconductor slab during thermal recording. If the pulse width, $t$, is 50 ns, the light spot has a diameter, $d_m$, of 2 μm, and the transmission efficiency, $\eta_{tr}$, of the laser is 30%, compute the temperature rise. The semiconductor thickness, $t$, is 400 nm, the specific heat, $\sigma_{sp}$, is 750 J/kg·K, and the density, $\rho_0$, is $2.4 \times 10^3$ kg/m³.

***Solution:***    The energy delivered to the semiconductor is $E = Pt = 15 \times 10^{-3}$ W $\times 0.3 \times 50 \times 10^{-9}$ s $= 0.225 \times 10^{-9}$ J. The mass of the semiconductor within the light spot is $m = (\pi/4)d_m^2 t\rho_0 = (3.14/4) \times (2 \times 10^{-6} \text{ m})^2 \times 4 \times 10^{-7} \text{ m} \times 2.4 \times 10^3 \text{ kg/m}^3 = 3 \times 10^{-15} \text{ kg}$. The temperature rise is $\Delta T = E/(\sigma_{sp}m) = 0.225 \times 10^{-9} \text{ J}/(750 \text{ J/kg·K} \times 3 \times 10^{-15} \text{ kg}) = 100 \text{ K}$. §

## Liquid Crystals

**Liquid crystals** are a suspension of rodlike organic molecules that have different structural arrangements under different bias voltages. These structural arrangements affect the optical properties, including the index of refraction and the permittivity. Liquid crystals, when placed in a capsule, have been used to form devices that act like electrically controlled light switches. Under normal conditions, liquid crystals have a structural arrangement that is dominated by the mutual interactions of the molecules and by the properties of the surface on the inside of the capsule. Under no bias, liquid crystals are transparent to light. A dc or low-frequency electric field will change the orientation of the liquid crystals within the capsule. When properly designed, this can make the liquid crystals opaque to light. A liquid-crystal display with a given electrode pattern therefore produces an image similar to the pattern appearing in the electrodes. The most common use of liquid crystals is in alphanumeric displays.

Fig. 4.37 shows a typical liquid-crystal molecule of cyanobiphenyl compounds. As mentioned earlier, liquid crystals exist in different phases, depending on the operating conditions. These can be in (1) a **nematic phase,** when the molecules are all parallel to each other but are otherwise randomly placed (this is the most common state); (2) a **cholesteric phase,** when the molecules are arranged in planes, with the molecules in each plane parallel to one another but in successive planes, twisted with a well-defined pitch; or (3) a **smectic phase,** in which the molecules are all lined up parallel to each other and arranged in layers. Each of these phases has different polarization properties that can be used to control the light reflection.

An understanding of the operation of a liquid-crystal display requires a knowledge of the construction of the capsule. Fig. 4.38 shows a liquid-crystal-display capsule made of transparent electrodes, polarizers, and a substrate reflector. The two transparent electrodes form the outer shell of the capsule and are covered on the inside with thin polymer layers. These polymer layers have line grooves on the inside that are at an angle 90° to each other. The two sets of grooves force the liquid crystals to align in their directions and impose a 90° twist (cholesteric phase). Light passing through these liquid crystals will also suffer the same 90° phase shift due to polarization. Thus, by deliberately using polarizers with crossed directions of polarization over the transparent electrodes, only light with the right polarization entering the

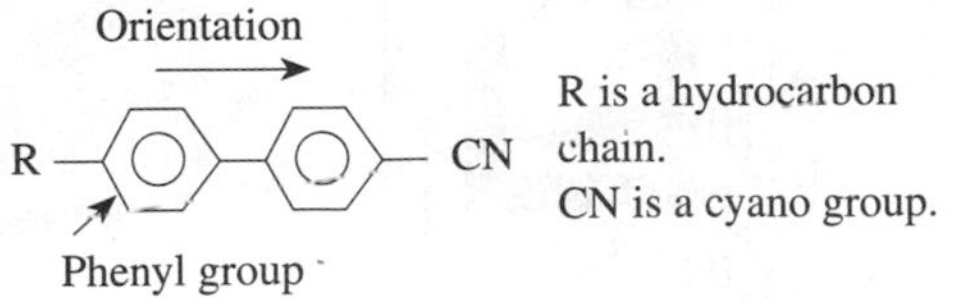

**Fig. 4.37**

A typical liquid crystal molecule

Orientation

R is a hydrocarbon chain.
CN is a cyano group.

Phenyl group

**Fig. 4.38**
LCD display capsule

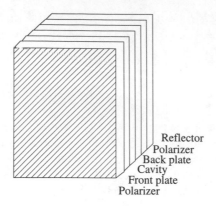

Reflector
Polarizer
Back plate
Cavity
Front plate
Polarizer

capsule will be transmitted. The reflector at the back of the capsule allows the transmitted light to be reflected back to the front of the capsule, and this generates the "light" pattern on the display (after a total of 360° phase shift). To generate the "dark" pattern, an electric field is applied to the transparent electrodes and the liquid crystals turn nematic. This blocks the light path due to random light scattering. Based on this principle, display patterns can be formed using electrodes of different shapes. Fig. 4.39 shows a typical seven-segment alphanumeric pattern for a liquid crystal display (LCD). Since there is no dc current involved, the power dissipation is very small (mainly that needed to align the liquid crystals). However, because of the inertia of the molecules, the response time of liquid crystals is rather slow (of the order of several milliseconds). This is one of the main handicaps in the use of LCDs.

Another shortcoming of the LCD is its spatial resolution. The transition between a light pattern and a dark pattern takes place over several molecules and does not allow for high-resolution imaging. An improvement is found by using a **super-twist structure** that has a greater angle of rotation in the liquid crystals (270° instead of 90°). Liquid crystals will exhibit colors through the attachment of suitable dye molecules. These dye molecules are selected to filter out a part of the incoming light during the cholesteric phase (transmission) and become ineffective in the nematic phase (when the image turns dark).

**Fig. 4.39**
Seven-segment
alphanumeric LCD pattern

Metallization

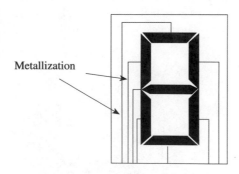

**Example 4.27**   What is the minimum number of layers of liquid crystals required to give a 90° twist if the maximum angle between the layers is 12°?

*Solution:*   The number of layers required is $90/12 = 7.5$, or 8 layers.   §

## Display Drivers

Most large-scale pixel-type displays require an efficient addressing mechanism, and drivers for the display elements are essential. This is especially true with LCD displays, where the display resolution is only moderately good. One highly suitable device for the making of drivers is thin-film transistors. Semiconductors such as CdSe and α-Si have been used to make thin-film transistors; α-Si has the advantage that the processing temperature can be less than 300 °C. Fig. 4.40 shows the cross-sectional view of an inverted α-Si FET used as a display driver. The gate is made of Cr deposited on the glass substrate. $Si_3N_4$ is then deposited as the insulator. This is followed by the deposition of a layer of α-Si, which forms the channel, the drain, and the source. Al contacts are deposited on the drain region and the source region by evaporation. Direct X-Y address is used, and when the pixel is illuminated, the FET will charge up a storage capacitor, which maintains a fixed bias voltage across the LCD. The turn-on time is set at 63.5 μs for standard TV display. The addressing sequence is repeated until all the pixels are covered. One major requirement for thin-film transistors is that the on-resistance must be small enough to adequately charge up the storage capacitor. Normally, the on-resistance must be less than 6 MΩ; the off-resistance must be at least 300 times larger so that charge can be retained within the refresh time in each cycle.

**Example 4.28**   A thin film transistor has the following geometry: channel length $L = 40$ μm, channel width $W_c = 500$ μm, and channel thickness $t_c = 300$ nm. If the channel mobility is $\mu = 1 \times 10^{-4}$ m²/V·s and the carrier density, $n$, is $1 \times 10^{25}$/m³, compute the channel resistance.

*Solution:*   The channel resistance is

$$R = L/(W_c t_c nq\mu)$$
$$= 40 \times 10^{-6} \text{ m}/(1 \times 10^{25}/\text{m}^3 \times 1.6 \times 10^{-19} \text{ C} \times 1 \times 10^{-4} \text{ m}^2/\text{V·s} \times 500 \times 10^{-6} \text{ m} \times 300 \times 10^{-9} \text{ m})$$
$$= 1.67 \text{ k}\Omega \quad §$$

Fig. 4.40
Inverted α-Si FET

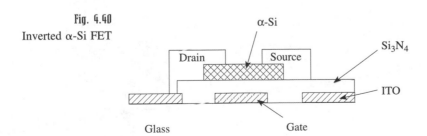

## 4.8  POWER GENERATION

**Photovoltaic diodes,** or **solar cells,** are used to harness electricity from sunlight. These devices are *p-n* junction diodes or MIS (metal-insulator-semiconductor) diodes (see Chapter 3); the solar energy is transformed into electrical energy via the generation of electrons and holes, as shown in Fig. 4.41.

In a *p-n* junction and a MIS diode, photoexcitation occurs in the depletion region, and the electrons and holes are separated by the internal electric field. Because of the direction of the internal electric field, the electrons move to the *n*-side and the holes move to the *p*-side (in a MIS diode, if the semiconductor is *p*-type, the holes will move to the semiconductor and the electrons will move to the metal electrode). As long as the light is on, the carrier densities in the *n*-side and the *p*-side remain above the equilibrium values, which gives rise to an **open-circuit voltage,** $V_{oc}$, across the device (the open-circuit condition implies that no load is placed across the device). If, however, a load is attached to the device, a current will flow into the load. Normally, $V_{oc}$ is less than the energy gap of the semiconductor. Typical values of $V_{oc}$ are in the neighborhood of 0.5 to 1 V (compared with an energy of 1.1 to 2.3 eV). Photovoltaic diodes have a limited power output, due to the nature of the solar spectrum and its low power density ($\approx 1$ kW/m$^2$). Very large panels are needed to provide a comparatively small amount of energy. Nevertheless, the energy is renewable and has a minimum environmental effect. Photovoltaic **conversion efficiency** is measured by the electrical power output divided by the input solar power. The latter can be at either air mass 0 (1353 W/m$^2$), air mass 1 (925 W/m$^2$), or air mass 2 (691 W/m$^2$), depending on the location on Earth where the solar power is measured. In addition, a fill factor, FF, related to the loss within the photovoltaic diode should be added. Thus, the conversion efficiency, $\eta_{\mathrm{conv}}$, is defined as

$$\eta_{\mathrm{conv}} = \mathrm{FF} \cdot V_{oc} I_{sc} / P_{\mathrm{in}} \tag{43}$$

where    FF is the fill factor (the fraction of usable output power),

$V_{oc}$ is the open-circuit voltage in V,

$I_{sc}$ is the short-circuit current in A, and

$P_{\mathrm{in}}$ is the input power in W.

**Fig. 4.41**
Carrier generation in a
*p-n* junction

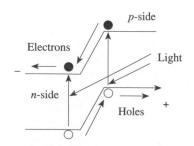

The best measured power-conversion efficiency for a conventional solar cell is in the neighborhood of 25%, or, effectively, an output of about 25 W/m$^2$ (nonconventional devices operating under special conditions can have a higher efficiency). The total amount of light received by a solar panel depends on the active area and on the absorption efficiency of the diodes. This effectively restricts the amount of metallization that is allowed to be placed on the solar cells. Even with a metal-film thickness of 30 to 50 nm, at least 50% of the sunlight is reflected. Finger electrodes are therefore used; these can reduce the reflection loss to less than 5%. An alternative method is to use transparent electrodes. Heavily doped SnO$_2$, In$_2$O$_3$, and ITO can achieve a resistivity as low as $1 \times 10^{-4}$ $\Omega \cdot$m and a transmittance of 80 to 95%. Although widely used, these transparent electrodes are somehow sensitive to preparation conditions. In order to minimize the reflective loss, **antireflection coatings** are deposited on top of the solar cells. For a semiconductor with an index of refraction $\eta_s$, the optimal index of refraction for the coating is $\sqrt{\eta_s}$. The coating thickness should also be one-fourth of the wavelength of the incident light. For instance, a Si solar cell with $\eta_{Si} = 3.5$ requires an ideal antireflection coating with $\eta_s = 1.87$. Assuming sunlight has a peak wavelength of 0.6 μm, the antireflection coating should then have a thickness of around 140 nm. Table 4.3 lists the values of the index of refraction, $\eta_s$, for common antireflection coatings. In addition to antireflection coatings, surface protection from cosmic particles can be achieved using a polyimide layer and a cover slide.

The cost of fabrication is one of the most important factors if photovoltaic diodes are to be economically competitive with conventional power sources. In principle, the ideal semiconductor material required for a good photovoltaic diode should have a direct energy gap in the neighborhood of 1.2 to 1.8 eV (the energy where the solar spectrum peaks is around 1.4 eV) and an extended junction region to increase the area for light absorption (this of course can increase the internal resistance and reduce the fill factor). A number of different materials have been seriously considered, including crystalline and noncrystalline semiconductors. Single-crystal Si is widely used because of its low cost (both material cost and processing cost), although GaAs, CdS, and InP single crystals are also good contenders. α-Si prepared by glow discharge provides a low-cost alternative; the peak efficiency is in the neighborhood of 7% to 10%. Very large amorphous solar-cell panels have been reported with an area up to 0.3 m$^2$. Fig. 4.42 shows the cross section of a typical α-Si solar cell. As

**Table 4.3**
Values of the index of refraction for the common antireflection coatings

| Antireflection Material | Index of Refraction |
| --- | --- |
| Ta$_2$O$_5$ | 2.2 |
| SiO$_2$ | 1.4 |
| SiO | 1.9 |
| SnO$_2$ | 1.9–2.1 |
| Si$_3$N$_4$ | 2.0 |
| In$_2$O$_3$ | 2.1 |
| Al$_2$O$_3$ | 1.7 |
| TiO$_2$ | 2.6 |

Fig. 4.42
Cross section of an α-Si
solar cell

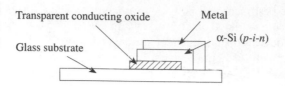

in other solar cells, the series resistance must be kept low to minimize internal loss. In some specific applications where a high output voltage (2 V or more) is required, several solar cells can be connected in series to give the desired output voltage (in this case, the internal resistance of the individual device must be kept very low).

Other techniques to improve the efficiency include the use of concentrated light from reflectors and the stacking of several thin-film solar cells together to form a **tandem cell.** In GaAs solar cells built with light reflectors to increase the solar intensity, a conversion efficiency above 30% has been measured. Stacking several solar cells with different absorption peaks together has also resulted in a higher conversion efficiency. In this case, a much larger portion of the solar spectrum is utilized. Such tandem cells normally would require the **short-circuit current,** $I_{sc}$, in each of the cells to be roughly the same. Direct current generated from a solar panel is sometimes a problem, since the majority of the conventional power devices, such as water pumps and dryers, operate using an ac current. Dc-to-ac current conversion is often required if the solar cell is used to supplement another power source. Up to now, terrestrial applications of photovoltaic cells have been restricted mainly to remote areas where other sources of power are not readily available.

**Example 4.29**      If 10% of the active area of a square solar cell is covered by Al finger electrodes, determine the series resistance of the solar cell if there are 20 fingers connected in parallel and the cell area is 20 cm × 20 cm. The resistivity of Al is $0.27 \times 10^{-7}$ $\Omega \cdot m$, and the Al thickness is 200 μm.

***Solution:***    The dimensions of the fingers are 0.1 cm × 200 μm × 20 cm. The resistance associated with each finger $R_f = \rho L/A_{cs} = 0.27 \times 10^{-7}$ $\Omega \cdot m$ × 0.2 m/(0.001 m × $200 \times 10^{-6}$ m) = 2.7 $\Omega$. Since the fingers are connected in parallel, the total series resistance $R = R_f/20 = 2.7 \, \Omega/20 = 0.13 \, \Omega$.   §

**Example 4.30**      For air mass 1, compute the conversion efficiency of a solar cell with an area of 1 cm², $V_{oc} = 0.92$ V, and $I_{sc} = 18$ mA/cm². The fill factor, FF, is 0.7.

***Solution:***    Based on Eq. (43), $\eta_{conv} = FF \cdot V_{oc} I_{sc}/P_{in} = 0.7 \times 0.92$ V × 18 × $10^{-3}$ A/cm²/0.925 W = 0.125, or 12.5%.   §

**Example 4.31**      Compute the index of refraction of α-Si if the reflectance between $Al_2O_3$ and α-Si is 0.7 and $\eta_{Al_2O_3} = 1.7$. (*Hint:* Replace $\eta_s$ by $\eta_2/\eta_1$ in Eq. (14).)

*Solution:*   Based on Eq. (14), replacing $\eta_s$ by $\eta_2/\eta_1$,

$$R^* = (\eta_2 - \eta_1)^2/(\eta_2 + \eta_1)^2$$

Or,

$$\begin{aligned}
\eta_2 &= \eta_1[(1 + R^{*2})/(1 - R^{*2})] \\
&= 1.7 \times [1 + (0.7)^2]/[1 - (0.7)^2] \\
&= 1.7 \times 1.47/0.51 = 4.9 \quad \S
\end{aligned}$$

# 4.9   SIGNAL-TRANSMISSION MEDIA

## Guided-Lightwave Transmission

Light transmission in a confined medium is becoming increasingly important in view of the very high data transmission rate that can be achieved. To transmit a light signal, a low-cost, low-loss medium is required. These criteria are met in silica fibers, more often called **optical fibers.** Optical fibers were first looked at seriously when it was discovered that the signal-propagation loss over a kilometer of fiber can be as low as 20 dB (a factor of 10), a reduction of ten times over conventional transmission media, such as Cu cables. This potential poses the possibility of replacing Cu cables in a large scale. Silica fibers are mostly silica doped with $GeO_2$, although dopants such as $B_2O_3$ and $P_2O_5$ are also used. In the simplest form, an optical fiber consists of an inner glass core and an outer cladding with a lower index of refraction, as shown in Fig. 4.43. The optical signals are transmitted through the core of the fiber, and the graded index of refraction provides a form of light confinement. Optical fibers are drawn from silica preforms, which are prepared by CVD on a fused-silica tube. The addition of dopants to the fiber increases the index of refraction, except for B and F, which have the opposite effect. Light propagation through an optical fiber is determined by the modes of propagation. A single-mode

**Fig. 4.43**
Cross section of a graded optical fiber

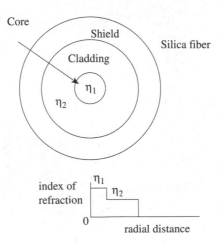

fiber will support only one waveform, whereas a multiple-mode fiber will support many (hundreds to thousands) waveforms.

Loss in optical fibers is usually the primary concern, and it can be related to the presence of heavy metal ions, such as $Fe^{2+}$, $Cu^{2+}$, $Cr^{3+}$, and $V^{3+}$ and other hydroxyl ($OH^-$) groups. These impurities generate strong absorption peaks at frequencies of 0.95 μm, 1.25 μm, and 1.38 μm, as listed in Table 4.4.

Low-absorption windows are also present at frequencies of 0.85 μm, 1.1 μm, and 1.5 μm. More recent development in fiber technology has shown that fibers can be produced with a loss less than 0.5 dB/km (6% loss/km) for a diameter around 10 μm, which occurs at the window near 1.5 μm. A second form of loss in optical fibers is the **Rayleigh scattering,** which is due to the microscopic fluctuations in the index of refraction of the fibers. Rayleigh scattering decreases proportionally to the fourth power of the wavelength. Most fibers do not transmit light of different wavelengths uniformly; this phenomenon is called **dispersion.** Due to dispersion, light at nearby wavelengths will travel at slightly different speed and can distort the output waveform. Such a problem is, however, minimized at the transmission wavelength of 1.3 μm, when dispersion in the fiber is at a minimum. The ideal low-loss fiber is supposed to have a loss of no more than 0.001 dB/km at the wavelength of 2.5 μm.

In addition to the above-mentioned problems with materials, light transmission in optical fibers also requires a suitable high-intensity light source and a sensitive light detector. The search for a suitable light source has extended the research of laser materials to the ternary and quaternary compound semiconductors. Compound semiconductors of the form ABCD, where A and B are the Group III elements and C and D are the Group V elements, have been explored extensively. The more popular systems include the GaAs/GaP system and the InSb/GaSb system. In the case of the GaAs/GaP system, there is a gradual transition from an energy gap of 1.43 eV for GaAs to an energy gap of 2.26 eV for GaP. The ternary compound also changes from direct gap to indirect gap at a composition of roughly 45% of GaAs. The lattice mismatch between the quaternary semiconductors and their substrates is another important factor governing the nonradiative recombination process. Misfit

**Table 4.4**

Absorption peaks due to metal ions present in optical fibers

| Ion | Peak (μm) |
|-----|-----------|
| $Cu^{2+}$ | 0.8 |
| $Fe^{2+}$ | 1.1 |
| $Ni^{2+}$ | 0.65 |
| $V^{3+}$ | 0.48 |
| $Cr^{3+}$ | 0.68 |
| $Mn^{3+}$ | 0.5 |

(*Source:* D. Marcuse, *Principles of Optical Fiber Measurement.* (New York: Academic Press, 1981). Reprinted by permission of AT&T Bell Laboratories Inc.)

layers are frequently limited to a thickness of 1 to 2 $\mu$m to minimize interface states. The more common Group III–Group V substrates are GaAs, InP, InAs, and GaSb. GaInAsP has a good lattice matching with InP substrates and will luminesce effectively in the range of 0.9 to 1.7 $\mu$m. Similar effects are observed in GaAs substrates, with an emission peak near 0.6 to 0.7 $\mu$m. Semiconductor lasers with an output wavelength between 1.3 and 1.5 $\mu$m have been developed using a InP/InGaAsP double heterostructure and can provide an output power of 1 to several milliwatts. The fibers are normally designed for use at a transmission speed of more than $10^9$ bits/s. To complement these light sources, solid-state detectors used in low-noise systems can have a signal-to-noise ratio of better than 40 dB (a factor of 100) at a transmission rate of 1 GHz. The noise level in these high-performance detectors still remains a key problem. Finally, optical amplifiers have also been invented for use in repeaters, and these can overcome some of the noise problems associated with long-distance data transmission. These amplifiers are fibers doped with Er ions (EDFA) and have been found to be capable of amplifying light signals through pumping with a high-power laser diode. An amplifier gain of the order of 30 dB (60 times) or better has already been achieved, and further improvements are expected.

**Example 4.32**   Compute the absorption coefficient $\alpha^*$ in optical fibers with a loss of 6% per kilometer.

*Solution:*   Based on Eq. (18), $\phi_L(x) = \phi_{L0} \exp(-\alpha^* x)$, or $\alpha^* = \ln[\phi_{L0}/\phi_L(x)]/x = \ln[1/(1 - 0.6)]/1000 \text{ m} = 6.19 \times 10^{-5}/\text{m}$.   **§**

**Example 4.33**   Compute the difference in the index of refraction of optical fibers that causes a light dispersion of 0.01 rad.

*Solution:*   Based on Eq. (13), $\eta_1/\eta_2 = \sin(\theta + 0.01 \text{ rad})/\sin(\theta) = 0.99991$. This implies $\Delta\eta_s \approx 1.07 \times 10^{-4}$.   **§**

**Example 4.34**   Compute the energy loss due to Rayleigh scattering at a wavelength of 1.4 $\mu$m if the loss of a fiber is 5 dB/km at 0.7 $\mu$m. Rayleigh scattering generates a loss that varies inversely with the fourth power of the wavelength.

*Solution:*   The energy loss due to Rayleigh scattering is 5 dB/km $\times$ $(0.7 \times 10^{-6} \text{ m}/1.4 \times 10^{-6} \text{ m})^4 = 0.31$ dB/km.   **§**

## Holography

**Holography** is not a solid-state phenomenon. Nevertheless, the development of a **hologram** requires the use of a laser (often a solid-state laser), and the image from a hologram is a powerful tool for studying solid surfaces and the dynamics of fine features on these surfaces. The idea of using holograms to amplify a light signal was first developed by D. Gabor, although the proper light source for its use was not available at that time. The idea was to use an X-ray laser source to produce a hologram, a photographic recording of the amplitudes and the phase of the reflected wavefronts from an object. When the image is reproduced by another coherent light of a longer wavelength, the resolution of the reconstructed image is magnified by the ratio of the long-wavelength light to the wavelength of the X ray. Although the concept is perfectly workable, so far an X-ray laser has not been fully developed. Holography, on the other hand, has found applications in other areas.

An ordinary photograph captures the light amplitudes reflected from an object but discards the phase information. With a hologram, both the light amplitudes and the phase are recorded. Holography involves two major steps:

1. The recording of an interference pattern produced by the superposition of two coherent light beams, with one of them reflected from an object and the other directly projected onto the hologram. A simple setup for the recording of a hologram is shown in Fig. 4.44. The interference pattern is then captured on a photographic film.
2. The reconstruction of the image from the interference pattern. Reconstruction is done by illuminating the hologram with a coherent light source identical to the one that produces the hologram. When done properly, the reconstructed image will reproduce the object in full depth and in three dimensions.

As shown in Fig. 4.44, other than the optics involved, the key component of the holographic setup is the laser source. During recording, the laser beam is split into an object beam and a reference beam. The object beam is directed at the object, giving rise to a reflected beam, which contains information on its image. The hologram is the interference pattern of the reference beam and the reflected beam.

To illustrate this concept, Fig. 4.45 shows the interference pattern arising from a point source, which—when superimposed with the pattern from the other points on the object—gives the full holographic image of the object. The hologram as it is recorded has a number of fuzzy fringes, and the image of the object is not recognizable. The reconstruction of the image is shown in Fig. 4.46. Two images appear, one a virtual image at the original location of the object and the other a real image refracted from the hologram in a different direction. Figure 4.47 shows the reconstruction of a point source through a hologram. Again, when superimposed

**Fig. 4.44**
A setup for recording a hologram

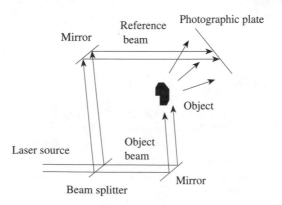

Fig. 4.45
Hologram from a point
source

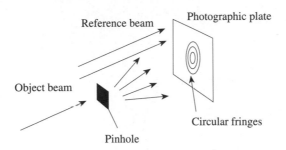

Fig. 4.46
Reconstruction of a
hologram

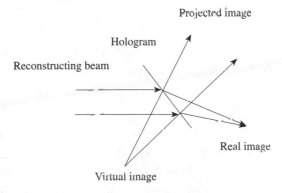

Fig. 4.47
Images created for a point
source

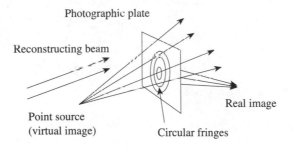

with holograms from other points in the object, a three-dimensional image is reconstructed.

A simple treatment of the interference pattern of a hologram will reveal how amplitude and phase information are recorded and restored. We shall assume a thin, transparent object, as shown in Fig. 4.48. The object beam (field strength) and the reference beam field strength can be represented by

$$S_b(x, y) = S_0 \exp(j(\omega t + 2\pi z/\lambda_L + \phi(x, y))) \qquad \text{(44)}$$

$$R_b(x, y) = R_0 \exp(j(\omega t + 2\pi x \cos(\theta_i)/\lambda_L + 2\pi z \sin(\theta_i)/\lambda_L) \qquad \text{(45)}$$

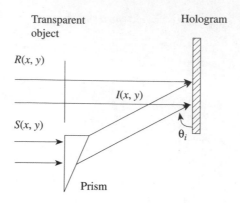

Fig. 4.48
Recording of a hologram

where    $S_0$ and $R_0$ are constants and are position dependent,

  $j$ is the imaginary number $\sqrt{-1}$,

  $\omega$ is the angular frequency of the light beam in rad/s,

  $z$ is the direction of propagation in m,

  $\theta_i$ is the angle of incidence of the object beam on the hologram in rad,

  $\phi$ is the phase angle in rad,

  $\lambda_L$ is the wavelength of the light beam in m, and

  $x$ and $y$ are the coordinates in the transverse plane in m.

The interference beam (intensity), $I_0(x, y)$, recorded on the hologram will simply be the magnitude $|S_b(x, y) + R_b(x, y)|^2$. Averaged over a period of time and for $z = 0$, $I_0(x, y)$ becomes

$$I_0(x, y) = R_0^2 + S_0^2 + 2R_0 S_0 (\cos(\alpha_h x) - \phi(x, y)) \tag{46}$$

where    $\alpha_h = 2\pi \cos(\theta_i)/\lambda_L$.

The first two terms of $I_0(x, y)$ contain the amplitude information of the beams, whereas the last term contains both the amplitude and the phase information. When $I_0(x, y)$ is used to develop a photographic plate (hologram), this information is recorded on the hologram.

Image retrieval requires a laser beam falling on the hologram. Let us assume that the conversion efficiency from $I_0(x, y)$ to the reconstructed image is $\eta_{conv}^*$; then the output light beam (intensity), $I_0'(x, y)$, is given by

$$I_0'(x, y) = \eta_{conv}^* R_b(x, y)(S_0^2 + R_0^2) + \eta_{conv}^* R_0^2 S_0 \exp(j\phi(x, y)) \tag{47}$$
$$+ \eta_{conv}^* R_0^2 S_0 \exp[j(2\alpha_h x - \phi(x, y))]$$

The first term of $I_0'(x, y)$ is directly proportional to the reconstructing beam $R_b(x, y)$ and represents direct transmission from the incident beam. The second term corresponds to an emerging (virtual) image and contains both the amplitude and the phase information of the object. The last term has a negative phase and is a converging beam capable of forming a real image. We have so far considered only a thin hologram. Thicker holograms are often used, and they are superior in terms of light-transformation efficiency; i.e., a great portion of the incoming light is used to produce the image.

There are numerous possible applications for holograms. One common application is the holographic interferometer, where objects having small vibrations or movements can be detected on a hologram. The movement changes the reflected light and produces fringe patterns in the reconstructed image. This application has been used to study the wall movement of magnetic domains. The resolution of a holographic interferometer can be close to the wavelength of the reconstructing light beam, which allows holography to be used in microscopy, whereby the image resolution can be improved using different lasers for the incident beam and the reconstructing beam. Optical Fourier transform performed on the reflected beam produces Fourier holograms, in which signal-processing techniques can be used to realize pattern recognition. A hologram storing a bit pattern has also been used as a computer memory. The data can be retrieved through reconstruction of the holographic image.

---

## GLOSSARY

| | |
|---|---|
| **light receptor** | an organ in the eye that responds to light. |
| **visible light** | light visible to the human eye. |
| **microwave radiation** | radiation with a wavelength of the order of $10^6$–$10^8$ nm. |
| **infrared light** | radiation with a wavelength between $10^3$ and $10^6$ nm. |
| **ultraviolet light** | radiation with a wavelength between 1 and $10^3$ nm. |
| **X rays and gamma rays** | radiation with very short wavelengths. |
| **photon** | a light particle. |
| **light intensity** | light energy, or power, delivered per unit area. |
| **light flux** | the number of photons passing through a unit cross-sectional area in unit time. |
| **wavelength** | the distance between two consecutive peaks of a sinusoidal waveform. |
| **permittivity** | a measure of polarization in a solid. |
| **index of refraction** | of a solid, the ratio of the light velocity in air over the light velocity in the solid. |
| **light refraction** | the change in the light path as it passes from one medium into another. |
| **Fermat's principle** | a principle relating the change in the light path as light travels from air into a solid to the angle of refraction. |
| **Snell's law** | a law relating the angle of incidence and the angle of refraction as light travels from one medium to another to the refractive indices. |
| **light transmission** | light passage through a transmitting medium. |
| **reflection** | at an interface, the rejection of a portion of the incoming light at the interface. |
| **reflectance** | the ratio of the reflected light energy over the incident light energy. |
| **Fresnel's formula** | an expression of the reflectance at an interface to the indices of refraction. |

| | |
|---|---|
| **brightness** | light intensity per unit area projected normal to the light path. |
| **absorption** | the process whereby light energy is used to promote an electron or lattice vibrations to a higher energy state. |
| **vibrational state** | the energy and momentum associated with the lattice. |
| **absorption coefficient** | the fractional light energy absorbed by a solid per unit distance along the light path. |
| **oscillator strength** | a measure of the coupling between two different energy states during light absorption. |
| **Restrahlen absorption** | light absorption that enhances lattice vibrations. |
| **photoemission** | the process by which light absorption in a solid results in the emission of electrons. |
| **work function** | the energy required for an electron to overcome the surface potential barrier of a solid. |
| **photoconductivity** | the increase in conductivity due to light absorption in a solid. |
| **exciton** | a weakly bound electron-hole pair in a semiconductor with a binding energy much less than the value of the energy gap. |
| **Franz-Keldysh effect** | the change in the absorption properties of a semiconductor in the presence of a strong electric field. |
| **Stark effect** | an effect involving a shift in the energy of an electronic state, such as an exciton state in a semiconductor, in the presence of an electric field. |
| **self-electro-optic device** | a device whose operation depends on the bistable absorption property in the presence of the Stark effect. |
| **extrinsic absorption** | light absorption in the presence of impurities in a semiconductor. |
| **band-to-band absorption** | the promotion of an electron from the valence band to the conduction band due to light absorption. |
| **sensitizing center** | an impurity whose presence will affect the lifetime of minority carriers in a semiconductor. |
| **blackbody radiation** | light emitted from a solid when the solid is at a temperature above that of its surroundings. |
| **spontaneous emission** | light emission in a solid that occurs naturally after an excitation. |
| **radiative emission** | an emission process that results in light generation. |
| **nonradiative emission** | an emission process that results in heat generation. |
| **direct transition** | an excitation or relaxation process in a solid that does not involve momentum change. |
| **radiative lifetime** | the lifetime of a minority carrier involved in a light-emission process. |
| **indirect transition** | an excitation or relaxation process in a solid that requires an exchange of momentum. |
| **direct gap semiconductor** | a semiconductor whose excitation and relaxation processes are dominated by direct transitions. |
| **indirect gap semiconductor** | a semiconductor whose excitation and relaxation processes are dominated by indirect transitions. |
| **recombination center** | an impurity that enhances the recombination process in a semiconductor. |
| **capture cross section** | the effective area for an impurity to capture an electron or a hole in a semiconductor. |
| **isoelectronic impurity** | an impurity having the same valence as the host semiconductor. |
| **Auger recombination** | the recombination process in a solid in which the excess energy is transferred from one carrier to another, with a small amount of heat generated in each step. |
| **luminescence** | light emission that occurs naturally. |
| **photoluminescence** | light emission after a solid has been illuminated with energetic light. |
| **electroluminescence (or cathodoluminescence)** | light emission in a solid due to electron injection and recombination. |

| | |
|---|---|
| **fluorescence** | light emission that takes place concurrently with the light-absorption process. |
| **phosphorescence** | light emission that persists for a long time after the excitation has terminated. |
| **bioluminiscence** | light emission in living organisms. |
| **activator** | an impurity that assists in the photoluminiscence process. |
| **coactivator** | an impurity that helps to maintain charge neutrality in a semiconductor. |
| **junction luminescence** | a form of electroluminescence in which light emission is due to carrier injection and recombination in a *p-n* junction. |
| **light-emitting diode** | a device whose operation depends on junction luminescence. |
| **bandtailing** | the interactions between the energy states in a heavily doped semiconductor, which result in the extension of the band edge into the energy gap. |
| **stimulated emission** | light emission in a solid in which an unstable condition (such as population inversion) is created and the emission process is initiated by an external source. |
| **population inversion** | the unstable condition when there are more electrons in the upper energy level in a solid than in the lower energy level. |
| **laser** | a device whose operation depends on stimulated light emission. |
| **Boltzmann statistics** | the statistical energy distribution of electrons at a finite temperature. |
| **pumping** | the mechanism applied in the excitation of electrons from the lower energy level to the higher energy level using an external energy source. |
| **resonant cavity** | a region of exact dimensions in which electromagnetic waves (energy) are able to oscillate back and forth with little or no losses. |
| **laser diode** | a laser built into a *p-n* junction. |
| **threshold current density** | the minimum current density needed in a laser to initiate stimulated emission. |
| **light confinement** | the use of a semiconductor with a lower index of refraction near the junction region of a laser diode to enhance light reflection at the interface. |
| **double heterostructure** | a graded structure built with epitaxial layers to increase light confinement in a laser diode. |
| **pulse mode** | the operation of a laser driven by a single-step current. |
| **continuous mode** | the operation of a laser driven by a sinusoidal current. |
| **solid-state imager** | a device that captures an image using an array of solid-state light sensors. |
| **photodiode** | a light-sensitive *p-n* junction designed so that its output current is proportional to the light input. |
| **phototransistor** | a light-sensitive transistor designed so that its output current is proportional to the light input and so that it also retains its gain characteristics. |
| **charge-coupled device imager** | an imager consisting of an array of light-sensitive pixels and a set of clock pulses that can facilitate the transfer of the photoexcited charges in each of the pixels to the output. |
| **electrophotography** | refers to the process whereby an image is captured by a light-sensitive thin film and then transferred to a piece of paper using an electrostatically charged toner. |
| **dark current** | the current due to thermally generated electrons and holes. |
| **photodetector** | a device capable of detecting the presence of light. |
| **transit time** | the time needed for an electron to travel across the length of a photoconductor. |
| **bandwidth** | the range of light energy that can induce a response from a photoconductor. |

| | |
|---|---|
| **avalanche photodiode** | a photodiode that operates at a large reverse-bias voltage, so carrier multiplication occurs within the depletion region leading to an internal gain in the device. |
| **noise-equivalent power** | the ideal input power at which the photocurrent is exactly equal to the noise current, corresponding to the minimum detectable signal level. |
| **detectivity** | a parameter used to compare the performance of different photodetectors. |
| **vidicon** | an imager built with photodiodes and phototransistors. |
| **MOS capacitor** | a device made of a metal-oxide-semiconductor structure capable of storing photoexcited charges after illumination. |
| **pixel** | a set of electrodes working together to make up a single point in the image. |
| **cathode-ray tube** | a device that displays an electric signal on a phosphorescence screen. |
| **electron gun** | a metal electrode that acts as a source of electrons through thermal emission. |
| **persistence** | prolonged luminescence in a solid. |
| **shadow mask** | a part of a color television screen consisting of a metal sheet with a large number of drilled holes, each aligned to a set of three different phosphor dots deposited on the screen. |
| **photocrystallization** | crystal regrowth in the presence of an intense light. |
| **silicon bubble** | a void generated in an $\alpha$-Si:H thin film due to the effusion of $H_2$ by laser heating; a bubble has a reflectivity of only a few percent (compared with 40% in $\alpha$-Si). |
| **liquid crystals** | a suspension of rodlike organic molecules that have different structural arrangements under different voltages. |
| **nematic phase** | a phase of liquid crystals that occurs when the molecules are all parallel to each other but are otherwise randomly placed. |
| **cholesteric phase** | a phase of liquid crystals that occurs when the molecules are arranged in planes, with the molecules in each plane parallel to one another but in successive planes, twisted with a well-defined pitch. |
| **smectic phase** | a phase of liquid crystals that occurs when the molecules are all lined up parallel to each other and arranged in layers. |
| **super-twist structure** | liquid crystals that have a greater angle of rotation (270° instead of 90°). |
| **photovoltaic diode,** or **solar cell** | a $p$-$n$ junction or MIS diode that converts solar energy into electrical energy. |
| **open-circuit voltage** | the voltage observed across a solar cell when the output is open-circuited (unconnected). |
| **conversion efficiency** | the ratio of the output power to the input power. |
| **antireflection coating** | a thin film deposited on top of a device to minimize light reflection at the air-device interface. |
| **tandem cell** | solar cells stacked one on top of another to increase the light-conversion efficiency. |
| **short-circuit current** | the current measured in a solar cell when the output is short-circuited (connected through a low resistance path). |
| **optical fiber** | a silica fiber with an inner glass core and an outer cladding of a lower index of refraction; used as a medium for light transmission. |
| **Rayleigh scattering** | a form of light scattering that decreases proportionally to the fourth power of the wavelength. |
| **dispersion** | the distortion of the light waveform in a medium due to the different speed of light at nearby wavelengths. |
| **holography** | a technique using coherent light to study objects and to enhance image resolution. |
| **hologram** | a photographic plate used to store the interference pattern produced by coherent light beams. |

**REFERENCES**   Jenkins, F. A, and H. E. White. *Fundamentals of Optics*. New York: McGraw-Hill Book Company, 1957.

Grovenor, C. R. M. *Electronic Materials*. Bristol and Philadelphia: Adam Hilger, 1989.

Madan, A. and M. Shaw. *The Physics and Applications of α-semiconductors*. New York: Academic Press, 1988.

Marcuse, D. *Principles of Optical Fiber Measurements*. New York: Academic Press, 1981.

Ready, J. F. *Industrial Applications of Laser*. New York: Academic Press, 1978.

Shur, M. *Physics of Semiconductor Devices*. Upper Saddle River, N. J.: Prentice-Hall, Inc., 1990.

Siegman, A. E. *An Introduction to Lasers and Masers*. New York: McGraw-Hill Book Company, 1968.

Sze, S. M. *Physics of Semiconductor Devices*, 2d ed. New York: John Wiley & Sons, Inc. 1981.

Wolfe C. M., N. Holonyak, Jr., and G. E. Stillman. *Physical Properties of Semiconductors*. Upper Saddle River, N. J.: Prentice-Hall, Inc., 1989.

**EXERCISES**   **4.2   Review of the Properties of Light**

1. What is meant by monochromatic light? Identify a light source that you feel should be close to monochromatic.
2. Identify the range of frequencies associated with the visible light. Suggest why two different colors when mixed together will form a third color to the human eye.
3. Verify that $hc \approx 1.24$ eV/μm (see Eq. (2)).
4. A device emits monochromatic green light (0.5 μm) at a rate of $1 \times 10^{16}$ photons/s. If the input to the device is 100 mW, suggest a value for the power-conversion efficiency.

**4.3   Light and Solid Interactions**

5. Give the equations for energy conservation and momentum conservation. Consider a hydrogen atom. Compute the energy needed to promote the electron from the ground state to the next-higher energy state, assuming energy conservation holds.
6. During light transmission from air into a solid, compute the light velocity and the angle of refraction inside the solid if the latter has an index of refraction of $\eta_s = 4$, and the angle of incidence is $\theta_i = 40°$. Also compute the critical angle.
7. Based on Eq. (14) and Table 4.1, what fraction of light (from air) will be reflected from the surface of a Si sample? Compare this with the value if light originates from water instead of air. Comment on the difference.
8. Verify that *brightness* as defined by Eq. (15) is indeed independent of the viewing angle.

**4.4   The Absorption Process**

9. Calculate the distance light has to travel within a solid if its intensity is to be reduced by a factor of 100, assuming an absorption coefficient $\alpha^* = 1 \times 10^{-4}$/cm.
10. What is the wavelength of light needed to cause electron emission from a Cu electrode?

11. Compute the fractional change in conductivity if intrinsic Si is illuminated to give a carrier density of $1 \times 10^{20}/m^3$ and $T = 300$ K. The intrinsic carrier density of Si at 300 K is $1.4 \times 10^{16}/m^3$. Assume no changes in the carrier mobilities.

12. From the following data, compute $A_{ab}$ and $n$ based on Eq. (20) in a given semiconductor. Given: $E_g = 0.7$ eV. The absorption coefficient is $\alpha^* = 5 \times 10^3/cm$ when $E_{photon} = 2$ eV, and $\alpha^* = 800/cm$ when $E_{photon} = 1$ eV.

13. Explain the Franz-Keldysh effect related to light absorption in semiconductors in the presence of a strong electric field.

14. The operation of a SEED device depends on the strength of the electric field in the intrinsic region of a Si $p$-$i$-$n$ diode. Assuming that the carrier generation lifetime is 1 ns, compute the light intensity needed to remove the intrinsic region (1 μm thick) entirely if the dopant densities on both sides of the junction are $1 \times 10^{22}/m^3$. The intrinsic carrier density of Si is $1.45 \times 10^{16}/m^3$.

## 4.5   The Emission Process

15. Using Eq. (23), estimate the radiative recombination lifetime, $\tau_r$, in a semiconductor at 300 K if the capture cross section is $1 \times 10^{-28}/cm^2$ and the density of the recombination centers is $1 \times 10^{17}/cm^3$. Assume $v_{th} = 1 \times 10^5$ cm/s.

16. Describe the various types of nonradiative recombination processes in semiconductors.

17. Describe the different types of radiative recombination processes in semiconductors.

18. What are the advantages of using ternary and quaternary compound semiconductors in light-emitting diodes? Name one example in which an impurity will improve radiative recombination in a semiconductor.

19. Describe a three-level system that can operate as a laser and explain what is meant by *pumping*. *Population inversion* is sometimes considered to be represented by a *negative-temperature effect*. Assume a three-level laser system has the same density of states in each of the three levels and the lower two levels are separated by an energy difference of 0.5 eV. Estimate the negative temperature if the middle energy level has one-tenth of the number of electrons in the ground level.

20. Using Eq. (31), compute the threshold current density, $J_T$, in a light-emitting diode if $n_0' = 1 \times 10^{21}/m^3$, $\tau_{eff} = 0.1$ μs, and $t_c = 1$ μm. What will be the photon density if $J = 10J_T$ and $\tau_p = 1$ ns?

21. Using Eq. (32), compute and plot $E_{trans}/E_{inc}$ for different values of the wavelength $\lambda_L$. Choose arbitrary parameters for the other constants. See if it has the same shape as Fig. 4.22.

## 4.6   Imagers

22. Explain briefly the steps used in electrophotography.

23. If the surface of a photosensitive Se thin film has dimensions of 20 cm × 28 cm × 0.25 cm (8 in. × 11 in. × 0.1 in.) and there is an electric field of $1 \times 10^6$ V/m across it, what will be the surface-charge density if the dielectric constant of Se is 10? What will be the peak-discharge current if the output load is 1 kΩ?

24. **(a)** Compute the photon-induced current in Si if the generation rate is $1 \times 10^{15}/m^3 \cdot s$ and the carrier lifetime is 10 μs. The electric field is 20 V/cm, and the cross-sectional area of the photoconductor is 3 mm × 3 mm. The length of the photoconductor is 1 mm.

**(b)** Compute the photoconductivity gain.

**(c)** What is the signal bandwidth for the photoconductor? The electron mobility for Si is 0.135 m²/V·s and for holes is 0.045 m²/V·s.

**25.** Describe how a charge-coupled device (CCD) operates.

**26.** What is the number of pixels per centimeter required in a two-phase CCD to give a spatial resolution of 20 μm if the device is to be used as a line scanner?

**27.** **(a)** If a CCD electrode has dimensions of 10 μm × 100 μm and the potential well to store charge is triangular in shape with a depth of 2 μm, compute the total number of electrons/holes that may be stored in the entire well if the substrate dopant density is $1 \times 10^{22}$/m³.

**(b)** Assuming that the carrier recombination lifetime is 100 ns, how long does it take for 90% of the charge to be drained away?

**28.** In a CCD, how long will it take the charge in a full well with the dimensions listed in Exercise 27 to be optically generated if the carrier generation rate is $1 \times 10^{23}$/m³·s?

**29.** What will be the series resistance of a single ITO electrode with dimensions of 10 μm × 50 μm × 20 μm? The resistivity of ITO is assumed to be $1 \times 10^{-6}$ Ω-m.

## 4.7 Displays

**30.** If the electrode of an electron gun is heated to a temperature of 1200 °C and the work function is 3 eV, what will be the electron current density emitted from the electrode? Assume that the thermionic emission current, $I_{th}$, is given by (see Chapter 3) $I_{th} = 120 \times A_{cs}T^2\exp(-\Phi_B/(kT))$ A, where $\Phi_B$ is the work function and $A_{cs}$ is the electrode surface area in square centimeters. $A_{cs} = 0.3$ cm × 0.3 cm.

**31.** Discuss the advantages of using a shadow mask in the design and construction of a television tube.

**32.** Explain the differences in the current-injection mechanisms between a rectifying contact and a p-n junction.

**33.** Name four different semiconductor materials that can produce red, yellow, green, and blue LEDs.

**34.** Suggest how ferroelectric solids may be used in light switching.

**35.** Explain the transformation cycle and the photocrystallization process in $Sb_2Se_3$, as shown in Fig. 4.35.

**36.** Describe the three common liquid-crystal structures.

**37.** Explain the operation of a seven-segment alphanumeric liquid-crystal display.

## 4.8 Power Generation

**38.** The solar spectrum gives a power density of 1 kW/m². For conversion efficiency of 20%, what is the collection area required to generate 1 kW?

**39.** What will be the minimum collection area needed to generate a current of 1 mA in a solar cell if the power conversion efficiency is 10%, the open-circuit voltage is $V_{oc} = 0.5$ V, and FF = 0.5?

**40.** Assuming $Ta_2O_5$ is used as an antireflection coating for a Si solar cell, what will be the optimal thickness, assuming the solar spectrum has an average wavelength of 500 nm? What is the series resistance if the solar cell has a surface area of 10 cm × 10 cm and a thickness of 100 μm? The electrode is comb-shaped, with fingers having widths of

10 $\mu$m and an interelectrode spacing of 90 $\mu$m. The resistivity of the surface electrode is $1 \times 10^{-6}$ $\Omega \cdot$m.

**41.** What are the advantages and disadvantages in using $\alpha$-Si in solar cells?

**42.** If we want to construct a tandem cell using CdS/CuS, GaAs, and Si thin films, suggest in what order we should deposit these thin films if illumination comes in from the upper surface.

**43.** A solar cell is modeled as a current source with a short-circuit current of 0.1 mA and an internal shunt resistance of 10 k$\Omega$. What is the short-circuit current if the resistive load is

(a) 10 k$\Omega$, and

(b) 100 k$\Omega$?

## 4.9 Signal-Transmission Media

**44.** Describe how optical fibers work in signal transmission. What are the advantages of using fibers over materials like Cu? If a fiber can achieve a maximum transmission loss of 0.5 dB/km, show that this corresponds to 6% loss per km. What will be the overall loss of the fiber in 100 km?

**45.** List the wavelengths of the transmission windows and the absorption peaks in conventional silica fibers.

**46.** A fiber transmits at a wavelength of 2.5 $\mu$m. Suggest how many channels can ideally be installed if the bandwidth of each channel is 250 MHz.

**47.** Briefly describe the operation of holographic imaging. How can it be used to achieve signal amplification?

**48.** If the hologram of a defect structure in a solid is generated using X rays with a wavelength of 0.5 nm, what will be the gain factor if the image is reproduced using green light with a wavelength 500 nm?

# CHAPTER 5

# Magnetic and Superconducting Properties of Solids

## 5.1   INTRODUCTION

Magnetism is one of the few areas in physics that has been studied for a long time, but it was not fully understood until quantum theory was applied. Nevertheless, as far as applications are concerned, magnetism is very important, and magnetic materials are widely used in different kinds of devices and systems. Like most other areas in solid-state physics, magnetism is closely associated with the electrons and their motion. Magnetism arises naturally from the magnetic dipoles present in a solid. These may be due to the circular motion of the orbiting electrons, the nuclear spins, or the electron spins. In fact, all solids have some form of magnetization (magnetic properties). Strong magnetization, however, is present only in elements and compounds of the transition metals and the rare-earth metals.

Magnetic solids can be classified as either weakly magnetic or strongly magnetic. Weakly magnetic solids are less useful in applications, but an understanding of their physics has provided important information on the structures of the atoms and their magnetic interactions. The study of paramagnetism and diamagnetism falls into this category. Strongly magnetic solids belong to a class of solids called *ferromagnets*. Their magnetization is mainly the result of spin interactions among the atoms and within the crystal as a whole. Ferromagnetism is found primarily in solids where domains of strong magnetization are present. In addition to elemental magnetic solids, magnetic alloys are also very important. The ability to selectively combine the properties of different magnetic solids in an alloy and maximize the magnetization is a very important area of study. The use of magnetic solids has also changed considerably over the years, and magnetic solids are no longer exclusively used in the making of permanent magnets and the cores of transformers. Information technology has opened up many new uses for magnetic materials. These include applications as the medium for data storage and in devices used for magnetic amplification, imaging, magneto-optics, and data processing.

Closely associated with magnetism is a phenomenon called *superconductivity*. Superconductivity occurs only at an extremely low temperature, when the electrons behave differently from the normal electrons. In the superconducting state, electrons move about collectively as a single entity, without energy loss or scattering by the lattice. A superconducting current will, therefore, flow indefinitely. Superconductivity exists in many metals and metallic compounds. Its presence is also affected by magnetic field, the current passing through it, and the temperature. The basic theory of superconductivity, BCS (Bardeen-Cooper-Schrieffer) theory, is reasonably well understood, and there are many practical devices made with superconductors. Widespread usage of conventional superconductors, however, has been impeded by the need for a liquid He-based cooling system and the expense of maintaining a low temperature close to absolute zero. Nevertheless, superconducting devices have demonstrated properties of extreme high sensitivity, high speed, and low noise, all of which are features essential to the development of high-speed electronics and telecommunication systems. The recent discovery of high-temperature (HT) superconductors is an important breakthrough in superconductivity and offers a strong incentive for further development in the emerging market of low-temperature electronics.

This chapter is divided into two parts. The first part is on magnetism in solids, its physical origin, the different kinds of magnetic solids, and their methods of preparation. In connection with these topics, we also examine magnetic phenomena, such as Zeeman splitting, exchange energy, magnetic domains, hysteresis, and magnetic anisotropy. A section is devoted to a study on magnetic devices. The second part of this chapter discusses superconductivity. We introduce the basic theory of superconductivity, the magnetic and temperature effects, Type I, Type II, and high-temperature (HT) superconductors and their physical properties. The last section studies superconducting devices and their applications.

## 5.2    MAGNETIC PROPERTIES OF SOLIDS

Magnetism is closely related to electricity, and the two make up a very diverse subject in physics. We shall focus only on static magnetic effects—i.e., **magnetization** that does not change with time. Static magnetic effects, unlike static electricity, are associated with **magnetic dipoles** and magnetic field rather than electric charges and electric field. Magnetic dipoles are sources of magnetization. In a solid, they are present in the atoms, the nuclear spins, the electron spins, and the circular motion

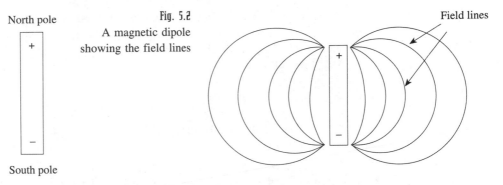

**Fig. 5.1**
A magnetic dipole

North pole

**Fig. 5.2**
A magnetic dipole
showing the field lines

Field lines

South pole

of the electrons. Magnetic dipoles give rise to **magnetic moments** and a magnetic field. The latter can be visualized as field lines emerging from the hypothetical poles of a magnetic dipole. Such a dipole representation of magnetic moment is illustrated in Fig. 5.1. The emerging magnetic field lines are called **magnetic flux,** and their density is measured in terms of a parameter called the **magnetic induction, B** in teslas (T), or webers/m$^2$ (Wb/m$^2$). The overall flux pattern emerging from a magnetic dipole looks somewhat like the groove pattern appearing on the shells of a bivalve. The magnetic field lines can be traced out using Fe filings, which fall automatically along the field lines. Figure 5.2 shows the field pattern in the neighborhood of a bar magnet. Although it is well known that human beings cannot sense magnetic field lines, there are suggestions that in some animals, biological organs do exist in the brain that are sensitive to magnetic field lines, such as those generated by the earth.

The simplest way to quantify magnetism is to associate the magnetic dipoles with their moments, $\mathbf{\mu}_m$ (A·m$^2$). In the presence of a magnetic induction **B,** a magnetic dipole will experience a torque, $\mathbf{t}_q$ (a turning force), which is given by

$$\mathbf{t}_q = \mathbf{\mu}_m \times \mathbf{B} \qquad \qquad (1)$$

where   $\mathbf{t}_q$ is in N·m.

The cross-product sign suggests that the torque will be in a direction perpendicular to the magnetic moment $\mathbf{\mu}_m$ and the magnetic induction **B,** as illustrated in Fig. 5.3.

**Fig. 5.3**
A torque acting on a
magnetic dipole

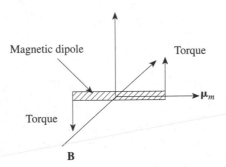

Magnetic dipole

Torque

Torque

$\mathbf{\mu}_m$

**B**

The effect of the torque is to align the magnetic dipole in the direction of **B**. The value of the magnetic moment depends on the nature of the dipole. For instance, a current $I$ (A) passing through a circular loop produces a dipole moment, $\boldsymbol{\mu}_m$ (A·m$^2$), which is given by

$$\boldsymbol{\mu}_m = IA_{\text{loop}} \tag{2}$$

where $A_{\text{loop}}$ is the area of the loop in m$^2$.

This type of magnetic effect is referred to as **electromagnetism.** The magnetic field lines in the electromagnet pass through the inside of the circular loop and return externally in a closed path. Magnetic moments can be generated in a current loop; they also originate from the self-gyration, or **spins,** of the electrons. In a solid, spin magnetic moment is most frequently measured in Bohr magneton $\boldsymbol{\mu}_\mathbf{B}$, which has a value of $9.27 \times 10^{-24}$ A·m$^2$. Physically, $\boldsymbol{\mu}_\mathbf{B}$ is the smallest difference in the orbital magnetic moment of an atom. Although $\boldsymbol{\mu}_\mathbf{B}$ is frequently used, the value of the magnetic moments of solids in $\boldsymbol{\mu}_\mathbf{B}$ are rarely integers (see Table 5.1). The potential energy of a magnetic dipole $\Phi_{md}$ (J) associated with a magnetic induction **B** is given by

$$\Phi_{md} = -\boldsymbol{\mu}_m\mathbf{B} \cos(\Theta) \tag{3}$$

where $\Theta$ is the angle between the directions of **B** and $\boldsymbol{\mu}_m$.

Here, once again, Eq. (3) suggests that the lowest energy of the dipole occurs when $\Theta = 0$, i.e., when the magnetic dipole is aligned with the magnetic induction.

Magnetization depends on the atomic structure, the placement of the electrons in the different energy orbits, and the crystal environment. In a solid, magnetism is due primarily to the electron spins and the circular motion of the electrons in the energy orbits. The latter gives rise to angular momentum magnetic moment because of the rotational motion. **Nuclear spin** magnetic moment is much weaker than electron spin magnetic moment and is not discussed here. In many ways, electron

Table 5.1    Electron spin magnetic moments in the atoms of the transition metals

| Solid | Spin Magnetic Moment ($\mu_B$) |
|---|---|
| Fe | 2.22 |
| Co | 1.72 |
| Ni | 0.61 |
| Gd | 7.10 |
| Tb | 9.34 |

(*Source:* C. Kittel, *Introduction to Solid State Physics,* 3d ed. © 1966 John Wiley & Sons, Inc. Reprinted by permission of John Wiley & Sons, Inc.)

spins are like small magnets with unique but interchangeable poles. The values of the spin magnetic moments in the transition metals are listed in Table 5.1.

Magnetism related to the orbiting electrons is quite different, because the electrons behave like small current loops. These current loops give rise to magnetic moments, which are weaker than the spin magnetic moments; their presence cannot be ignored, especially in those solids that have little or no spin magnetic moments. Magnetization due to the orbiting electrons is called **diamagnetism.** Diamagnetism exists in all solids and can be explained by the change in the orbital motion of the electrons in response to an applied magnetic field. Strong magnetization, on the other hand, is due to electron spins and their mutual interactions at the atomic level and at the interatomic level within the magnetic domains. **Magnetic domains** are found in the **transition metals,** their alloys, and some of the **rare-earth metals.** These solids have a different arrangement of the electrons in the energy orbits, which results in the parallel alignment of the electron spins. They are called ferromagnets. **Ferromagnetic solids** were discovered a long time ago and have been used as materials for making permanent magnets and compasses (a compass is used in the detection of the earth's magnetic poles during navigation or travel). An understanding of ferromagnetism requires the concept of **quantum mechanics,** which explains the mutual interactions between the electron spins. Quantum mechanics also explains the interactions between different magnetic domains and the forces involved in the formation of the domain wall.

## Magnetic Field

The magnetic field is linked to electron motion (we shall delay the study of electron spins for the time being). If we have a current passing through a circular loop, a magnetic dipole is created, as shown in Fig. 5.4.

According to Ampere's law, the **magnetic field intensity** $H$ (A/m) at the center of the loop is given by

$$H = I_{\text{loop}}/(2r_{\text{loop}}) \qquad (4)$$

where     $I_{\text{loop}}$ is the current through the loop in A, and

        $r_{\text{loop}}$ is the radius of the loop in m.

**Fig. 5.4**
A current loop forming a magnetic dipole

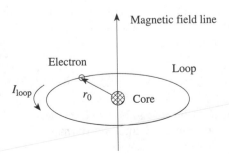

In addition to $H$, the magnetic field intensity, magnetic field lines (per unit area) are measured by the magnetic induction $\mathbf{B}$ (in T). $\mathbf{B}$ is related to $H$ by

$$\mathbf{B} = \mu H \tag{5}$$

where    $\mu$ is the permeability of the medium in H/m (henries/m).

Physically, $\mathbf{B}$ is a measure of the density of the magnetic field lines. In handling the magnetic field lines, it is often convenient to separate the free space (vacuum) contribution ($\mathbf{B} = \mu_0 H$), and the materials contribution ($\mathbf{B} = \mu_0 M$), i.e.,

$$\mathbf{B} = \mu_0(H + M) \tag{6}$$

where    $M$ is the magnetization in A/m, and

$\mu_0$ ($= 4\pi \times 10^{-7}$ H/m) is the vacuum **permeability.**

In terms of the field lines, Eq. (6) may be viewed as the algebraic sum of the contributions from the (external) magnetic field $H$ and the (internal) magnetic sources $M$. In many solids, $M$ is proportional to $H$, which gives

$$\mathbf{B} = \mu_0(1 + \chi_e)H \tag{7}$$

where    $\chi_e$ is the **magnetic susceptibility.**

Magnetization depends on the magnetic moments in the solid. The latter arise from both the orbiting electrons and from the electron spins. If the orbiting electrons are responsible, the magnetic susceptibility, $\chi_e$, of the solid is given by

$$\chi_e = Nn_a\Delta\boldsymbol{\mu}_m/H \tag{8}$$

where    $N$ is the atomic density in /m$^3$,

$n_a$ is the number of orbiting electrons in each atom,

$\Delta\boldsymbol{\mu}_m$ is the change in magnetic moment per atom in A·m$^2$, and

$H$ is the magnetic field intensity in A/m.

In a paramagnetic solid, the magnetic susceptibility is primarily due to the electron spins, and $\Delta\boldsymbol{\mu}_m$ in Eq. (8) is replaced by a term dependent on the spin magnetic moments. Spin paramagnetic moments vanish at a high temperature. The relative permeability, $\mu_r$, of a solid is defined as

$$\mu_r = 1 + \chi_e \tag{9}$$

For a ferromagnetic solid, $\mu_r$ can be much greater than 1. This implies that there are many more magnetic field lines inside the magnetic solid than outside, or, effectively, energy is stored in the solid in the form of magnetic energy. The relative

Table 5.2 Relative permeabilities of the different magnetic materials

| Material | Solid | $\mu_r$ |
|---|---|---|
| Paramagnetic | Al | 1.00002 |
| | Mn | 1.00098 |
| Diamagnetic | Cu | 1.00001 |
| | Au | 1.00004 |
| Magnetic | Fe | $2 \times 10^2$ |
| | Fe/Si | $1 \times 10^3$ |
| | Co/Fe/B/Si | $4 \times 10^6$ |
| | Ni/Fe alloy | $1 \times 10^5$ |

permeability is an important parameter in characterizing a magnetic solid. Table 5.2 lists the relative permeabilities of the different magnetic materials.

## Magnetostatic Energy

Theoretically, **magnetostatic energy** is the energy required to bring the magnetic dipoles together to form a solid. This is expressed in terms of $E_m$, the energy per unit volume (J/m$^3$). Over a unit volume, $E_m$ can be written as the sum of the energies of the individual dipoles:

$$E_m = \Sigma_i q_i \phi_{id}/2 \qquad (10)$$

where $\Sigma_i$ is summation over the $i$th magnetic dipoles per unit volume in /m$^3$,

$q_i$ is the dipole strength in C, and

$\phi_{id}$ is the potential at the location of the $i$th dipole in V.

It can also be shown that

$$\Sigma_i q_i \phi_{id} = \mu_0 \Sigma_i \mu_{mi} H_i \qquad (11)$$

where $\mu_{mi}$ is the magnetic moment of the $i$th dipole in A·m$^2$, and

$H_i$ is the magnetic field intensity at the location of the $i$th dipole in A/m.

Assuming that the average magnetic field intensity is $H_{av}$, the magnetic energy per unit volume of the solid is given by

$$E_m = \mu_0 M H_{av}/2 \qquad (12)$$

where $M (= \Sigma_i \mu_{mi})$ is the magnetization in A/m.

**Example 5.1** (a) Determine the magnetic field intensity at the center of a loop with radius 0.5 cm when a current of 1 mA passes through the loop in vacuum, and (b) find the magnetic induction if the medium has a magnetic susceptibility of 0.2.

*Solution:*
    **(a)** Based on Eq. (4), $H = I_{loop}/(2r_{loop}) = 1 \times 10^{-3}$ A/$(2 \times 5 \times 10^{-3}$ m) $=$ 0.1 A/m.
    **(b)** Based on Eq. (5), $\mathbf{B} = \mu_0 \mu_r H = 1.25 \times 10^{-6}$ H/m $\times (1 + 0.2) \times$ 0.1 A/m $= 1.50 \times 10^{-7}$ H·A/m$^2$ (T).  §

**Example 5.2**   Compute the magnetic field due to a dipole with a magnetic moment $\boldsymbol{\mu}_m$

*Solution:*   We assume the magnetic pole strength is given by $q$, and $d$ is the separation of the dipole. Since $\boldsymbol{\mu}_m = qd\cos(\Theta)$, the magnetic potential $\phi$ at a distance $r$ (when $r \gg d$) is given by $qd\cos(\Theta)/(4\pi\mu r_{loop}^2) = \boldsymbol{\mu}_m\cos(\Theta)/(4\pi\mu r_{loop}^2)$. The field strength $H$ in the radial direction is then given by $H = -\Delta\phi/\Delta r_{loop} = 2\boldsymbol{\mu}_m\cos(\Theta)/(4\pi\mu r_{loop}^2)$.  §

**Example 5.3**   Compute the magnetostatic energy of Fe if $H_{av} = 0.1$ A/m.

*Solution:*   Based on Eq. (12), $E_m = \mu_0 M H_{av}/2 = \mu_0 \mu_r H_{av}^2/2 = 1.25 \times 10^6$ H/m $\times 2 \times 10^2 \times (0.1$ A/m$)^2/2 = 0.63 \times 10^6$ J/m$^3$.  §

**Example 5.4**   What is the magnetic induction **B** if the magnetic field intensity is 1 A/m?

*Solution:*   Based on Eq. (9) and Table 5.2, $\mathbf{B} = \mu_0\mu_r H = 1.25 \times 10^{-6}$ H/m $\times 200 \times 1$ A/m $= 0.25 \times 10^{-3}$ H·A/m$^2$ (T).  §

## 5.3   SOURCES OF MAGNETIZATION

In magnetic solids, there are two sources of magnetization, those originating from the electron spins and those from the circular motion of the orbiting electrons. In the following paragraphs we discuss the magnetization arising from these sources.

### Diamagnetism and Paramagnetism

Diamagnetism and paramagnetism are relatively weak forms of magnetization found in solids. They arise from the response of the orbiting electrons and their spins to an external magnetic field. In diamagnetism, the magnetic force slows down the orbiting electrons in a manner that repels the external magnetic field lines. The result is a small but negative magnetization in the solid. A simple mathematical model useful for understanding diamagnetism is one created by setting up an equation of the forces acting on an orbiting electron, as shown in Fig. 5.5.

**Fig. 5.5**
Forces acting on an electron

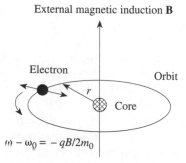

External magnetic induction **B**

Electron

Orbit

$r$

Core

$\omega - \omega_0 = -qB/2m_0$

   In the absence of magnetic induction, the attractive force between the positively
charged nucleus and the electron will be exactly balanced by the centrifugal force
of the orbiting electron. If an external magnetic induction, $\mathbf{B}$ (in T), is present,
however, the orbiting electron will experience an additional force (the Lorentz force
is $qr\omega\mathbf{B}$), where $r$ (m) is the radius of the orbit and $\omega$ (rad/s) is the angular frequency.
To offset the Lorentz force, $\omega$ will change according to the following equation,
which represents the balance of the (additional) forces:

$$m_0\omega^2 r + qr\omega\mathbf{B} = m_0\omega_i^2\mathrm{r} \tag{13}$$

where    $m_0$ is the mass of the electron in kg,

   $q$ is the electron charge in C,

   $\omega$ is the instantaneous angular frequency of the orbiting electron in rad/s,
   and

   $\omega_i$ is the **angular frequency** of the orbiting electron in rad/s prior to the
   application of $\mathbf{B}$.

For a small $\mathbf{B}$, Eq. (13) becomes

$$\Delta\omega = \omega - \omega_i \approx -q\mathbf{B}/(2m_0) \tag{14}$$

$\Delta\omega$ is the change in the angular frequency due to $\mathbf{B}$. The negative sign suggests a
slowdown of the orbiting electron. The reduction in the magnetic moment, $\Delta\mu_m$,
is given by

$$\Delta\mu_m = qr^2(\omega - \omega_i)/2 \tag{15}$$

By putting realistic values for the parameters into Eqs. (15) and (8), it can be shown
that diamagnetic permeability is quite small (the change in $\mu_r$ (i.e., $\Delta\mu_r$) is about
$10^{-5}$ in most solids).

   **Paramagnetism** is found in solids with energy orbits that are only partially
occupied by electrons. In these orbits, electron spins are not totally aligned but give
rise to a small net magnetic moment. In the presence of an external magnetic field,
the magnetic dipoles will have a lower energy if they are aligned in the direction
of the magnetic field. Magnetization resulting from the alignment of these dipoles
is given by

$$M = Nn_a\mu_0\mu'^2 H/(3kT) \tag{16}$$

where    $N$ is the atomic density in /m$^3$

   $n_a$ is the number magnetic dipoles per atom,

   $\mu'$ is the spin magnetic dipole moment in A·m$^2$,

   $H$ is the magnetic field intensity in A/m,

   $k$ is the Boltzmann constant, and

   $T$ is the absolute temperature in K.

At a finite temperature, paramagnetism is quite weak ($\Delta\mu_r \approx 10^{-5}$) in most solids and will disappear if the thermal energy ($E_{th} \approx kT$) exceeds the magnetization energy. This temperature is called the **Curie temperature, $T_c$.** Different from diamagnetism, paramagnetism gives a positive magnetization, i.e., a small increase in the density of the field lines passing through the solid. As expected, elements with completely filled energy orbits, such as the inert gases, do not exhibit paramagnetism.

**Example 5.5**    Based on Eq. (15), estimate the change in the magnetization for a single electron in a hydrogen atom. The radius of the hydrogen atom is $0.53 \times 10^{-10}$ m.

*Solution:*    Based on Eqs. (14) and (15), $\Delta\mu_m = -q^2 r^2 \mathbf{B}/(4m_0)$. This leads to a magnetization $\Delta M = -q^2 r^2/4m_0 = (1.6 \times 10^{-19} \text{ C} \times 0.53 \times 10^{-10} \text{ m})^2/ (4 \times 0.91 \times 10^{-30} \text{ kg}) = 0.197 \times 10^{-28}$ A/m. In terms of the magnetic susceptibility, $\Delta\chi_e = 0.197 \times 10^{-28}$ A/m/$(1.25 \times 10^{-6}$ H/m$) = 0.157 \times 10^{-22}$.    §

**Example 5.6**    Estimate the magnetization of a hydrogen atom if $H = 0.01$ A/m and $T = 300$ K. Assume $\mu' = \mu_B$.

*Solution:*    Based on Eq. (16), $M = Nn_a\mu_0\mu'^2 H/(3kT) = 1 \times 1/\text{m}^3 \times 12.65 \times 10^{-7}$ H/m $\times (9.27 \times 10^{-24}$ A/m$)^2 \times 0.01$ A/m/ $(3 \times 1.38 \times 10^{-23}$ J/K $\times 300$ K$) = 8.69 \times 10^{-55}$ A/m.    §

**Example 5.7**    Compute the magnetic susceptibility due to diamagnetism in a solid with an atomic density of $10^{29}/\text{m}^3$. Each atom has 10 orbiting electrons. Assume a small magnetic induction of 0.1 T is applied and the average radius of the orbit is 0.1 nm.

*Solution:*    Based on Eqs. (14) and (15), $\Delta\mu_m = q(\omega - \omega_i)r^2/2 = -q^2 r^2 \mathbf{B}/4m_0 = -(1.6 \times 10^{-19} \text{ C} \times 10^{-10} \text{ m})^2 \times 0.1 \text{ T}/(4 \times 9.1 \times 10^{-31} \text{ kg}) = 0.70 \times 10^{-29}$ A·m$^2$. Based on Eq. (8), $\chi_e = \mu_0 Nn_a\Delta\mu_m/\mathbf{B} = 1.25 \times 10^{-6}$ H/m $\times 1 \times 10^{-29}/\text{m}^3 \times 10 \times 0.70 \times 10^{-29}$ A/m$^2$/0.1 T $= 0.87 \times 10^{-5}$.    §

## Magnetic Dipole Moments in Solids

Our understanding of the more complex magnetic properties in a solid will require an understanding of the atomic orbits and the arrangement of the electrons. We first consider a one-electron atom with the electron circulating the nucleus in a loop. The magnetic moment, $\mu_m$, associated with this current loop is given by the product of the current through the loop and the loop area:

$$\mu_m = \pi r^2 qv/(2\pi r) \tag{17}$$

where    $q$ is the electron charge in C,

$v$ is the electron velocity in the loop in m/s, and

$r$ is the radius of the loop in m.

But, according to classical electromagnetic theory such an orbiting electron will be radiating electromagnetic energy, and it cannot stay in the orbit indefinitely. The problem was resolved by N. Bohr, who proposed that an electron is restricted to specific energy orbits—i.e., its **angular momentum** has discrete values. The angular momentum of the electron is therefore given by

$$m_0 v r = n' h / 2\pi \tag{18}$$

where   $m_0$ is the electron mass in kg,

$h$ is Planck's constant, and

$n'$ is an integer.

By eliminating $v$ in Eqs. (17) and (18), it can be shown that $\mu_m$ has a value of

$$\mu_m = n' q h / (4 \pi m_0) = n' \mu_B \tag{19}$$

where   $\mu_B \ (= q h / (4 \pi m_0))$ is the **Bohr magneton.**

This simple model of the atom suggests that the magnetic moment of an electron in an atomic orbit is discrete (quantized), i.e., restricted to integer multiples of $\mu_B$. In a real atom, there can be more than one electron, and so the total magnetic moment will be the sum of the magnetic moments of the individual electrons. Equation (19) can be further refined to take into account the magnetic moments due to other angular momentum contributions. For instance, the magnetic moment in the $z$ direction is given by $m \mu_B$ (where $m$ is a positive or negative integer or 0) and the magnetic moment due to the total angular momentum of an electron is given by

$$\mu_l = \sqrt{l(l + 1)} \mu_B \tag{20}$$

where   $l$ is an integer with a magnitude always greater than $m$ but less than $\tilde{n} - 1$, $\tilde{n}$ being a different integer called the principal quantum number.

The integers $\tilde{n}$, $l$, and $m$ are called the **quantum numbers.** The principal quantum number, $\tilde{n}$, defines the energy orbits (see Chapter 1). For $\tilde{n} = 1$, both $l$ and $m$ are 0, and the first energy orbit can have two electrons with different spins; for $\tilde{n} = 2$, $l = 0$ and 1, and $m = 0$ and $\pm 1$. There are eight electrons in this second orbit with positive and negative spins. The same process can go on for the higher values of $\tilde{n}$ as we build up the energy orbits for all the atoms in the periodic table. Although $\tilde{n}$, $l$, and $m$ seem to define adequately the different energy orbits, in the higher-energy orbits, the effect of $l$ on the total energy of the atom is more complicated, and it is not always true that the lower-energy orbits are completely filled before the next energy orbit begins to be occupied. This happens with the $3d$ ($\tilde{n} = 3$ and $l = 2$) orbits and the $4s$ ($\tilde{n} = 4$, $l = 0$) orbits and is responsible for the strong magnetization found in the transition metal ions. In addition to the angular momentum magnetic moment, there is the magnetic moment due to the electron spins. Spin magnetic moment, $\mu_s$, also has fixed values in terms of the Bohr magnetons and is given by

$$\mu_s = 2 \mu_B \sqrt{s(s + 1)} \tag{21}$$

where   $s = \pm \frac{1}{2}$.

The direction of $\mu_s$ can only be in the direction of the applied magnetic field or opposite to it. The energy associated with electron spins will be modified by the presence of an applied magnetic field. The spin energy will split up into closely spaced energy levels according to an effect called Zeeman splitting. **Zeeman splitting** can be used to measure a small magnetic field very accurately by measuring the spin energy absorption. Spin magnetic moments also interact with the angular momentum magnetic moments to generate mixed magnetic moments. In most solids, however, the mutual interactions among the same type of magnetic moment always prevail over mixed types. In interactions involving the same type of magnetic moment, the individual magnetic moments add up algebraically. Finally, it is important to point out that the placement of the electrons in the energy orbits follows what are called *Hund's rules*. Hund's rules may by summarized as follows:

1. Both $l$ and $s$ will take on their maximum values allowed by the Pauli exclusion principle.
2. The total angular momentum of the atom is given by $|l - s|$ when the shell (energy level) is less than half full and by $|l + s|$ when the shell (energy level) is more than half full.

Thus, based on Hund's rules, the electrons in a given orbit will prefer to have parallel spins and aligned magnetic moments over antiparallel spins and unaligned magnetic moments. This leads to strong magnetization in atoms with partially filled energy orbits and zero magnetization in the full orbits.

In addition to the magnetic moments found in the individual atoms, the overall magnetic properties of a solid are largely determined by the interactions of the magnetic moments between the atoms. This interaction depends on the electron distribution. In an ionic solid, the electrons are localized near the ion cores, whereas in a metal, the electrons are distributed over many atomic distances. We first examine the ionic solids, which in many respects are similar to the isolated ions. Table 5.3

**Table 5.3**    Predicted spin magnetic moments associated with the ions of the transitional metals

| Ion | Magnetic Moment ($\mu_B$) |
| --- | --- |
| $Ca^{2+}$ | 0 |
| $Sc^{2+}$ | 1.55 |
| $Ti^{2+}$ | 1.63 |
| $Cr^{2+}$ | 0 |
| $Mn^{2+}$ | 5.92 |
| $Fe^{2+}$ | 6.70 |
| $Co^{2+}$ | 6.64 |
| $Ni^{2+}$ | 5.59 |
| $Cu^{2+}$ | 3.55 |
| $Zn^{2+}$ | 0 |

(*Source:* D. J. Craik, *Structure and Properties of Magnetic Materials,* 1971, p. 80. Reprinted with permission of Pion Limited, London.)

shows the predicted spin magnetic moments associated with the ions of the transition metals.

With the differences in spin magnetic moments shown in Table 5.3, it is quite obvious that some ionic solids will have larger magnetic moments than others. Ionic solids that exhibit a strong magnetization effect are called *ferromagnets*. **Ferromagnetism** is due primarily to spin magnetic moments, since the angular momentum magnetic moments in ferromagnets can be shown to be almost nonexistent. As in ions, strong magnetization is found in solids where the ions contain a large number of unpaired electron spins and partially filled (3$d$ and 4$s$) energy orbits. The number of unpaired spins essentially determines the magnetization. $Fe^{2+}$, for instance, has the predicted equivalence of 6.7 $\mu_B$ per ion. Even though the individual ions have large magnetic moments, it is not always necessary that the neighboring ions in an ionic solid will have their magnetic moments pointing in the same direction. In fact, in the case of a $H_2$ molecule, as shown in Fig. 5.6, the preference is to have the spins of the neighboring ions aligned opposite to each other. Fortunately, in many of these solids, the electron orbits of the neighboring atoms overlap. Physically, overlapping allows the neighboring electrons to have parallel spins, provided that the overall electron distribution of the orbits (spin and angular momentum magnetic moments combined) are antisymmetric ($s$ is half-integral). The energy related to the overlapping of the electrons is called the **exchange energy.** Whether exchange energy favors a parallel spin configuration or an antiparallel spin configuration depends on the energy of the system and on the sign of an interaction parameter called the **exchange integral, $J'$**. If $J'$ is positive, parallel spin configuration is favored. Otherwise, there will be antiparallel spins.

Exchange interactions are observed in ionic solids that also have large magnetic moments (see Table 5.3). The presence of nonmagnetic ions in a magnetic oxide affects the exchange energy; often, the nonmagnetic ions will also participate in the exchange interactions. This usually results in an antiparallel spin configuration, as in most of the spinel ferrites except $CrO_2$ and CrTe (where the neighboring spins are actually parallel to each other). Table 5.4 lists the magnetic moments found in the transition metal ferrites. Note that the magnetic moments are essentially due to those of the first ions.

Unlike the ionic solids, transition metals are modeled as a collection of free

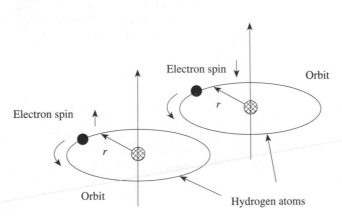

**Fig. 5.6**
A hydrogen molecule having atoms with different spins

Table 5.4    Magnetic moments found in the transition metal ferrites

| Ferrite | Magnetic Moment ($\mu_B$) | Ideal Value ($\mu_B$) |
|---|---|---|
| $MnFe_2O_4$ | 4.5 | 5 |
| $FeFe_2O_4$ | 4.1 | 4 |
| $CoFe_2O_4$ | 3.9 | 3 |
| $NiFe_2O_4$ | 2.3 | 2 |
| $CuFe_2O_4$ | 1.3 | 1 |
| $ZnFe_2O_4$ | 0 | 0 |

(*Source:* D. J. Craik, *Structure and Properties of Magnetic Materials,* 1971, p. 107. Reprinted with permission of Pion Limited, London.)

electrons, each with a given spin. The distributed electrons in a metal no longer have the strong magnetization effect observed in the isolated atoms or ions but behave like a collection of magnetic dipoles similar to the case of paramagnetism. For a metal with a parabolic energy band structure (i.e., the energy of the energy states varies as the square of the (crystal) momentum), the energy states are partially filled by electrons of positive spins and partially filled by electrons of negative spins, as shown in Fig. 5.7.

The presence of magnetic field shifts the energy band in favor of the aligned spins and results in a net spin magnetic moment. This result is called **Pauli paramagnetism.** Only the most energetic electrons near the vacant energy states will take part in Pauli paramagnetism, since the other electrons are all paired up in the lower energy states. The resulting magnetization is, nevertheless, quite small, but it agrees with the magnetization found in monovalent metals such as Li, Na, and K and the fact that the observed magnetic susceptibilities are temperature independent. Pauli paramagnetism cannot explain the large magnetic moments observed in the transition metals, as listed in Table 5.1. Obviously more complex energy interactions are needed to explain these effects, and some form of exchange interactions between the electrons and the ion cores has to exist. To achieve the observed scale of magnetization, it appears that the electrons in the $3d$ energy orbits are only partially distributed, since they make up the majority of the electrons in the transition metals. Some localized properties in these electrons will allow for the existence of exchange interactions similar to what is found in the ionic solids. The detailed study of these interactions is very complicated and is not considered here.

Fig. 5.7
Energy shift in Pauli magnetism

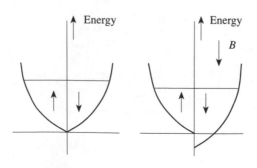

**Example 5.8**   Compute the value of the Bohr magnetron.

**Solution:**   The Bohr magneton is $\mu_B = qh/(4\pi m_0) = 1.6 \times 10^{-19}\,C \times 6.62 \times 10^{-34}\,J\cdot s/(4 \times 3.14 \times 0.91 \times 10^{-30}\,kg) = 0.927 \times 10^{-23}\,A\cdot m^2$.   §

**Example 5.9**   How many energy states are there in an atom when the principal quantum number, $\tilde{n}$, is 3?

**Solution:**   For $\tilde{n} = 3$, $l = 2, 1, 0$ and $m = \pm 2, \pm 1$, and 0. Remember that if $m < 1$, there should be a total of 18 states after accounting for the different electron spins.   §

**Example 5.10**   What is the value of the spin magnetic moment?

**Solution:**   The spin magnetic moment is $\mu_s = 2\mu_B\sqrt{s(s + 1)} = \pm 2\mu_B\sqrt{1/2 \cdot 3/2} = \pm 1.73\,\mu_B$.   §

## Ferromagnetism, Antiferromagnetism, and Ferrimagnetism

Ferromagnetism refers to the spontaneous magnetization found in solids due to spin alignment between the neighboring atoms. Parallel alignment is possible only in the presence of exchange interactions in both the ionic solids and in the transition metals. Physically, exchange interaction is found to fall off rapidly with the interatomic distance, which is why ferromagnetism exists only in certain types of crystal structures and not in others. For instance, bcc $\alpha$-Fe is ferromagnetic, and fcc $\gamma$-Fe is not. Spin alignment due to exchange interactions is sensitive to temperature change. When thermal energy exceeds exchange energy, interactions cease to exist, and the solid becomes paramagnetic. This takes place at the Curie temperature, $T_c$. Table 5.5 shows the Curie temperatures, $T_c$, for the common ferromagnetic solids.

Table 5.5   Values of the Curie temperature for the common magnetic solids

| Solid | $T_c$ (°C) |
| --- | --- |
| Fe | 770 |
| Co | 1120 |
| Ni | 358 |
| Gd | 20 |
| Tb | −52 |

(Source: D. J. Craik, *Structure and Properties of Magnetic Materials*, 1971, p. 107. Reprinted with permission of Pion Limited, London.)

Another parameter that appears to affect the alignment of the spin magnetic moments in ferromagnets is the ratio of the interatomic distance, $d_0'$, divided by the orbital radius of the electrons, $r_0'$. Alignment of the spin magnetic moments is favored if $d_0'/r_0'$ is large. This happens to the $3d$ electrons in the transition metals and in some of the rare-earth metals. Figure 5.8 shows how the exchange integral, $J'$, varies with $d_0'/r_0'$ for several magnetic solids. Alloys of magnetic solids behave differently, depending on the type of magnetic ions present. For instance, Ni—with the addition of nonmagnetic ions—will behave like a different solid with some intermediate magnetic properties. This is illustrated in Fig. 5.9 by a plot of the effective number of magnetons present in the Ni alloys as the Ni concentration changes. In the case of an alloy of Fe ($\alpha$-Fe), however, the magnetization effect decreases progressively with an increase in the fraction of nonmagnetic ions.

In a ferromagnetic solid, atoms with parallel spins will group together to form magnetic domains, as shown in Fig. 5.10. Each domain possesses some significant magnetic moments. In the natural state these domains are randomly oriented, and the overall magnetic moment of the solid is zero. This is the nonmagnetic state of the ferromagnetic solid. In the presence of an applied magnetic field $H$, the domains will orientate in the direction of the magnetic field. If the magnetic field is sufficiently strong, all the domains will align in the same direction, resulting in a strong magnetization. The progressive orientation of the magnetic domains in the direction of the magnetic field is illustrated in Fig. 5.11.

**Fig. 5.8**
Exchange integral $J'$
versus $d_0'/r_0'$

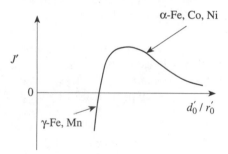

**Fig. 5.9**
Effects of impurities on
magnetization

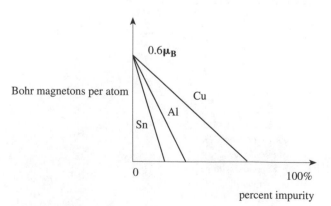

**Fig. 5.10**
Magnetic domains in a
cubic crystal

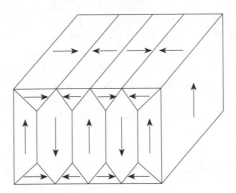

**Fig. 5.11**
Progressive magnetization
in a ferromagnet

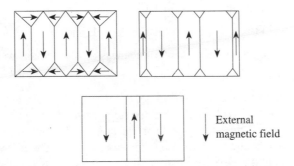

|   | External
magnetic field |

The extent of alignment and magnetization depends on the magnetic field intensity, *H,* but will saturate when *H* is high enough and all the domains are in the same direction. This condition is called *saturation.* A strong ferromagnet usually has a high saturation field. Depending on the crystal structures, most ferromagnetic solids have some preferred crystal axes that will allow for an easy alignment of the magnetic domains. This is called **magnetic anisotropy.**

The change in the orientation of the magnetic domains depends effectively on the movement of the **domain wall,** which separates the different regions of spin magnetization. Figure 5.12 shows a schematic of the spin orientations in a domain wall separating two domains. During magnetization, the domain wall shifts in favor of those domains that are already aligned to the external magnetic field. This shift

**Fig. 5.12**
180° spin flip at a domain
wall

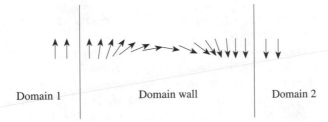

Domain 1                    Domain wall                    Domain 2

is achieved by the rotation of the spin magnetic moments in the neighborhood of the domain wall. The energy involved in the formation of the domain wall depends on the exchange energy and the anisotropy energy (energy related to spin alignment in the preferred direction of the crystal), both of which are a function of the angle of rotation $\Theta$—i.e., the angle between the spins in the domains and the external magnetic field. The energy density, $E_w$ (J/m$^2$), in the domain wall for a crystal with uniaxial (single-direction) anisotropy is given by

$$E_w = A_m \int_{-\infty}^{\infty} (\partial\Theta/\partial z)^2 \, dz + \int_{-\infty}^{\infty} K_m \cos^2(\Theta) \, dz \qquad (22)$$

where    $A_m$ is a constant in J/m,

$K_m$ is a constant in J/m$^3$,

$\Theta$ is the angle of rotation in rad, and

$dz$ is the incremental thickness of the domain wall in m.

Equation (22) can be used to solve for the minimum energy $E_w$(min) to form the wall. For $\Theta = 180°$—i.e., when the spins are totally flipped over across the wall—the domain wall thickness is $\delta = \pi\sqrt{A_m/K_m}$ and $E_w$(min) $= 4\sqrt{(A_m K_m)}$.

The movement of the domain wall is also affected by the impurities present, especially if the size of the impurities is of the same dimension as the wall thickness $\delta$. In this case, there is a tendency for the domain wall to be partitioned at the location of the impurities, and this reduces the average size of the domains. Small impurities usually do not impede the movement of the domain wall, and they may be ignored. In an ac magnetic field, the domain wall will require a finite time delay to move; this is called the **switching time, $\tau_{sw}$**. Theoretically, $\tau_{sw}$ can be estimated by assuming a frictional force that retards the wall movement, and $\tau_{sw}$ (s) is then given by

$$\tau_{sw} = 1/S_m(H - H_c) \qquad (23)$$

where    $H$ is the magnetic field intensity in A/m,

$S_m$ is a constant in m/A·s, and

$H_c$ is the coercive field strength in A/m.

For most magnetic solids, $\tau_{sw}$ at a moderate field intensity is of the order of tens of microseconds.

In a ferromagnet, the process of repeated magnetization and demagnetization in an ac magnetic field is only partially reversible. In fact, a remnant magnetic induction $B_0$ will result after the magnetization field is removed. This effect, called **hysteresis,** is associated with the energy consumed in changing the domain wall. Hysteresis is responsible for the observed magnetization found in permanent magnets. A typical hysteresis curve is shown in Fig. 5.13. The area enclosed in the hysteresis loop is a measure of the energy loss per unit volume during the magnetization and demagnetization process. A larger hysteresis loop implies that more energy is consumed to move the domain wall, and so it will be more difficult to demagnetize the solid. Permanent magnets can be demagnetized if the temperature is raised above the Curie temperature, $T_c$. At this temperature, the spin magnetic moments become randomized within the domains and result in negligible overall magnetic moments in the solid (similar to paramagnetism). Thus, when $T$ falls below $T_c$, the domains are randomly oriented. The Curie temperatures for most ferromagnetic solids are below 1000°C (see Table 5.4). As mentioned earlier, spin magnetic moments are antiparallel if the exchange integral, $J'$, is negative. This can result in an **antiferromagnetic solid;** a typical example is MnO. Some antiferromagnetic solids will become ferromagnetic at a sufficiently high temperature.

Ferrimagnetism occurs in compounds of Fe and O that have the inverse spinel crystal structure, as shown in Fig. 5.14. For the inverse spinel crystal structure, the Bravais lattice is fcc. In many **ferrimagnetic solids,** even though the spin magnetic moments are antiparallel, a residual magnetic moment remains in the solids, because the antiparallel spin magnetic moments are not equal. Just as in the ferrimagnetic solids, there are magnetic domains in the ferrimagnetic solids, but the spin magnetic

**Fig. 5.13**
Magnetization loop of a
ferromagnet

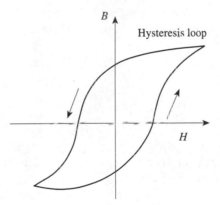

**Fig. 5.14**
The inverse spinel crystal
structure

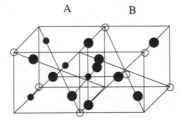  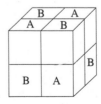

●   Metal ions on octohedral sites

○   Metal ions on tetrahedral sites

⬤   Oxygen ions

moments are much weaker. Ferrimagnetic solids include the oxides of Mn, Ni, and Zn. The main advantage of ferrimagnetic solids is that they are poor current conductors, and any rf loss due to the presence of an induced current (in waveguides, mixers, etc.) is quite low.

**Example 5.11**   Estimate the values of $A_m$ and $K_m$ appearing in Eq. (22) if the domain wall thickness, $\delta$, is 1 $\mu$m and $E_w$ is $1 \times 10^{-3}$ J/m$^2$.

*Solution:*   Since $\delta = \pi\sqrt{A_m/K_m}$ and $E_w(\min) = 4\sqrt{A_m K_m}$, $A_m = \delta E/4\pi = 1 \times 10^{-6}$ m $\times 1 \times 10^{-3}$ J/m$^2$/(4 $\times$ 3.14) = $8.0 \times 10^{-11}$ J/m. $K_m = (\pi/4)(E/\delta) = (3.14/4) \times 1 \times 10^{-3}$ J/m$^2$/1 $\times 10^{-6}$ m = $0.785 \times 10^2$ J/m$^3$.   §

**Example 5.12**   If the switching time in a ferromagnetic solid is 20 $\mu$s, estimate the constant $S_m$ in Eq. (23) if $H - H_c = 0.1$ A/m.

*Solution:*   Based on Eq. (23), $S_m = 1/(\tau_{sw}(H - H_c)) = 1/(20 \times 10^{-6}$ s $\times 0.1$ A/m) = $5 \times 10^5$ m/A·s.   §

**Example 5.13**   If the exchange energy in a ferromagnetic solid is the same as the thermal energy at the Curie temperature, estimate the exchange energy for Ni.

*Solution:*   The Curie temperature for Ni is 358°C, or 631 K. The exchange energy is $J' \approx kT = 1.38 \times 10^{-23}$ J/K $\times 631$ K = $8.7 \times 10^{-21}$ J.   §

## 5.4   MAGNETIC ANISOTROPY AND INVAR ALLOYS

### Magnetic Anisotropy

Many magnetic solids have complex crystal structures, and the latter give rise to directional effects. For instance, Co has a hcp (hexagonal close-packed) structure, and the direction perpendicular to the hexagonal $c$ plane is an easy axis of magnetization. Thus, Co is an anisotropic magnetic solid. For a similar reason, it is found that Fe has an easy axis of magnetization along the $<1\,0\,0>$ direction; for Ni, it is in the $<1\,1\,1>$ direction. The tendency for the magnetic moments to align with the easy axis is determined by a parameter called the **anisotropy energy (per unit volume)**, $E_{ai}$. Perfect alignment will result if $E_{ai}$ is at a minimum or zero. For a cubic crystal, the anisotropy energy, $E_{ai}$ (J/m$^3$), is given by

$$E_{ai} = K_1(\sigma_1^2\sigma_2^2 + \sigma_2^2\sigma_3^2 + \sigma_3^2\sigma_1^2) + K_2\sigma_1^2\sigma_2^2\sigma_3^2 + \cdots \tag{24}$$

where     $K_1$ and $K_2$ are constants specific to the crystal structures in J/m$^3$, and

$\sigma_1$, $\sigma_2$, and $\sigma_3$ are the cosines of the angles between the principal axes of the cubic crystal and the direction of the magnetic field.

Based on Eq. (24), it is possible to show that geometrically, for a positive $K_1$, $E_{ai}$ is a minimum if the field direction is in the $<1\,0\,0>$ direction. On the other hand, if $K_l$ is negative, $E_{ai}$ will be a minimum if the field direction is in the $<1\ 1\ 1>$ direction. Table 5.6 shows the values of $K_1$ for the more common transition metals.

The negative value of $K_1$ in Ni indicates that it has an easy axis in the $<1\ 1\ 1>$ direction. In a hexagonal crystal (hcp) structure such as Co, $E_{ai}$ (in J/m$^3$) is given by

$$E_{ai} = K_{c0} + K_{c1}\sin^2(\Theta) + K_{c2}\sin^4(\Theta) + \cdots + K_{c4}\sin^6(\Theta)\cos(\phi) + \cdots \quad (25)$$

where   $K_{c0}$, $K_{c1}$, $K_{c2}$, and $K_{c4}$ are constants specific to the crystal structures in J/m$^3$,

$\Theta$ is the angle between the $c$ axis and the direction of spontaneous magnetization, and

$\phi$ is the component of spontaneous magnetization in the $c$ plane.

Magnetization is also affected by the built-in strain. Any applied stress will interact with the existing magnetic moments in a way that changes the direction of magnetization. This is called **magnetostriction.** In the presence of an applied stress, $\sigma_s$ (N/m$^2$), $K_1$ in Eq. (24) will have to be replaced by $K_1 + K_\sigma$, where $K_\sigma$ is given by

$$K_\sigma = 3\epsilon\sigma_s/2 \quad (26)$$

where   $\epsilon$ is the observed strain ($= \Delta l/l$).

In principle, it is possible that strain anisotropy may override crystal anisotropy (i.e., $K_\sigma > K_1$) and cause a change in the direction of magnetization. In such a case, the magnetic solid is strain-sensitive. Ferromagnetic properties are also affected by the size of the magnetic domains. Large magnetic domains have lower energy since the amount of energy needed to form the domain wall is smaller. However, this increases the magnetostatic energy in the domain since the magnetic field lines within a domain do not close (the magnetic field lines is closed outside the domain). The trade-off in most ferromagnetic solids in fact occurs at a domain size of the order of 50 nm, when **single-domain particles** are known to be formed. A single-domain particle is the smallest magnetic domain.

**Table 5.6**   Values of $K_1$ in the common transition metals

| Metal | $K_1$ (J/m$^3$) |
| --- | --- |
| Fe | $4.8 \times 10^4$ |
| Co | $5.3 \times 10^5$ |
| Ni | $-4.5 \times 10^3$ |

*(Source:* D. J. Craik, *Structure and Properties of Magnetic Materials,* 1971, p. 43. Reprinted with permission of Pion Limited, London.)

Remnant ferromagnetic effect is optimized by magnetization along an easy axis and suppressed when the magnetization is perpendicular to the easy axis. Magnetization along the easy axis in an anisotropic ferromagnetic solid results in a large **remanence** (residual magnetization), whereas a similar magnetization perpendicular to the easy axis results in negligible remanence. Often, elongated single-domain particles with the easy axis along the longitudinal axis are prepared to achieve significant anisotropy in the solid. **Demagnetization** in single-domain particles involves either reversing the spins in the magnetic domains or moving the domain wall. A lesser energy is required in the latter, since the coercivity is quite high with single-domain particles. Low **coercivity** is usually found in an isotropic solid. Nonmagnetic additions such as Si and B can reduce the anisotropy of a ferromagnetic solid and increase its resistivity. Based on this principle, strong ferromagnetic solids can be prepared with little or no anisotropy, as in some of the Ni–Fe alloys. These solids usually have high purity and large domains with flexible domain wall. Further reduction in anisotropy is achieved in the amorphous alloys. These magnetic solids are particularly useful in the making of transformers when large resistance, high magnetic-field saturation, and low coercivity are required. In a solid such as Co, anisotropy can be affected by the addition of other ferromagnetic solids. The addition of 1% Fe to Co will change the easy axis from the $c$ direction to a direction in the $c$ plane. The reason for this behavior is found to be related to a transformation in the crystal structure of Co.

**Example 5.14**    Fe saturates at a magnetic induction of 2.18 T. How many electrons per Fe atom are involved? $\mu_B = 9.27 \times 10^{-24}$ A·m$^2$.

*Solution:*    Magnetic induction due to each electron is $\mu_0 \mu_B = 1.25 \times 10^{-7}$ H/m $\times 9.27 \times 10^{-24}$ A·m$^2 = 1.16 \times 10^{-30}$ T·m$^3$. Since there are $0.85 \times 10^{29}$ atoms/m$^3$ in Fe, the number of electrons involved per atom in the magnetization process is 2.18 T/($1.16 \times 10^{-30}$ T·m$^3 \times 0.85 \times 10^{29}$/m$^3$) = 2.19.   §

**Example 5.15**    Estimate the value of applied stress necessary to offset the negative value of $K_1$ in Ni. Assume $\epsilon_{st} = 34 \times 10^{-6}$.

*Solution:*    From Table 5.6, $K_1 = -4.5 \times 10^3$ J/m$^3$. For $K_1 = K_\sigma = 3\epsilon\sigma_s/2$, $\sigma_s = 2K_1/3\epsilon_{st} = 2 \times 4.5 \times 10^3$ J/m$^3$/($3 \times 34 \times 10^{-6}$) = $8.8 \times 10^7$ J/m$^3$.        §

## Invar Alloys and Magnetovolume Effect

The Fe–Ni alloy, when formed in a composition in the neighborhood of 65% Fe and 35% Ni, exhibits a negligible thermal expansion coefficient at room temperature. Such a solid is called an **Invar alloy** because of its invariable thermal expansion coefficient. The Fe–Ni Invar alloy has a bcc structure. The volume expansion coefficient as a function of temperature is shown in Fig. 5.15. Invar alloys also exist in other ferromagnetic solids, such as Zr–Zn$_2$ and Cr–Mn. The near-zero expansion coefficient found in these solids is closely related to the volume magneto-striction (or magnetovolume effect).

The **magnetovolume effect** is best explained by the spontaneous magnetization

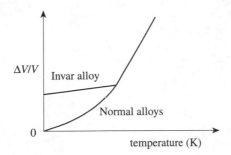

Fig. 5.15
Volume expansion
coefficient versus $T$ in an
Invar alloy

observed in an Invar alloy as it is cooled below the Curie temperature. The energy
band structure of an Invar alloy is shown in Fig. 5.16. Magnetization can be achieved
by increasing the number of electrons with one type of spin at the expense of a
higher kinetic energy. The increase in kinetic energy, however, is minimized if there
is a simultaneous expansion in the density of states or a flattening in the energy
bands. This increases the interatomic distance and results in a volume expansion.

Thus, spontaneous magnetization in the Invar alloy generates a volume expansion
in the crystal lattice. The associated volume expansion will offset any lattice shrinkage due to cooling and gives rise to a minimal temperature dependence in the
thermal expansion coefficient. This is commonly called the magnetovolume effect.

We assume that the energy of the Invar alloy (J) is given by

$$E = \Delta E + k_v \Delta V^2 / 2 \qquad (27)$$

where   $\Delta E$ is the change in the kinetic energy of the electrons in the Invar alloy
in J,

$k_v$ is the compressibility in J/m$^6$, and

$\Delta V$ is the volume change in the alloy in m$^3$.

To a first order of approximation, if $\Delta E / \Delta V$ is proportional to $-M^2$, the square of
the magnetization, then the fractional change in the volume, $\Delta V/V$, of the Invar
alloy is given by

$$\Delta V/V = A_v M^2 \qquad (28)$$

where   $A_v$ is a constant in m$^2$/A$^2$.

Fig. 5.16
Flattening of the energy
bands in an Invar alloy

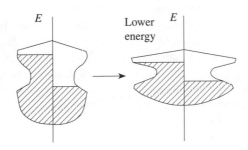

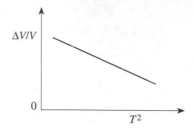

Fig. 5.17
Thermal expansion
coefficient versus $T^2$ in
$ZrZn_2$

Equation (28) suggests that an increase in the magnetization of the Invar alloy also increases its volume. The reverse is also true, and a reduction in the volume of the Invar alloy is observed if the magnetization is somehow reduced. In addition to the magnetic Invar alloys, other magnetic alloys, such as $Zr–Zn_2$, also exhibit a negative temperature expansion coefficient, as shown in Fig. 5.17. Invar alloys have many different applications. In particular, they are used in precision scales, transfer tubes for liquefied gases that frequently undergo temperature changes, and in applications where dimensional tolerance due to thermal effect is small.

**Example 5.16**    In an Invar alloy, if the thermal expansion coefficient is $30 \times 10^{-6}$, estimate the magnetization required to have zero thermal expansion coefficient. $A_v = 1 \times 10^{-22} m^2/A^2$.

*Solution:*    For volume expansion, the thermal expansion coefficient is approximately three times the linear value or, in the given Invar alloy, it is $90 \times 10^{-6}$. Based on Eq. (28), $\Delta V/V = A_v M^2$ or

$$M = \sqrt{\Delta V/(V A_v)}$$
$$= \sqrt{90 \times 10^{-6}/(1 \times 10^{-22} \ m^2/A^2)}$$
$$= 9.5 \times 10^{-8} \ A/m \quad \S$$

## 5.5    AMORPHOUS MAGNETIC MATERIALS

Amorphous magnetic solids are usually prepared as a mixture of (1) a transition metal and a nontransition metal; (2) a transition metal and a rare-earth metal, or (3) a transition metal oxide and a metal oxide. Most amorphous magnetic solids are unstable at room temperature, and they usually contain 15% to 25% metalloid elements, such as B or P, which give the stability and the magnetization. Magnetization found in amorphous magnetic solids is lower than that found in the crystalline form, but otherwise the magnetic properties are quite similar, except that magnetocrystalline anisotropy is absent. Some amorphous solids, such as GdCo, have been reported to have strong anisotropy, although it is possible that they may possess some form of microscopic crystalline structures.

Most commercial amorphous magnetic solids are prepared by a continuous rapid-cooling technique. During preparation, the molten magnetic alloy is ejected from a nozzle onto a roller and quickly solidified at a quenching rate of $10^5$ to $10^6$ °C/s. The roller draws the molten liquid into a ribbon with a thickness usually less than 100 μm. A simple setup is shown in Fig. 5.18. The as-prepared solids require heat treatment at a temperature of 200 °C to 450 °C to stabilize and develop the magnetic properties. Most Fe-based amorphous solids will require the lower heat-treatment temperature. Although amorphous magnetic solids do not possess magnetocrystalline anisotropy, they are subject to recrystallization during heat treatment and induced magnetic anisotropy. Amorphous magnetic solids have high resistivities and stiffness, surpassing the standard ferrimagnetic solids. They have been used extensively in transformer cores, recording heads, and various types of magnetic cores.

**Example 5.17**  Compare the resistivity of magnetic solids in the crystalline state, as oxides, and in the amorphous state.

*Solution:*  Crystalline metals have a resistivity of about $10^{-7}$ Ω·m. Metallic amorphous Fe-related solids have a value of about $10^{-6}$ Ω·m, whereas a ferrite has a value of about $10^2$ Ω·m.  **§**

**Example 5.18**  Estimate the resistance per unit length of a magnetic amorphous solid with the following dimensions: width = 10 cm, thickness = 100 μm, and resistivity = $1 \times 10^{-6}$ Ω·m.

*Solution:*  Resistance per unit length = $\rho/A_{cs}$ = $1 \times 10^{-6}$ Ω·m/ $(0.1 \times 10^{-4}$ m$^2$) = 0.1 Ω/m.  **§**

## 5.6  HARD AND SOFT MAGNETS

Metallic ferromagnets can be soft or hard, depending on the difficulty in moving the magnetic domains. This is called coercivity. **Soft magnets** are less coercive and can be easily magnetized or demagnetized by a reversal of the external magnetic field. Because of the reversible properties, soft magnets are frequently used to make magnetic cores for transformers. A **transformer** is a current coupling device and

**Fig. 5.18**

Setup to form amorphous ribbons

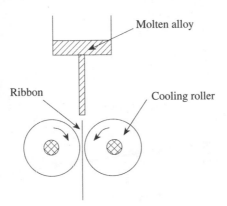

it consists of two coils coupled to each other through a magnetic core. Both the primary coil and the secondary coil have a large number of windings, and any current change in the primary coil will be reflected in the secondary coil. Transformers are primarily used for voltage step-up or step-down. The efficiency of a transformer depends critically on the properties of the magnetic core. A magnetic core is a metal block made of Fe or an Fe alloy. Some cores are made with a stack of metallic plates. The design of the magnetic cores requires (1) a small hysteresis loop in the B-H plot, (2) a high electrical resistance, and (3) a high magnetic permeability. The last item is, of course, needed to increase the energy-storage capability. Soft magnets are frequently made with amorphous ferromagnets in order to lower the anisotropy and, hence, the coercivity. Amorphous ferromagnets and ferrites also have high resistivities, which lower the induced (eddy) current in the magnetic cores. Resistance in ferromagnets can be increased with the addition of an impurity, such as Si. The more popular soft magnets are made of Ni–Fe alloys.

**Hard magnets** are used as permanent magnets since they cannot easily be demagnetized. A figure of merit of the magnetic strength of a ferromagnet is the value of $(BH)_{max}$—i.e., the peak value in a plot of the magnetic induction versus the magnetic field intensity. $(BH)_{max}$ in fact is a measure of the peak magnetic induction and the peak demagnetization field. A typical plot is shown in Fig. 5.19.

High-quality magnetic alloys (NdFeB) have a $(BH)_{max}$ value in excess of $1 \times 10^5$ T·A/m. Hard magnets often have impurities such as Ti to increase their coercivity. In addition to the more common metallic ferromagnets, ceramics can also form permanent magnets. They exist in the form of **ferrites,** or **garnets,** which are $Fe_2O_3$ mixed with other metallic ions. Many ceramic ferrimagnets possess significant magnetic moments. In addition to their strong magnetization, ceramic magnets are also good current insulators. Most ceramic magnets have an inverse spinel structure, as shown in Fig. 5.14. The Bravais lattice is fcc. Some outstanding use of ceramic magnets are in the coating of magnetic tapes and disks. The key material is $\gamma\text{-}Fe_2O_3$, either in grains or a single-domain (acicular) particles. In addition to the requirement of a large magnetization effect, the power-conversion efficiency of a permanent magnet is also very important. In the case of an **electric motor,** the

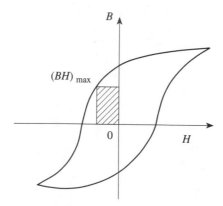

Fig. 5.19
Hysteresis plot showing
$(BH)_{max}$

power, in watts, delivered by moving the shaft at a velocity $v$ in a given magnetic induction **B** (in T) is given by

$$P_m = \mathbf{B}Ilv \qquad (29)$$

where  $l$ is the length of the loop for the core in m,

  $I$ is the current in A, and

  $v$ is the velocity of the shaft in m/s.

The electrical power supplied to the motor (W) is given by

$$P_{in} = lI^2/(A_{cs}\sigma) \qquad (30)$$

where  $A_{cs}$ is the cross section of the wire in m$^2$, and

  $\sigma$ is the electrical conductivity in S/m.

The **power efficiency**, $\eta_E$, given by the ratio of $P_m/P_{in}$, is

$$\eta_E = P_m/(P_m + P_{in}) = 1/(l + II/(BvA_{cs}\sigma)) \qquad (31)$$

Based on Eq. (28), in order to maximize $\eta_E$ both $B$ and $\sigma$ should be large in an electric motor. This requirement can only be partially met by the use of a metallic core.

Ferromagnets are also used as an energy-storage medium, as in the case of a transformer. Unlike a motor, there is no mechanical movement involved, and the losses are mainly electrical. Losses include the energy dissipated in moving through the hysteresis loop of the B-H curve and resistive loss, in the form of an eddy current. An eddy current is induced in the magnetic core whenever there is a change in the magnetic field. For a rectangular wire loop with a changing magnetic flux density $dB/dt$, the eddy current, $I_{ed}$ (A), is given by

$$I_{ed} = hl(dB/dt)/R \qquad (32)$$

where  $h$ and $l$ are the dimensions of the loop in m, and

  $R$ is the wire resistance in $\Omega$.

To minimize $I_{ed}$, either a more resistive transformer core is used or there must be a smaller cross section in the current loop. The latter can be achieved by using a laminated transformer core made with thin isolated magnetic plates, as shown in Fig. 5.20.

**Fig. 5.20**

Eddy currents in a metal block and in a laminated core

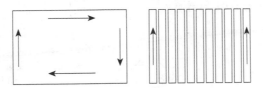

**Example 5.19**    If the $B$-$H$ plot of a ferromagnetic solid is given by $B = H_0 - H$, determine $(BH)_{max}$.

*Solution:*    To obtain the peak value, $\partial(BH)/\partial H = 0$. Since $BH = H_0 H - H^2$, $H_0 - 2H = 0$. This result gives $H = H_0/2$. Therefore, $(BH)_{max} = (H_0 - H_0/2)H_0/2 = H_0/4$.    §

**Example 5.20**    Estimate $(BH)_{max}$ if the $B$-$H$ curve is governed by the following relationship: $B = \tanh(C_1(H + H_0))$, where $C_1$ and $H_0$ are constants. $H_0$ can be positive or negative. Obtain an analytical solution if $C_1$ is small.

*Solution:*    $HB = H \tanh(C_1(H + H_0))$. To obtain $(BH)_{max}$, we set $\partial(HB)/\partial H = 0$, or $\partial(H \tanh(C_1(H + H_0)))/\partial H = 0$. The derivative gives a complex function in $H$. But, if we assume $C_1$ is small, it can be shown that $H \approx H_0/2$.    §

**Example 5.21**    Si is known to increase the resistivity of Fe. What will be the reduction in eddy current loss if the Si content of an Fe transformer core is increased from 3% to 10% (see Section 2.6)? $\beta$ for Si in Fe is 117.

*Solution:*    The fractional increase in resistivity in the core is $[(1 + 0.1\beta) - (1 + 0.03\beta)]/(1 + 0.03\beta) = [(1 + 117 \times 0.1) - (1 + 117 \times 0.03)]/(1 + 117 \times 0.03) = (8 - 3.5)/4.5 = 1.81$. Note that if nothing else changes, the eddy loss will be reduced by 1.81 times.    §

**Example 5.22**    Which of the following materials are more suitable for a transformer core: (1) sheets of SiFe; (2) MnZn ferrite; or (3) dry air?

*Solution:*    Both (2) and (3) have low permeability and are not suitable. Their principal advantage would be low voltage and probably a high breakdown voltage in the case of dry air.    §

**Example 5.23**    Determine the value of the eddy current in a transformer if the current loop is 5 cm $\times$ 5 cm, the wire is made of Cu and has a diameter of 0.050 cm, and the magnetic induction is 0.01 T at 1 kHz. The resistivity of Cu is $1.6 \times 10^{-8}$ $\Omega$·m.

*Solution:*    The resistance of the loop is $R = \rho L/A_{cs} = 1.6 \times 10^{-8}$ $\Omega$·m $\times 4 \times 0.05$ m$/(3.14 \times (2.5 \times 10^{-4}$ m$)^2) = 16.3 \times 10^{-3}$ $\Omega$. Based on Eq. (32), $I_{ed} = hl(dB/dt)/R = 0.05$ m $\times 0.05$ m $\times 0.01$ T$/1000/$s $\times 1/(16.3 \times 10^{-3}$ $\Omega) = 1.53 \times 10^6$ A $= 1.53$ μA.    §

## 5.7    MAGNETIC DEVICES AND APPLICATIONS

### Magnetic Recording

**Magnetic recording** involves the storage of data using a magnetic medium. The data may be a magnetization pattern, as in the case of analog recording, or a sequence of binary magnetization states, as in the case of recording digital data. The density of magnetic storage has increased significantly over the years, and so have the recording technologies. High-density recording depends on the **recording wave-**

**length** and the width of the medium. Feature size down to 1 μm by 20 μm can be achieved in **video-tape recording** (VTR), giving a recording time as long as 8 h. The most popular form of magnetic recording uses either a tape or a disc. The magnetic materials are simply deposited on a polymeric sheet or an Al disc in the form of powder, grains, or needlelike (acicular) particles. Early magnetic materials were mainly powdered $\gamma$-$Fe_2O_3$; more recently, metallic powder and evaporated thin films have been used. The recording density has also increased to $10^6$ bits/cm$^2$ or higher. Table 5.7 lists the magnetic properties of some of the more popular recording materials.

Although the materials used in magnetic recordings has changed from the isotropic Fe-based powders to the Co-based thin films, the data-storage format also increasingly has changed from analog to digital. Digital data storage requires a much shorter recording wavelength (down to 0.1 μm) when the bits are stored vertically rather than laterally, and this can yield a tenfold increase in the storage capacity. In addition, thin-film metallic tapes also have a smaller film thickness ($\approx$0.1 μm in the thin films compared with 5 μm in the conventional tapes), which allows for a more uniform magnetic induction during recording.

The principle of recording on tapes or discs requires a coil carrying a current placed in close proximity to the magnetic medium. The current required for straightforward writing of data is usually quite large, and the magnetic induction may require using a coil of many windings around a magnetized armature, as shown in Fig. 5.21. This basic configuration is used in a **recorder head.** The number of coil windings and the permeability of the armature determine the strength of magnetization, and induction effect is achieved through the fringing field at the edge of the armature. Because of the lower saturation field at the edge of the armature, the head is often coated with a (high-permeability) metallic layer to focus the field lines.

Table 5.7   Magnetic properties of the more popular recording materials

| Recording Material | $B_{out}$ (T) | Coercivity (A/m) | $T_c$ (°C) |
|---|---|---|---|
| $\gamma$-$Fe_2O_3$ | 0.5 | 1,600 | 675 |
| (Co-coated) | 0.5 | 4,000 | 400 |
| Fe powder | 2.0 | 80,000 | 770 |
| Co–Vi thin film | 1.6 | 4,000 | 1,000 |
| Co–Cr thin film | 0.5 | | 500 |

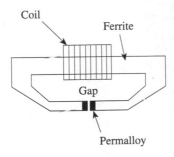

Fig. 5.21
A typical recorder head

As a rule the **magnetic gap length**—i.e., the distance between the ends of the armature—should be about one-third of the recording wavelength. This implies a magnetic gap length of around 0.3 μm and a tolerance of one-tenth of that value. During reading, the magnetic head is subject to wear and tear when it is in contact with the fast moving tape or disc. Materials such as MnZn, which have a good mechanical hardness, are becoming more widely used to extend the lifetime of the recording head.

Reading a **magnetic disk** or tape involves the reverse to the write process. The changing flux on the magnetic medium as the tape or disk passes through the head induces a voltage in the head, resulting in a small voltage change corresponding to the strength of the magnetization present in the medium. The induced voltage, $V_{ind}$ (V), is given by

$$V_{ind} = nv \, d\phi/dx \tag{33}$$

where    $n$ is the number of windings in the coil,

$v$ is the speed of the tape in m/s, and

$d\phi/dx$ is the gradient of the magnetic flux in T·m.

In addition to the magnetic tape and disk, an older computer memory technology known as *core memory* has been used for data storage. Computer core memory is a different form of magnetic recording. The **magnetic cores** of the memory are arranged in a matrix interlaced through fine metal wires both horizontally and vertically, as shown in Fig. 5.22. A change in the state only occurs during reinforced magnetization, i.e., when both the horizontal current and the vertical current pass through the core in the same direction (activation in one of the wires will not induce a change in the magnetization of the core). This form of direct read-write address is frequently used in semiconductor memories. Reading of the magnetic core is achieved using a third sense wire threaded through the core. It will pick up an induced voltage if the core changes state (a current pulse has to be sent through the sense wire to test the polarity of the existing state). To facilitate a fast response for a high-speed memory, soft magnets are always used in the cores.

**Fig. 5.22**
A computer core memory

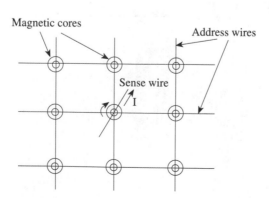

**Example 5.24**   If a magnetic tape contains granular particles of diameter 0.1 μm, determine the speed of the rotor if a full cycle of the recording requires at least 10 grains and the audio range is 2 kHz.

*Solution:*   The shortest wavelength will be 0.1 μm $\times$ 10 = 1 μm. Speed = wavelength $\times$ frequency = $1 \times 10^{-6}$ μm $\times 2 \times 10^3/s = 2 \times 10^{-3}$ m/s, or 0.2 cm/s.  §

**Example 5.25**   Determine the induced signal voltage if a flux reversal occurs over a 1-μm distance along the tape when the flux density is 0.01 T. The reading head has a coil with 10 windings and is 1 μm above the tape (with a width of 50 μm), moving at a velocity of 40 m/s.

*Solution:*   Based on Eq. (33), the induced voltage is $V_{ind} = nv\ d\phi/dx = 10 \times$ 40 m/s $\times (2 \times 0.01 \times 10^{-5})$ T $\times (1 \times 10^{-6}$ m$)^2/(1 \times 10^{-6}$ m$) = 8 \times$ $10^{-5}$ V = 80 μV.  §

## Magneto-optic Effect

A **magneto-optic effect** is observed when polarized light passes through the domains in a magnetic medium. The reflected light and the transmitted light will experience a finite rotation with a polarization angle, $\Theta_p$, of a fraction of a degree for a given direction of magnetization perpendicular to the surface of the recording medium. The angle of rotation is reversed if the magnetization is in the opposite direction. This result allows us to use the reflected or transmitted light beam from a polarized source such as a laser to directly read the magnetization pattern in the recording medium. A light **polarizer** will reveal the pattern with different contrasts when there are domains with more than one direction of magnetization. A typical pattern is shown in Fig. 5.23. When the reflected light is used to study magnetization, it is called the **Kerr effect.** The **Faraday effect,** on the other hand, refers to the case when the transmitted light is used. Most magneto-optic systems use the Kerr effect. The advantage of a magneto-optic system is that it allows us to read or write a magnetic storage medium without making direct contact, thus eliminating the wear and tear associated with moving parts, as in the case of a magnetic head. Table 5.8 lists the values of magneto-optical properties of some common recording media.

**Fig. 5.23**
A magneto-optical pattern
on a recording medium

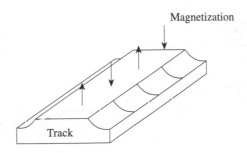

Magnetization

Track

Table 5.8    Values of the magneto-optical properties of some common recording media

| Material | Recording Temperature (°C) | $\Theta_p$ (degrees) | $H_c$ (A/m) |
|---|---|---|---|
| GdCo | 600 | 0.28 | 1,600 |
| DyFe | 70 | 0.24 | 6,400 |
| MnBi | 360 | 0.7 | 20,000 |
| TbFeCo | | 0.63 | 800,000 |
| TbFe | 140 | 0.24 | 400,000 |
| GdFe | 220 | 0.25 | 8,000 |

Writing data onto a magnetic medium involves a thermomagnetic step. The magnetic medium is first heated using a focused laser beam to a temperature above the Curie temperature in order to remove any residual magnetization. It is then allowed to cool in the presence of a magnetic field. The magnetic field can be generated using a coil (or several coils) carrying a current driven by an input signal placed close to the recording medium. In this way, the input signal is transferred directly into a magnetic pattern.

**Example 5.26**    A 5-mW laser is focused on a region for 100 ns in a magneto-optic medium of area $1 \times 10^{-13}$ m$^2$ and thickness 1 μm. What will be the temperature rise if the thermal capacity, $C_{th}$, is 1000 J/°C·kg and the density, $\rho_o$, is $1 \times 10^4$ kg/m$^3$?

*Solution:*    The energy dissipated is $E = Pt = 0.005 \times 10^{-7}$ J $=$ $5 \times 10^{-10}$ J. The corresponding temperature rise is $E/(A_{cs}t\rho_0 C_{th}) =$ $5 \times 10^{-10}$ J/($10^{-13}$ m$^2 \times 10^{-6}$ m $\times 1 \times 10^4$ kg/m$^3 \times 10^3$ J/°C·kg) $= 500$°C.   §

**Example 5.27**    A magneto-optic read system relies on the Faraday effect to read the magnetic domains. Suggest what the attenuation would be if the film thickness is 100 nm and the absorption coefficient, $\alpha^*$, is $2 \times 10^6$/m.

*Solution:*    The percent attenuation is $(1 - \exp(-\alpha^* t)) \times 100\% =$ $(1 - 2 \times 10^6$/m $\times 10^{-7}$ m$) = 80\%$.   §

## Magnetic Bubble Memories

**Magnetic bubbles** are used in the making of nonvolatile computer memories. The bubbles are very minute magnetic domains generated during the magnetization of a thin strip of an anisotropic ferromagnet with the easy axis perpendicular to the strip, as shown in Fig. 5.24. As the ferromagnetic strip is increasingly magnetized by the external field, the antiparallel domains diminish in size until they are the size of a bubble (between 1 and 3 μm). These bubbles will vanish if the bias field

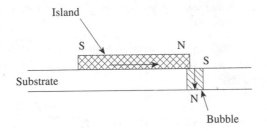

Fig. 5.24
The formation of a
magnetic bubble

increases further. The magnetic bubbles, once formed, can be moved freely from one location to another, even though they tend to congregate in regions of low magnetic-field intensity. Because of their magnetization (and the same polarity), magnetic bubbles repel each other. The manipulation of magnetic bubbles allows us to store digital data nondestructively. This storage is accomplished by generating a surface magnetic field on high-permeability islands deposited on a magnetic substrate, allowing the bubbles to reside in the low-field regions. A bubble memory circuit formed by depositing high-permeability materials in the form of I-shaped or T-shaped islands is shown in Fig. 5.25. By changing the magnetic polarization on these islands, the bubbles can be steered and moved from one island to another. It is also possible to sense the presence of a bubble using a sense coil (similar to the case of a recorder head). Magnetic-bubble memories are available in the market for large-scale nonvolatile storage (1–4-Mbit device), and the main obstacle to high-speed operation has been the relatively long transit time required to move the bubbles. At present, 0.35-$\mu$m bubbles are being investigated; they will require narrow permalloy line tracks of the order of 0.5 $\mu$m. This requirement reduces the bit period from 4 $\mu$m to 2 $\mu$m and increases the memory density to 16 to 64 Mbit.

**Example 5.28**   If the velocity of magnetic bubbles is given by $v = \mu_w(\Delta H - 8H_c/\pi)$, determine the magnetic field, $\Delta H$, when $v = 30$ m/s. Assume $H_c = 72$ A/m and $\mu_w = 2$ m$^2$/A·s.

***Solution:***   The magnetic field is $H = v/\mu_w + 8H_c/\pi = 30$ m/s/2 m$^2$/A·s + $8 \times 72$ A/m/3.14 = 233 A/m.   §

Fig. 5.25
Islands in a bubble
memory

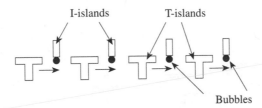

**Example 5.29**    Estimate the average spacing between magnetic bubbles if the density of the bubble memory is 1 terabit/m³.

*Solution:*    Average spacing between bubbles is $\sqrt{10^{-12}}$ m = $1 \times 10^{-6}$ m, or 1 μm. Note that this situation implies a bit period of 1 μm and requires submicron bubbles.    §

## Nuclear Magnetic Resonance (NMR) Device

Both protons and neutrons in an atom possess spins, just like the electrons. Their magnetic effect, however, is much weaker, because they are heavier and do not possess exchange interactions, as in the case of electron spins. Similar to other forms of dipoles, nuclear magnetic dipoles have different energy states corresponding to the extent of nuclear-spin magnetic interactions. A simple system such as a proton will have two energy states, corresponding to parallel and antiparallel spins in the presence of a dc magnetic field. The energy difference depends on the field intensity and the nuclear magnetic moment, $\mu_n$. The population of the higher-energy states (antiparallel spin) increases due to absorption of the ac magnetic field. Resonance (strong absorption) results if the ac field is at or near the resonant frequency corresponding to the spin transition. Fig. 5.26 shows a setup illustrating how **nuclear magnetic resonance** may be measured. The primary coil in the setup creates a vertical dc magnetic field (which sets up the resonant frequency for the energy levels in the sample), whereas the secondary coil produces an ac horizontal field (at or near the resonant frequency to excite the spin magnetic moments). If, at some later time, the horizontal field is terminated, the spin magnetic moments progressively return to their initial states, i.e., parallel to the vertical field. The relaxation process creates a decaying induction voltage in the secondary coil at the natural resonant frequency. This induced voltage is a measure of the nuclear magnetic resonance (NMR) effect. Its value is proportional to the spin density in the sample. NMR is particularly suitable for the detection of protons in living tissues. During a NMR measurement, a horizontal gradient in the vertical field is used to identify the location of the protons in the sample. The field gradient sets up a range of resonant frequencies at different points across the sample, and the response spectrum at different resonant frequencies identifies the approximate positions of the protons. The response intensity at a particular resonant frequency is proportional to the spin density in the sample. For physiological NMR measurements a dc magnetic induction of 1 T will give a proton resonant frequency at 42.5 MHz. The horizontal field gradient can be as small as 0.01 T/m.

**Fig. 5.26**
Setup for NMR
measurement

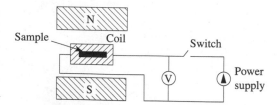

**Example 5.30**   Compute the energy of the nuclear magnetic resonance of a proton at a magnetic induction of 1 T.

*Solution:*   Since $E = h\nu$, the resonance energy is $6.62 \times 10^{-34}$ J·s $\times 4.25 \times 10^7$ s $= 2.8 \times 10^{-26}$ J.   §

**Example 5.31**   Estimate the current necessary to obtain a field strength of 1 T in NMR measurement if the coil has a diameter of 1 m and the number of turns in the coil is 50.

*Solution:*   The relationship between the magnetic induction **B** and the current through the coil is given by $NI = 2r_0B/\mu_0$. So, $I = 2r_0B/(\mu_0N) = 2 \times 1$ m $\times$ 1 T$/(50 \times 4\pi \times 10^{-7}$ H/m$) = 3.18 \times 10^4$ A $= 31.8$ kA.   §

## Metallic Magnetic Superlattices

A magnetic superlattice involves the deposition of alternating thin layers of solids of different chemical compositions, of which at least one is magnetic. The superlattice can be prepared to exact dimensions with a thickness down to a monolayer, as shown in Fig. 5.27. In general, a superlattice has a configuration given by $S[X(x)/Y(y)]_n$, where $S$ is the substrate material, $X$ is a ferromagnetic transition metal or a rare-earth ion, $Y$ is a nonmagnetic material, $n$ is the number of repetitive units in the structure, and $x$ and $y$ stand for the thicknesses of $X$ and $Y$, respectively. In a superlattice, $x$ and $y$ can vary between a few to several tens of nanometers. Table 5.9 (p. 326) lists the materials commonly used for $S$, $X$, and $Y$.

Most alternating magnetic and nonmagnetic layered structures have their magnetization in the plane of the layers, and the extent of coupling is an oscillatory function of the thickness of the spacer (nonmagnetic) layer. Furthermore, the electron spin in the magnetic layers can be either parallel or antiparallel. Electrically, when the spins in the layers are antiparallel, the resistance is high. The reverse is also true when the spins are parallel in the layers, as illustrated in Fig. 5.28.

However, when the antiparallel magnetic layers are subjected to a strong magnetic induction, the magnetization can be reversed; this process gives rise to a negative magnetoresistance effect. Materials possessing these properties include Co/Ru and Fe/Cr. The magnetoresistance of a magnetic superlattice is shown in Fig. 5.29. In general, the change in the magnetoresistance is independent of the direction of the applied magnetic field and the current flow. There is strong evidence that the change is related to the spin orientation. It appears that as the spin-up and spin-down electrons transverse through the different layers, the amount of scattering is dependent on the magnetization. With parallel spins, the scattering effect is substantially reduced.

**Fig. 5.27**
A magnetic superlattice

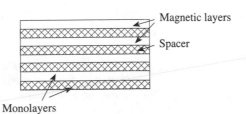

Magnetic layers

Spacer

Monolayers

Table 5.9     Materials used in metallic magnetic superlattices

| S | X | Y |
|---|---|---|
| Cu | Fe | Cu |
| Au | Co | Ag |
| Al | Ni | Au |
| Cr | Ni–Fe | Mg |
| Mo | Dy | Sn |
| LiF | Er | V |
| NaF | Gd | Nb |
| AlAs | Ho | Ta |
| GaAs | Tm | Cr |
| Ge | | Mo |
| ZnSe | | W |
| MgO | | Pd |
| Ni–P | | |

(*Source:* L. M. Falicov, "Metallic Magnetic Superlattices," *Physics Today,* October 1992, American Institute of Physics.)

Fig. 5.28
Magnetic superlattice showing parallel and antiparallel spins

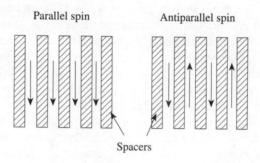

Fig. 5.29
Magnetoresistance versus applied magnetic induction in a magnetic superlattice

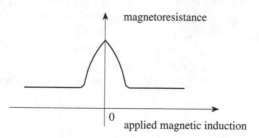

The maximum change due to the magnetization effect can be as great as 60%. Other techniques, including the surface magneto-optics effect, have also been used to study the coupling between the ferromagnetic layers.

One of the major applications of the negative magnetoresistance effect is in its use as a recorder head. In principle, by removing the need for an inductive coil (present in the conventional recorder head) and using the magnetoresistance effect, both the signal sensitivity and the dependence on the mechanical movements of the tape or disk can be reduced (inductive magnetization depends on the time rate of change of the magnetization).

# 5.8   PROPERTIES OF SUPERCONDUCTORS

## Superconducting Solids

The superconducting properties of a solid are closely associated with the properties of the superconducting electrons, although they are also affected by the current density in the superconductor, the applied magnetic field, and the temperature. A superconducting material is one that has a negligible resistance even when measured with the most sophisticated instrument. The resistivity of a superconducting wire has been measured to be less than $1 \times 10^{-26}$ $\Omega \cdot$m. This value is many orders of magnitude lower than what we can measure in the best conductor. Figure 5.30 shows the difference in resistivity between a normal metallic conductor and a **superconductor.** Note that the resistance in a superconductor changes abruptly at the **critical temperature** $T_{cr}$, whereas the resistance of a normal conductor will decrease progressively below that temperature (Matthiessen rule). Superconductors have many interesting properties not commonly found in other solids. The most important one of course is the fact that the resistance vanishes at the critical temperature $T_{cr}$. This feature has created many possible applications, in particular, the development of a lossless electromagnet. We shall first examine how a current $I$ (in A) may be induced in a superconducting coil.

Figure 5.31 shows a superconducting coil connected to a normal voltage source. If at $t = 0$ the switch S is closed, the current originally appearing in the coil will persist in the circuit as long as the energy loss remains negligible. This is one way a perpetual current can be generated. An interesting question arises in the current division if there is more than one—say, two—current branches in the superconducting coil. The superconducting current will somehow have to be divided between

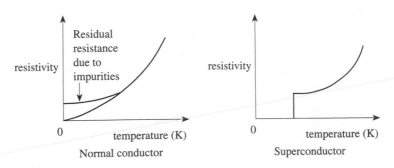

**Fig. 5.30**
Resistivities of a normal conductor and a superconductor

**Fig. 5.31**
A superconducting coil
connected to a voltage
supply

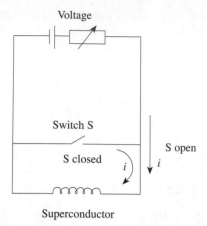

the two branches, as shown in Fig. 5.32. Since resistance is negligible in either coil, **Kirchhoff's law** cannot be applied.

The answer to this question comes from the initial conditions of the circuit—i.e., the current distribution when the voltage source just turns on. At this moment, the voltage across the two branches has to be the same, which implies that

$$L_1 d(i_1)/dt = L_2 d(i_2)/dt \qquad (34)$$

where    $L_1$ and $L_2$ are the respective inductances of the two branches in H, and

$i_1$ and $i_2$ are the respective currents in A.

Thus, the current ratio is given by the ratio of the inductances:

$$i_1/i_2 = L_2/L_1 \qquad (35)$$

Another question arises as to what may happen to the superconductor in the presence of an external magnetic field $H_x$. For a normal conductor, the magnetic

**Fig. 5.32**
Current division between
two superconducting coils

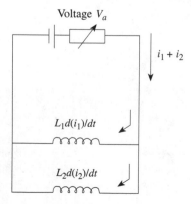

field lines will penetrate through the conductor, as shown in Fig. 5.33. When the resistance of the conductor goes to zero, there cannot be any current inside the (perfect) conductor. If the external magnetic field is switched off, a surface current will flow in the conductor so that the internal magnetic flux density remains the same. Since there is no resistance, the surface current will persist forever. This situation does not happen to a superconductor. Just the opposite of a conductor, a superconductor will repel any external magnetic field at all times through the presence of an induced surface current. The surface current will disappear once the external magnetic field is removed. This result is called the **Meissner effect** and is a useful way to distinguish between a normal conductor and a superconductor. Although a superconductor will reject a magnetic field, at its surface there is a thin transition region in which the external magnetic field, $H_x$, can have some finite penetration. This distance is called the **penetration depth** of the superconductor, $\lambda_L$. To estimate the penetration depth, we must use the Maxwell equations. We shall assume a superconductor with a carrier density $n_s$; each carrier has mass $m_s$, charge $q_s$, and velocity $v_s$. In one dimension, the spatial variation of the magnetic induction, $\mathbf{B}$ ($= \mu H$, see Eq. (5)) (T), inside a superconductor can be shown to have the following properties:

$$d^2\mathbf{B}/dx^2 = \mathbf{B}/\lambda_L^2 \qquad (36)$$

where   $\lambda_L$ ($= \sqrt{(m_s/(\mu_0 n_s q_s^2))}$ is the penetration depth in m, and

$\mu_0$ is the permeability in vacuum in F/m.

The solution to Eq. (36) is given by

$$\mathbf{B} = B_0 \exp(-x/\lambda_L) \qquad (37)$$

where   $B_0$ is the surface flux density in T.

Roughly speaking, the magnetic field lines will penetrate a distance of the order of $\lambda_L$ into a superconductor. Typical values of $\lambda_L$ are in the range of 10 to 1000 nm, as listed in Table 5.10.

Since $\lambda_L$ depends on the carrier density, $n_s$ (see Eq. (36)), it is temperature dependent. Experimentally, $\lambda_L$ has the following empirical temperature dependence:

$$\lambda_L(T)/\lambda_L(0) = 1/\sqrt{1 - (T/T_{cr})^4} \qquad (38)$$

**Fig. 5.33**

Field distribution near a conductor

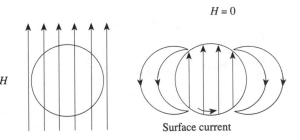

$H = 0$

$H$

Surface current

**Table 5.10**    Materials parameters of the common superconductors

| Superconductor | $T_{cr}$ (K) | $\lambda_L$ (nm) |
|---|---|---|
| Al | 1.2 | 50 |
| In | 3.4 | 64 |
| Sn | 3.7 | 51 |
| Pb | 7.2 | 39 |
| Nb | 9.2 | 47 |
| $YBa_2Cu_3O_7$ | 92 | 150 |

(*Source:* V. Z. Kresin and S. A. Wolf, *Fundamentals of Superconductivity* (New York: Plenum Publishing Corporation, 1992). Reprinted with permission of Plenum Publishing Corporation.)

where    $\lambda_L(0)$ is the penetration depth at 0 K (absolute zero) in m,

$T_{cr}$ is the critical temperature in K, and

$T$ is the temperature in K.

Based on Eq. (38), at $T = T_{cr}$ the penetration depth extends throughout the solid, and the superconductor becomes a normal conductor. Typical values of $T_{cr}$ are listed in Table 5.10.

Superconductivity in a solid is due to special electron pairs called **Cooper pairs.** Cooper pairs have the unique property that the pair momentum remains zero at all times in the superconducting state. What this means is that Cooper pairs will not experience any changes in momentum due to scattering by the lattice, and hence they will conduct a current persistently without suffering any losses. Physically, a Cooper pair consists of a pair of electrons with opposite spins kept at a distance of 0.1 to 1 $\mu$m apart. This distance is called the **coherence length,** and it is roughly equal to $10^4$ atomic spacing. In a superconductor, there are normal electrons in addition to the Cooper pairs. The major difference is that all the Cooper pairs exist as a single entity (as a single coherent electron wave) and respond collectively to an external force. In the ground state, all the electrons in a superconductor are bound to Cooper pairs, and the binding energy for a Cooper pair, $2\Delta^*$, has a value roughly between 2 to 20 meV. At the superconducting temperature, Cooper pairs are quite stable, but they will dissociate if there is a large current flowing. At the critical current, there will be enough energy in the superconductor to overcome the binding energy, $2\Delta^*$, and to break up the Cooper pairs. This result occurs at about 100 A for a wire diameter of 1 mm.

Raising the temperature of the superconductor reduces $\Delta^*$ according to the following empirical relationship:

$$\Delta^*(T)/\Delta^*(0) = \sqrt{\cos(\pi(T/T_{cr})^2/2)} \qquad (39)$$

where    $\Delta^*(0)$ is the binding energy at 0 K (absolute zero) in eV, and

$T_{cr}$ is the critical temperature in K, above which superconductivity ceases to exist.

Thus, at the critical temperature, $\Delta^*(T_{cr})$ vanishes, and the superconductor again becomes a normal conductor. It is possible to view the normal electrons as originated from the Cooper pairs; they are called the **quasiparticles.** In a superconductor, the quasiparticles are the excited states of the Cooper pairs, and they are separated from the ground states by an **energy gap** of $2\Delta^*$, as shown in Fig. 5.34. In this case, the energy gap is the binding energy of the Cooper pairs. As we shall see, the quasiparticles are responsible for current flow in a manner quite similar to the electrons in the energy bands of a semiconductor.

Heat loss due to a surface current is important in a superconductor. This loss is similar to the **skin effect** found in a normal conductor in the presence of an ac electric field. If we assume that the superconductor has a complex conductivity due to the normal electrons (or quasiparticles) and Cooper pairs, the **complex conductivity,** $\sigma_s$ (S/m), is given by

$$\sigma_s = \sigma_1 + j\sigma_2 \qquad (40)$$

where   $\sigma_1$ is the real part of the complex conductivity due to the normal electrons in S/m, and

$\sigma_2$ is the imaginary part of the complex conductivity in S/m.

With $\sigma_2 \gg \sigma_1$, the real part of the surface impedance, $R'_s$, is given by

$$R'_s \approx 0.5\mu_0^2\sigma_c\lambda_L^3\omega^2(T/T_{cr})^4/[1 - (T/T_{cr})^4]^{1.5} \qquad (41)$$

where   $\sigma_c$ is the conductivity in S/m when $T = T_{cr}$, and

$\omega$ is the angular frequency of the ac signal in rad/s.

Equation (41) shows that the heating effect due to resistive (joule) heating in a superconductor has a quadratic frequency dependence. This is quite different from

**Fig. 5.34**

An energy band diagram for quasiparticles

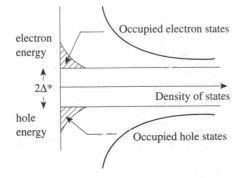

the case of a normal conductor, when $R_s = \sqrt{\omega\mu_0/(2\sigma)}$. A plot of $R_s$ for a supercon-ductor and a normal conductor is shown in Fig. 5.35.

It is possible to relate the critical current density, $J_{cr}$, that leads to the breakup of a Cooper pair to the energy gap $2\Delta^*$ (J). The critical current density, $J_{cr}$ (A/m$^2$), is given by

$$J_{cr} = qn_s\Delta^*/\sqrt{2m_sE_F} \tag{42}$$

where    $E_F$ is the Fermi energy in J.

Assuming that $J_{cr}$ is confined to the penetration depth, $\lambda_L$ (m), the **critical magnetic field intensity,** $H_{cr}$ (A·m), can be shown to be given by

$$H_{cr} = J_{cr}\lambda_L \tag{43}$$

For a circular wire with the radius $r$ (m) $>> \lambda_L$, the **critical current** (A) becomes

$$I_{cr} = 2\pi rH_{cr} \tag{44}$$

In some superconductors, it is possible that a finite number of magnetic field lines will penetrate the superconductors. However, because of the coherent properties of the Cooper pairs, the enclosed magnetic flux lines will be quantized (that is, they must have discrete values). One way to visualize this is to treat the Cooper pairs that make up the surface current around the flux lines as a standing wave (remember that there is no momentum change for the Cooper pairs). In a closed loop, the phase change can only be in multiples of $2\pi$. The same argument extends to perturbations caused by the external magnetic field lines. The unit of magnetic quantization in a superconductor is called the **fluxoid,** $\Phi_0$. It has a value of $h/2q$ (= $2.07 \times 10^{-15}$ Wb). Magnetic penetration occurs only in **Type II superconductors.** These

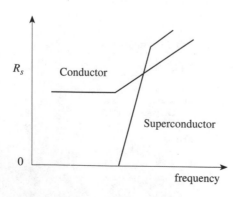

**Fig. 5.35**
Surface resistance versus frequency for a conductor and a superconductor

$R_s$

Conductor

Superconductor

0

frequency

superconductors are unique in the sense that above a certain magnetic field intensity, they possess **normal** (electron) **states** in addition to pure superconducting states. The magnetic field lines concentrate into small cylindrical volumes within the superconductors that have lost their superconducting properties. A schematic of the mixed states of a Type II superconductor is shown in Fig. 5.36.

**Example 5.32**   What is the current ratio between two superconducting coils connected in parallel, with one having three times as many windings?

*Solution:*   Based on Eq. (35), the coil with the smaller number of windings will carry three times as much current as the one with a larger number of windings.   §

**Example 5.33**   Compute the penetration depth of Pb at 3 K.

*Solution:*   Based on Eq. (38), $\lambda_L(T) = \lambda_L(0)/\sqrt{1 - (T/T_{cr})^4}$. From Table 5.9, $\lambda_L(0) = 39$ nm. Therefore, $\lambda_L(T) = 39$ nm$/\sqrt{1 - (3 \text{ K}/7.2 \text{ K})^4}$ nm $= 39.6$ nm.   §

**Example 5.34**   Compute the ratio of $\Delta^*/\Delta^*(0)$ in Pb at 3 K.

*Solution:*   Based on Eq. (39), $\Delta^*(T)/\Delta^*(0) =$
$\sqrt{\cos(\pi(T/T_{cr})^2/2)} = \sqrt{\cos(3.14/2 \times (3 \text{ K}/7.2 \text{ K})^2)} = 0.98$.   §

**Example 5.35**   Estimate $J_{cr}$ if a superconductor has $\Delta^* = 0.17$ mV. Assume $n_s = 1 \times 10^{28}/\text{m}^3$, $m_s = 0.91 \times 10^{-30}$ kg, and $E_F = 11.7$ eV.

*Solution:*   Based on Eq. (42), $J_{cr} = qn_s\Delta^*/\sqrt{2m_sE_F} = 1.6 \times 10^{-19}$ C $\times 1 \times$
$10^{28}/\text{m}^3 \times 0.17 \times 10^{-3} \times 1.6 \times 10^{-19}$ V$/$
$\sqrt{2 \times 0.91 \times 10^{-30} \text{ kg} \times 1.6 \times 10^{-19} \text{ V/eV} \times 11.7 \text{ eV}} = 2.35 \times 10^{10}$ A/m².
§

**Fig. 5.36**
The mixed states in a Type
II superconductor

Normal states

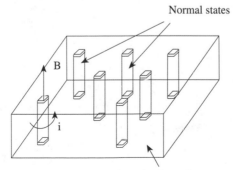

Superconducting substrate

## 5.9 APPLICATIONS OF SUPERCONDUCTORS

### Superconductor-Insulator-Superconductor (SIS) Junction

A **superconductor-insulator-superconductor** (SIS) **junction** has some very useful tunneling properties that can be exploited in device applications. A typical SIS junction is shown in Fig. 5.37. To achieve tunneling, the insulator layer has to be very thin, less than 10 nm. As we are aware, both normal electrons and Cooper pairs can take part in the current-conduction process. We shall for the time being restrict our attention to normal electrons. The Cooper pairs are assumed to be suppressed due to the presence of, say, a strong magnetic field (see Eq. (42)). In the SIS junction, we consider the normal electrons to be similar to the electrons in the valence band and the conduction band of a semiconductor with an energy gap equal to $2\Delta^*$ (see Fig. 5.34). As we observed in the figure, the main difference is that density of states of the normal electrons in a superconductor is highest near the band edge, decreasing away from the band edge. Similar to the case of a **tunneling diode,** when a bias is applied to the SIS junction, the energy bands shift proportionally. Current starts to flow abruptly once the filled conduction band states are opposite to the empty valence band states. This situation corresponds to a bias voltage of $2\Delta^*/q$. The current-voltage *(I-V)* characteristics of a tunneling SIS junction are shown in Fig. 5.38. For comparison, we also show the *I-V* characteristics of a conductor-insulator-conductor sandwich structure.

**Fig. 5.37**
A typical SIS junction

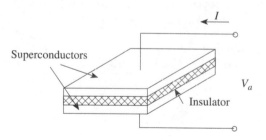

Superconductors

Insulator

$V_a$

$I$

**Fig. 5.38**
Tunneling *I-V*
characteristics of an SIS
junction

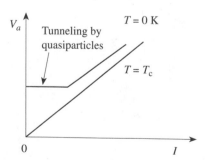

$V_a$

$T = 0$ K

Tunneling by
quasiparticles

$T = T_c$

0

$I$

Mathematically, the dc $I$-$V$ characteristics of the SIS junction can be expressed as

$$I_{dc} = (2G_n/q)\exp(-\Delta^*/(kT))\sqrt{2\Delta^*/(qV_a + 2\Delta^*)} \cdot \qquad \textbf{(45)}$$
$$(qV_a + \Delta^*)\sinh(qV_a/(2kT))K_0(qV_a/(2kT))$$

where   $G_n$ is the conductance of the tunneling junction in S,

   $2\Delta^*$ is the energy gap in J,

   $k$ is Boltzmann's constant in J/K,

   $T$ is the temperature in K,

   $V_a$ is the applied voltage in V, and

   $K_0(\;)$ is a special function.

In addition to the dc $I$-$V$ characteristics, an SIS junction will also respond to an ac (say, microwave) signal. Let us assume the SIS junction is placed in a microwave circuit with an input voltage, $V_{ac}(t)$ in volts, given by

$$V_{ac}(t) = V_0 + V_1\cos(\omega t) \qquad \textbf{(46)}$$

where   $V_0$ is the dc component of the input voltage in V,

   $V_1$ is the amplitude of the ac signal in V, and

   $\omega$ is the angular frequency in rad/s.

Because of the size of the energy gap ($\approx$ a few meV), **microwave radiation** will be able to assist in the tunneling process, as illustrated in Fig. 5.39.

Therefore, instead of an abrupt change, the **tunneling current** will increase in small steps proportional to the number of (microwave) photons that are participating

**Fig. 5.39**
Radio frequency radiation-assisted tunneling

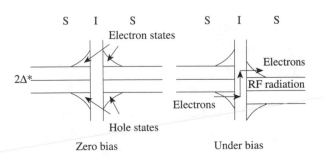

in the tunneling process. This situation is shown in Fig. 5.40. The voltage step is equal to $h\omega/(2\pi q)$.

In addition, the dc current component of the **photon-assisted tunneling** process is given by

$$I_0 = \sum_{n=-\infty}^{+\infty} J_n^2(2\pi V_1/h\omega) \cdot I'_{dc} \tag{47}$$

where $\Sigma$ is a summation sign,

$J_n(\ )$ is another special function, and

$I'_{dc}$ is the same as $I_{dc}$ in Eq. (45), except $V_a$ is replaced by $V_0 + nh\omega/(2\pi q)$ because of the ac effect.

Equation (47) illustrates how an SIS junction may be used as a microwave detector. For a small RF voltage, $J_n(\ )$ is proportional to its argument, and $\Delta I_0$ is proportional to $V_1^2$. Thus, the SIS junction is a **quadratic (microwave) signal detector.** The detector sensitivity can be shown to be proportional to the square of the curvature in the transition region (in fact, the sensitivity is approximately $(\partial^2 I/\partial V_a^2)/(\partial I/\partial V_a))$ of the $I$-$V$ curve (see Fig. 5.40). In the SIS junction, the sensitivity approaches the quantum limit of the detector, i.e., $2\pi q/(h\omega)$. This value corresponds to a tunneling electron for every RF photon absorbed. Assuming a detector sensitivity of $2 \times 10^4$/V for a typical SIS junction, this value translates into a frequency limit of 13 GHz. In this limit, the noise-equivalent power, NEP, can be as low as $1 \times 10^{-15}$ W/Hz (an exceptionally sensitive detector).

The construction of a typical SIS junction detector is shown in Fig. 5.41. The same SIS junction may also be used as a microwave mixer (a device that combines

**Fig. 5.40**

*I-V characteristics of an SIS junction with RF absorption*

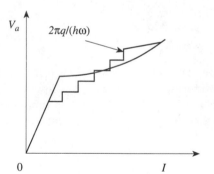

**Fig. 5.41**

*An SIS junction detector*

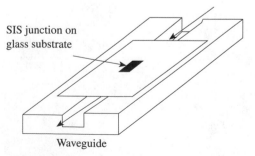

SIS junction on glass substrate

Waveguide

two microwave signals). In this application, the nonlinear $I$-$V$ characteristics of the SIS tunneling diode generate an output signal with a frequency that is the sum of the frequencies of the **local oscillator** (an ac source to be mixed with the input signal) and the harmonics of the input signal. Similar to the SIS signal detector, the noise level is extremely low in an SIS **mixer,** far superior to the ordinary Schottky diode (metal-semiconductor junction) mixers. SIS mixers can operate at a frequency in the range of hundreds of gigahertz.

**Example 5.36**    Show that for a small RF voltage, $\Delta I_0$ is proportional to $V_1^2$ if $J_n^2(2\pi V_1/h\omega)$ is proportional to its argument.

*Solution:*    Based on Eq. (47), $I_0 = \sum\limits_{n=-\infty}^{+\infty} J_n^2(2\pi V_1/h\omega)I_{dc}(V_0 + nh\omega/2\pi q)$.

When $n = 1$, $J_n(\ )$ is proportional to its argument, so $I_0$ is proportional to $V_1^2$. Note that this is a quadratic signal detector.    §

**Example 5.37**    Compute the quantum limit of sensitivity for an SIS junction if the device is operating at 100 GHz.

*Solution:*    The quantum limit is given by $q/(hv) = 1.6 \times 10^{-19}$ C/$(6.62 \times 10^{-34}$ J·s $\times 10^{11}$/s$) = 2.4 \times 10^3$ V.    §

## Josephson Junction

A **Josephson junction** is quite similar to an SIS junction; the main difference is that the carriers involved in the tunneling process are Cooper pairs. Because of phase coherence in the Cooper pairs, the phase has to be matched on either side of the insulator layer. If the phase difference is $\Delta\phi = \phi_1 - \phi_2$ (where $\phi_1$ and $\phi_2$ are the phases of the electron waves in the two superconducting regions), the result will be a tunneling current density $J_{dc}$ (*without any applied voltage*) of the form

$$J_{dc} = J_{cr}\sin(\Delta\phi) \tag{48}$$

where    $J_{cr}$ is the critical current density in A/m$^2$.

Let us look at the case when there is a dc voltage bias across the Josephson junction. Then $d\phi/dt$ will be proportional to the applied voltage, $V_a$ (V), which results in an ac current density, $J_{ac}$ (A/m$^2$), given by

$$J_{ac} = J_{cr}\sin(4\pi qV_at/h + \phi_0) \tag{49}$$

where    $\phi_0$ is a constant,

$q$ is the electron charge in C, and

$h$ is Planck's constant in J·s.

Equation (49) suggests that the Josephson junction will exhibit ac oscillation under a dc bias. The oscillation frequency, $\omega$(rad/s), is given by

$$\omega = 4\pi qV_a/h \tag{50}$$

The ac oscillation is superimposed on the dc *I-V* characteristics. Since the current flow in the Josephson junction is no different from that in the SIS junction, we have to include the current contribution from the normal electrons as well as from the Cooper pairs. At $T = 0$ K, the overall *I-V* characteristics of the Josephson junction are shown in Fig. 5.42. As in any other ac circuits, the Josephson junction under a dc bias will emit radiation at a frequency equal to $\omega$. This is normally in the far infrared range.

A magnetic field when applied to a Josephson junction will induce flux penetration in the insulator layer. Assuming a cross section of the interface, as illustrated in Fig. 5.43, it can be shown that the critical current, $I_{cr}$ (A), in the presence of a transverse magnetic field, $H_y$ (A/m), will have a value given by

$$I_{cr}(H_y) = I_{cr}(0)\left|\sin((\pi\Phi/\Phi_0)/(\pi\Phi/\Phi_0))\right| \tag{51}$$

where   $\Phi\ (= \mu_o H_y l(d + 2\lambda_L))$ is the applied flux in Wb (see Fig. 5.43 for the dimensions of the Josephson junction), and

$\Phi_0 = h/2q = 2.1 \times 10^{-15}$ Wb.

Figure 5.44 shows how the critical current, $I_{cr}(H_y)$, varies with the applied flux $\Phi$.

So far, our model of a Josephson junction applies only to a narrow junction. In practice, this can be a superconducting point contact when the narrow tip end of the contact has turned into a normal conductor or when there is a narrow constriction in a planar superconductor. Fig. 5.45 shows the equivalent-circuit model of a superconducting junction. The junction consists of a capacitor, $C_j$ (F), a resistor, $R_n$ ($\Omega$), and an

**Fig. 5.42**
*I-V* characteristics showing Josephson current and quasiparticle current

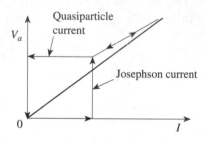

**Fig. 5.43**
Cross section of a Josephson junction

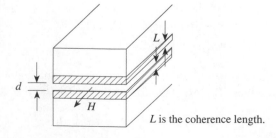

*L* is the coherence length.

**Fig. 5.44**
Critical current, $I_{cr}$, versus
normalized magnetic flux

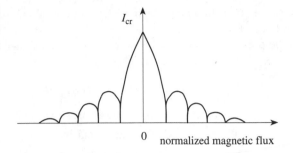

Critical current, $I_{cr}$, versus normalized magnetic flux

**Fig. 5.45**
An equivalent circuit
model for a Josephson
junction

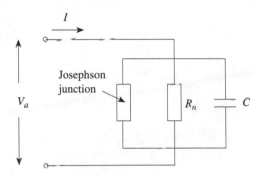

ideal Josephson junction connected in parallel. The total current, I(t) (A), passing through the device in the absence of any applied ac voltage is given by

$$I(t) = I_{cr}\sin(\phi) + V_a/R_n + C_j d(V_a)/dt \qquad (52)$$

where $V_a$ is the terminal voltage in V.

The dc solution of Eq. (52) for the case when $C_j = 0$ is relatively simple. When $I > I_{cr}$, the dc I-V characteristic is given by

$$V_a = R_n\sqrt{I^2 - I_{cr}^2} \qquad (53)$$

This situation is illustrated in Fig. 5.46. In the figure, the ac oscillation (resulting from Eq. (46)) is superimposed on the dc I-V characteristics. For the case of a finite

**Fig. 5.46**
I-V characteristics of a
Josephson junction
showing ac oscillation

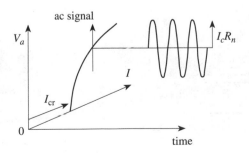

capacitance, $C$, a hysteresis loop will be observed in the $I$-$V$ characteristics, as illustrated in Fig. 5.47.

In the presence of an ac signal similar to the one shown in Eq. (46), the junction current, $i_j(t)$ (A), through the Josephson junction is given by

$$i_j(t) = I_{cr}\sin[4\pi qV_0t/h + 4\pi qV_1/(h\omega)\sin(\omega t) + \phi_0] \tag{54}$$

where     $V_0$ is the dc component of the input signal in V,

$V_1$ is the amplitude of the ac signal in V,

$\omega$ is the angular frequency in rad/s, and

$\phi_0$ is a constant.

Equation (54) can also be written as

$$i_j = I_{cr} \int_{m'=-\infty}^{+\infty} \{J_m(2\pi V_1/(h\omega))\sin(m'\omega + 4\pi qV_0/h)t + \phi_0\} \tag{55}$$

where     $J_m(\ )$ is a special function.

To obtain a dc solution from Eq. (55), we set $m'\omega = -4\pi qV_0/h$; this leads to quantization in the dc input voltage, which is given by

$$V_0 = m'\omega h/(4\pi q) \tag{56}$$

Based on Eq. (56), a range of dc currents will be observed for each value of $V_0$, as shown in Fig. 5.48. We can now see how a Josephson junction may be used as a

**Fig. 5.47**

*I*-*V* characteristics of a Josephson junction

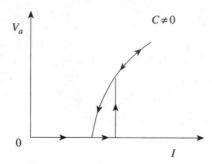

**Fig. 5.48**

Modulated dc current versus $V_0$

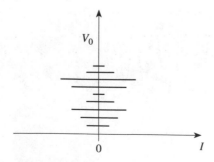

voltage standard. Equation (56) essentially describes the operation of a frequency-to-voltage converter. By choosing a suitable microwave signal with a frequency $\omega_0$ (which can be controlled very precisely), a particular value of $V_0$ will be established across the Josephson junction for a given $m'$. The maximum value of $V_0$ for a single Josephson junction cannot exceed the value of the energy gap because when $V_0$ is greater than $2\Delta^*$, the junction becomes rectifying. A series of Josephson junctions, however, will be able to provide a voltage value of the order of a volt or higher. In this way, the Josephson junctions can be used as a voltage standard similar to the **bandgap voltage reference (circuit),** which relies on the value of the energy gap in a *p-n* junction.

A Josephson junction can also be used to sense a small magnetic field. To achieve a very sensitive magnetic-field measurement, a device called a SQUID (**superconducting quantum interference device**) is used. A simple SQUID consists of two or more Josephson junctions connected in parallel to form a closed loop. Figure 5.49 shows a typical connection. During sensing, the external magnetic flux, $\Phi_F$ (Wb), penetrates the loop; because of phase coherence of the Cooper pairs, the total **phase** difference across the entire loop has to be equal to $n2\pi$, where $n$ is an integer. Assuming the phases in the Josephson junctions are $\phi_1$ and $\phi_2$, respectively, the quantized phase change due to $\Phi_F$ will be $2\pi\Phi_F/\Phi_0$. Thus, the overall phase difference is given by

$$\phi_1 = \phi_2 + 2\pi\Phi_F/\Phi_0 \tag{57}$$

The dc current, $I_g$ (A), is given by

$$I'_g = I_1 + I_2 = I_{c1}\sin(\phi_1) + I_{c2}\sin(\phi_2) \tag{58}$$

where $I_{c1}$ and $I_{c2}$ are the respective critical currents in A.

If $I_{c1} = I_{c2}$, we get

$$I'_g = I_{c1}(\sin(\phi_1) + \sin(\phi_2 + 2\pi\Phi_F/\Phi_0)) \tag{59}$$

**Fig. 5.49**
A typical connection of a SQUID

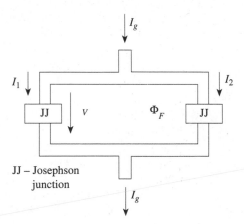

JJ – Josephson junction

The maximum value of $I'_g$ that doesn't generate a dc voltage drop across the device occurs when

$$I_{gc} = 2I_{c1}|\cos(\pi/\Phi_F/\Phi_0)| \tag{60}$$

Ideally, Eq. (60) has a peak value of $2I_{c1}$. Figure 5.50 shows the actual $I$-$V$ characteristics measured in a real device. For a given magnetic flux, $\Phi_F$ (varying between $n\Phi_0$ and $(n + 1)\Phi_0$), and an input current, $I_g$, the device will have the dynamic characteristics shown in the figure. For a properly designed device, the detectivity of such a magnetic-field sensor can be down to $1 \times 10^{-33}$ J/Hz, a value quite comparable to the best available magnetic-field sensor. Other applications of the Josephson junction include its use as a microwave detector, a signal mixer, or an amplifier. In general, a Josephson junction device has a much higher operating frequency (well into the hundreds of gigahertz range) when compared with an SIS junction device, even though the latter has less noise.

**Example 5.38**   Compute the ac oscillation in a Josephson junction at an applied voltage of 1 V.

*Solution:*   Based on Eq. (50), the oscillation frequency is $\omega = 4\pi q V_a /h = 4 \times 3.14 \times 1.6 \times 10^{-19}$ C $\times 1$ V/$(6.62 \times 10^{-34}$ J·s$) = 3 \times 10^{15}$/s (Hz).   §

**Example 5.39**   Compute the critical current in a Josephson junction if $I_{cr}(0) = 10$ μA and the magnetic induction in the $y$ direction, $\mathbf{B}_y$, is $8 \times 10^{-2}$ T. The gap spacing is 10 nm, $\lambda_L = 10$ nm, and the width of the junction is 10 μm. $\Phi_0 = h/2q = 2.1 \times 10^{-15}$ Wb.

*Solution:*   Based on Eq. (51), the applied flux is $\Phi = \mu_0 H_y l(d + 2\lambda_L) = 8 \times 10^{-2}$ T $\times 1 \times 10^{-5}$ m $\times (10^{-8}$ m $+ 2 \times 10^{-8}$ m$) = 2.4 \times 10^{-14}$ Wb. The critical current is $I_{cr}(H_y) = I_{cr}(0) \,| \sin((\pi\Phi/\Phi_0)/(\pi\Phi/\Phi_0)) \,| = 10 \times | \sin(\pi \times 2.4 \times 10^{-18}$ Wb/$(2.1 \times 10^{-15}$ Wb$))/(\pi \times 2.4 \times 10^{-14}$ Wb/$2.1 \times 10^{-15}$ Wb$) \,| = 0.83 \times 10^{-6}$ A $= 0.83$ μA.   §

**Example 5.40**   Compute the voltage drop across the Josephson junction when $I = 15$ μA. Use the equivalent circuit model when $C_j = 0$, $R_n = 0.02$ Ω, and $I_{cr} = 10$ μA.

**Fig. 5.50**
*I-V characteristics of a SQUID circuit showing the flux dependence*

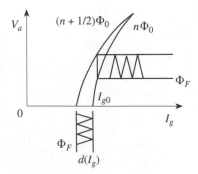

***Solution:***   Based on Eq. (53), $V_a = R_n \sqrt{I^2 - I_{cr}^2} = 0.02 \, \Omega \times$
$\sqrt{15^2 - 10^2} \times 10^{-6} \, A = 0.1 \times 10^{-6} \, V = 0.1 \, \mu V.$   **§**

**Example 5.41**   As a frequency-to-voltage converter, compute the dc voltage across a given Josephson junction if the input ac signal frequency is 10 GHz.

***Solution:***   Based on Eq. (56), the voltage across the Josephson junction is
$V_0 = m' h \omega/(4 \pi q) = 1 \times 6.62 \times 10^{-34} \, J \cdot s \times 10^{10}/s/(4 \times 3.14 \times 1.6 \times 10^{-19} \, C) = 3.3 \times 10^{-6} \, V = 3.3 \, \mu V.$   **§**

## High-Temperature Superconductors

Our understanding of **high-temperature** (HT) **superconductors** has improved significantly in the last few years, and great improvements have been made in the achievement of a high critical temperature and certain types of structural stability. HT superconductors have very complex structures; to harness their utility, there has to be a much better understanding of the superconducting mechanism. Many of the HT superconductors have a layered cuprate structure, and the oxygen content appears to play an extremely significantly role. The actual electronic states of the Cu ions also appear to be quite important. The $YBa_2Cu_3O_7$ HT superconductors have a critical temperature up to 92 K, which can be carefully manipulated through changing the oxygen content. $Tl_2Ba_2CuO_6$ HT superconductors increase their critical temperature when annealed in a pressurized atmosphere containing $H_2$ and Ar. Annealing can occur as low as 300 °C, and prolonged annealing raises the superconducting temperature to around 90 K. It is believed that the increase in the critical temperature is due to a change in the hole carrier density. Other HT superconductors, such as $Tl_2Ba_2Ca_2Cu_3O_{10}$, become superconductors only when annealed in an $O_2$ atmosphere. Thus, there is ample evidence that the presence or absence of oxygen vacancies plays a crucial role in eliciting the superconducting properties in the family of cuprate superconductors.

In general, HT superconductors can be classified as having the general formula $(ACuO_{3-x})_m(AO)_n$, where A stands for Ba, Tl, Ca, Th, Bi, Pb, and so on. The determining factor, however, seems to be $x$ in the oxygen content. The structure $(ACuO_{3-x})_m(AO)_n$ consists of both perovskite layers and rock salt layers $(AO)_n$ (with $n$ up to 3), as shown in Fig. 5.51. The perovskite layers are referred to as the

**Fig. 5.51**
The perovskite layer (I) and the rock salt layer (II) in an HT superconductor

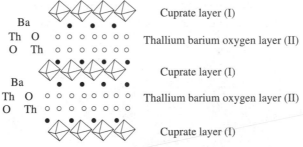

(*Source:* B. Raveau, "Defects and superconductivity in layered cuprates," *Physics Today* (October 1992): p. 54. Reprinted by permission of the American Institute of Physics, USA.)

superconducting layers and the rock salt layers as the normal layers. These layers can rearrange themselves so that superconductors are formed with the proper oxygen content.

Some of the major problems with HT superconductors appear to be the difficulty in stabilizing the magnetic flux vortices in the thin films and the fact that the critical current in single crystals and ceramics is very low. The latter problem is believed to be due to weak links (nonsuperconducting regions) present at the grain boundaries. Some form of melt-textured growth appears to be capable of removing the grain boundaries and is helpful in increasing the critical current. Magnetic flux pinning in thin films, on the other hand, is achieved through the addition of impurities and defects, which appear to be able to pin the vortices, as shown in Fig. 5.52. Implantation of heavy nonmagnetic ions such as Pb at several GeV can produce stable vortices, which show a critical current density of 3.5 A/m$^2$ at 77 K and 2.6 A/m$^2$ at the same temperature in the presence of a magnetic induction of 7 T. Thus, the research into HT superconductors involves an understanding of the role of the layered structure and the defect properties leading to the generation of the superconducting energy states.

HT superconductors have the advantages that they will become superconductors at a temperature above liquid N$_2$ temperature and in principle can be used to perform most of, if not all, the tasks of the low-temperature superconductors. These include use as cables in power transmission with a current density in excess of 10$^9$ A/m$^2$ and in power devices such as electromagnets, transformers, and generators. The use of HT superconductors in the generation of a high magnetic field (>2 T) is particularly appealing because of the high critical field, $H_{cr}$. At present, most of these devices are made using Nb–Ti superconductor wires clad with a thin Cu coating. There are, however, a number of technical problems with the use of HT superconductors. The principal one has to do with brittleness of these superconductors. Being similar to ceramics, HT superconductors cannot be easily made into thin wires, and most HT superconductors are deposited in the form of thin films.

A possible application of HT superconductors is in **magnetic levitation.** The idea is to use the magnetic repulsive force to achieve levitation of heavy objects. The same concept has been extended to develop magnetically levitated trains (**maglevs**). Magnetic levitation can be easily demonstrated in the lab by making a slab of HT superconductor to suspend over a permanent magnet. The expulsion of the flux lines in the HT superconductor will allow the slab to be suspended in midair, as shown in Fig. 5.53.

The following is a simple mathematical model that may be used to illustrate the effect of magnetic levitation. The total energy, $E$ (J), required to suspend a

**Fig. 5.52**
Defect pinning of the magnetic vortices in an HT superconductor

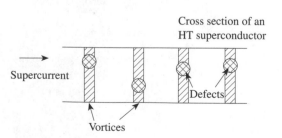

Fig. 5.53
A superconductor in
suspension

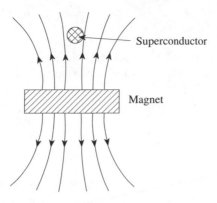

Superconductor

Magnet

superconducting slab in midair (above a permanent magnet) is given by the sum of the magnetic energy and the gravitational potential energy:

$$E = \mu_0 H^2 h_1^2 V_{sl}/(2z^2) + \rho_0 V_{sl} g_r z \qquad (61)$$

where    $H$ is the magnetic-field intensity in A/m,

$h_1$ is the thickness of the slab in m,

$z$ is the position of the slab in midair in m,

$\rho_0$ is the density of the slab in kg/m³,

$V_{sl}$ is the volume of the slab in m³, and

$g_r$ is the gravitational acceleration in m/s².

At equilibrium, when $dE/dz = 0$,

$$z^* = \mu_0^{1/3} H^{2/3} h_1^{2/3}/(\rho_0 g_r)^{1/3} \qquad (62)$$

where    $z^*$ is the height of the slab above the permanent magnet in m.

Thus, the HT superconductor slab will be suspended at a distance $z^*$ above the permanent magnet when it becomes superconducting. There are a number of suggestions as to how to achieve maglev trains. One suggestion is to use a magnetic guide in the track and a HT superconductor to generate a magnetic field of the same polarity on board the train. In principle, the repulsive force can generate a gap between the train and the guide as wide as 0.01 m. The lateral propulsion of the train is more complicated and can involve a form of magnetic-field gradient to generate the driving force.

**Example 5.42**    Compute the height of levitation for a HT superconducting slab with the following data: $H = 100{,}000$ A/m, $h_1 = 0.01$ m, $\rho_0 = 8000$ kg/m³, and $g_r = 9.8$ m/s².

***Solution:***    Based on Eq. (62), the suspension height is $z^* = \mu_0^{1/3} H^{2/3} h_1^{2/3}/(\rho_0 g_r)^{1/3} =$ $(4 \times 3.14 \times 10^{-7}$ H/m$)^{1/3} \times (100{,}000$ A/m $\times 0.0$ m$)^{2/3}/(8000$ kg/m³ $\times$ 9.8 m/s$)^{1/3} = 0.025$ m, or 2.5 cm.  **§**

## 5.10    SUPERCONDUCTING MATERIALS

The more common superconducting devices have a very simple superconductor-insulator-superconductor (SIS) structure that can be easily fabricated using standard planar technology. The key materials in these devices are the superconducting electrodes and a thin insulator film. In particular, the insulator film must be sufficiently thin for tunneling to take place. To achieve this, the layer thickness has to be between 5 to 10 nm. In addition, when many of these devices are placed in a superconducting circuit, the uniformity of the insulator layer over a large area must be ensured.

Superconducting electrodes are made with thin films of known superconductors. Two types of superconductors are frequently used, **soft superconductors** and **hard superconductors.** Soft superconductors include Pb, In, and Sb and can be easily deposited by vapor-phase deposition or evaporation at a reasonably low temperature. Hard superconductors include Nb and V, which are more difficult to deposit due to their higher melting temperatures. Table 5.11 lists the important parameters for the common superconducting electrodes. In general, soft superconductors are more popular. Notwithstanding the difficulty in deposition, hard superconductors have been found to give better stability and a longer lifetime. Alloys of superconductors, such as Pb–In–Au, are also used. Some of the problems related to the superconducting electrodes include surface roughness, the possibility of peel-off during temperature cycling, a lack of mechanical strength and chemical stability, gas absorption, and the presence of a leakage current. Of the different superconductors listed in Table 5.11, Nb is most widely used.

Normally, the insulator used in an SIS sandwich structure is either a native oxide of the superconductor used for the electrodes or a chemically deposited metal oxide, which forms an artificial barrier. Native oxide grown by thermal oxidation is preferred, since it has a better texture and is more compatible with the substrate. This has the effect of reducing the thermal stress and provides a better surface coverage. For instance, $Nb_2O_5/Nb$ can be grown at a temperature between 100 °C and 200 °C for a few minutes. In using a native oxide, some form of surface protection (such as coating with a silica film) is often needed to prevent further oxidation of the electrodes. As with other thin-film insulators, the most important

Table 5.11    A list of important parameters for the superconducting electrodes

| Superconductor | $T_{cr}$ (K) | $H_{cr}$ ($\times$ $10^4$ A/m) | $\Delta^*$ (mV) |
|---|---|---|---|
| In | 3.40 | 2.3 | |
| Sb | 3.72 | 2.4 | |
| Pb | 7.19 | 6.4 | 2.7 |
| Pb–In–Au | 7.0 | | 1.2 |
| Nb | 9.26 | 1.6 | 1.5 |
| Nb–Ge | 23.6 | | 3.9 |
| V | 5.30 | 8.1 | |

properties of the insulator are thickness uniformity and low leakage current. Techniques used to form oxides include **glow discharge** and **plasma discharge.** Nonnative oxides offer a much wider variety of choice, and in these thin films, interface adhesion is an important factor. Thin films of $Al_2O_3$ and $MgO$ are frequently used because of their low leakage. Most SIS structures have some special geometrical requirements (like a thin neck region), and a lithographic technique is usually needed to form the electrodes. For instance, in the microbridge structure shown in Fig. 5.54, in order for the neck region to be normal, its dimensions have to be less than the coherence length, which may be between 3 and 30 nm (sometimes, this requirement is not strictly adhered to, and even with a neck dimension of 100 to 300 nm in $Nb_3Ge$ films, reasonable SIS device characteristics can still be observed).

Another very important type of superconductors are the recently discovered high-temperature (HT) superconductors. These are mixed oxides of metals that become superconductors at a temperature above liquid $N_2$ temperature (77 K), making them extremely attractive as far as commercial superconducting products are concerned. A typical HT superconductor is $YBa_2Cu_3O_{7-\delta}$ ($\delta$ indicates O deficiency), which has a critical temperature, $T_{cr}$, of 92 K. The lattice structure for $YBa_2Cu_3O_7$ is shown in Fig. 5.51. It is very similar to the perovskite lattice, except that a few O atoms are missing. The $YBa_2Cu_3O_7$ structure consists of a Cu cubic lattice with either a Ba ion or a Y ion in the middle of the cube. Additional O atoms are also present on the edges of the cube. Like the low-temperature superconductors, an HT superconductor also exhibits Meissner effect and has a well-defined energy gap between the superconducting states and the normal (electron) states.

Bulk HT superconductors are prepared by solid-state reaction. The starting material is a cold-pressed mixture of $Y_2O_3$, $BaCO_3$, and $CuO$. It is annealed for many hours at 950 °C in air and then followed by a similar treatment at a lower temperature in an $O_2$ atmosphere. These bulk superconductors are essentially superconducting grains embedded in some insulating layers. The electrical and magnetic properties indicate that the HT superconductors act as a composite consisting of a network of many small SIS junctions. Relatively pure superconducting thin films can be formed by sputtering, evaporation, and MBE growth. The quality of the thin films improves with increasing axial orientation and larger grains. Large-scale production of superconducting thin films uses a mixture of metals and oxides annealed at a temperature above 900 °C in an $O_2$ atmosphere.

**Fig. 5.54**
A superconducting microbridge structure

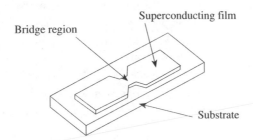

Bridge region

Superconducting film

Substrate

## GLOSSARY

| | |
|---|---|
| **magnetization** | a measure of the magnetic properties of a solid that gives rise to internal magnetic field. |
| **magnetic dipole** | a source for magnetization. |
| **magnetic moment** | a parameter that measures the strength of a magnetic dipole. |
| **magnetic flux** | the distribution of field lines (per unit area) originated from magnetic dipoles. |
| **magnetic induction** | a measure of magnetic flux. |
| **electromagnetism** | the phenomenon associated with an ac electric field and magnetic field. |
| **spin** | a parameter that describes the magnetic properties of an electron. |
| **nuclear spin** | the spin properties of the nucleus. |
| **diamagnetism** | the properties associated with the changes in the orbital motion of electrons in an atom in response to an applied magnetic field. |
| **magnetic domain** | a region within a solid where all the magnetic dipoles are aligned in the same direction. |
| **transition metal** | a class of metals with partially filled $3d$ and $4s$ orbits. |
| **rare-earth metals** | a special class of metals. |
| **ferromagnetic solid** | an ionic solid with magnetic domains of parallel spins. |
| **quantum mechanics** | a mathematical formulation of the properties of matter with very small dimensions. |
| **magnetic field intensity** | a measure of the intensity of the magnetic field strength. |
| **permeability** | a measure of magnetic properties in a solid. |
| **magnetic susceptibility** | the ratio of magnetization to the magnetic field intensity. |
| **magnetostatic energy** | the energy required to bring magnetic dipoles together to form a solid. |
| **angular frequency** | the frequency of a rotating object, measured in radians per second. |
| **paramagnetism** | magnetism due to net spin magnetic moment found in solids with partially filled energy orbits. |
| **Curie temperature** | the temperature when magnetization in a solid spontaneously vanishes. |
| **angular momentum** | a parameter that measures the circular motion of a rotating object. |
| **Bohr magnetron** | a unit of the magnetic moment. |
| **quantum numbers** | a set of numbers that describe the electronic states of an electron. |
| **Zeeman splitting** | the broadening of the energy levels (division into closely spaced energy levels) of a solid in the presence of a magnetic field. |
| **ferromagnetism** | the spontaneous magnetization in solids due to spin alignment between the neighboring atoms. |
| **exchange energy** | the energy related to the overlapping of the electrons in a solid. |
| **exchange integral** | a parameter governing whether the spin configuration of atoms is parallel or antiparallel. |
| **Pauli paramagnetism** | magnetism due to a collection of free electrons, each possessing a spin magnetic moment. |
| **magnetic anisotropy** | the property of solids having preferred crystal axes that allows for easy alignment of the magnetic domains. |
| **domain wall** | a boundary between magnetic domains. |
| **switching time** | the time required to move the domain wall in an ac magnetic field. |
| **hysteresis** | the partial reversibility of magnetization in an ac magnetic field. |
| **antiferromagnetic solid** | a solid with a small amount of net magnetization, since the spin magnetic moments in the domains are antiparallel. |
| **ferrimagnetic solids** | compounds of Fe and O that have a residual magnetic moment. |

| | |
|---|---|
| **anisotropy energy (per unit volume)** | the energy that determines whether the magnetic moments of the atoms in a solid will align with the easy axis of magnetization. |
| **magnetostriction** | the change in the direction of magnetization under an applied stress. |
| **single-domain particle** | the smallest magnetic domain. |
| **remanence** | residual magnetization in a solid. |
| **demagnetization** | the process used to remove magnetization in a solid. |
| **coercivity** | the ability of a magnetic solid to retain magnetization. |
| **Invar alloy** | a solid that exhibits a negligible thermal expansion coefficient near room temperature. |
| **magnetovolume effect** | the spontaneous magnetization observed in an Invar alloy as it is cooled below the Curie temperature. |
| **soft magnet** | a magnet that is less coercive and can be easily magnetized or demagnetized by a reversal of the external magnetic field. |
| **transformer** | a current-transfer device consisting of two coils coupled to each other through a magnetic core. |
| **hard magnet** | a permanent magnet that cannot easily be demagnetized. |
| $(BH)_{max}$ | a parameter that measures the peak magnetic energy stored in a transformer. |
| **ferrite (or garnet)** | a ferrimagnetic solid. |
| **electric motor** | a machine that transforms electrical power into mechanical power. |
| **power efficiency** | the ratio of the magnetic power stored in the magnet to the input power. |
| **magnetic recording** | the process by which data is stored in a magnetic medium. |
| **recording wavelength** | the distance on the magnetic medium between two digital data. |
| **video-tape recording** | refers to recording geared for capturing pictures dynamically. |
| **recorder head** | a device that writes and reads data from a magnetic storage medium. |
| **magnetic gap length** | the distance between the two ends of the armature of a magnetic head. |
| **magnetic disk** | a metal disk coated with magnetic particles for the purpose of storing magnetic signals and information. |
| **magnetic core** | a small circular ring made of magnetic materials used in magnetic memories. |
| **magneto-optic effect** | the changes observed when polarized light passes through magnetic domains in a solid. |
| **polarizer** | an optical device that selects only a particular polarized light to pass through. |
| **Kerr effect** | the use of reflected light to study magnetization. |
| **Faraday effect** | the use of transmitted light to study magnetization. |
| **magnetic bubbles** | small magnetic spheres used as storage elements in nonvolatile computer memories. |
| **nuclear magnetic resonance** | the absorption phenomenon associated with the magnetic spins of protons and neutrons. |
| **superconductor** | a solid that exhibits zero resistance below a critical temperature. |
| **critical temperature** | the temperature above which superconductivity fails to exist. |
| **Kirchhoff's law** | a law stating that the sum of the voltages in a closed circuit loop is zero. |
| **Meissner effect** | the repulsion of the magnetic field by a superconductor at all times through the presence of an induced surface current. |
| **penetration depth** | the distance an external magnetic field enters the surface of a superconductor. |
| **Cooper pair** | an electron pair with opposite spins (the pair momentum is zero at all times in the superconducting state). |

| | |
|---|---|
| coherence length | the distance separating a Cooper pair. |
| quasiparticle | the excited state of a Cooper pair, with properties similar to an electron and a hole in a semiconductor. |
| energy gap | the binding energy of the Cooper pairs. |
| skin effect | the phenomenon related to the confinement of the surface current in a conductor in the presence of an ac electric field. |
| complex conductivity | the real and imaginary components of conductivity. |
| critical magnetic field intensity | the maximum magnetic field intensity a superconductor can withstand without loosing its superconducting properties. |
| critical current | the maximum current a superconductor can carry before superconductivity disappears. |
| fluxoid | the unit of quantized magnetization in superconductors. |
| Type II superconductor | a superconductor with partial magnetic field penetration. |
| normal state | an electron state in a superconductor with nonsuperconducting properties. |
| superconductor-insulator-superconductor junction | a device that promotes the tunneling of superconducting electron states. |
| tunneling diode | a highly doped $p$-$n$ junction that conducts a tunneling current in the forward direction. |
| microwave radiation | an electromagnetic wave with a wavelength of the order of $10^6$ to $10^8$ nm. |
| tunneling current | a current generated by electrons moving through a barrier without having sufficient energy to surmount the barrier. |
| photon-assisted tunneling | electron tunneling involving the lattice vibrations. |
| quadratic signal detector | a detector whose output signal is proportional to the square of the input signal. |
| local oscillator | a device that produces a sinusoidal signal at a given frequency. |
| mixer | a device that combines two signals of different frequencies. |
| Josephson junction | an SIS junction with Cooper pairs involved in the tunneling process. |
| bandgap voltage reference (circuit) | a circuit that produces a fixed output voltage independent of temperature. |
| superconducting quantum interference device | a superconducting device that can measure magnetic field very accurately. |
| phase | the modulation part of a sinusoidal signal. |
| high-temperature superconductor | a mixed oxide of metals, which superconducts at a temperature above liquid $N_2$ temperature (77 K). |
| magnetic levitation | the use magnetic repulsive force to lift heavy objects. |
| maglevs | magnetically levitated trains. |
| soft superconductor | a metallic superconductor that can be easily deposited by vapor-phase deposition or evaporation at a reasonably low temperature. |
| hard superconductor | a metallic superconductor that is difficult to deposit by vapor-phase deposition or evaporation due to the higher melting temperature. |
| glow discharge | the luminous region created by the impact ionization of ions and electrons under a high dc electric field in a discharge tube. |
| plasma discharge | the region of positive and negative ions and electrons created under an ac electric field, analogous to the dc glow discharge. |

**REFERENCES**   Burns, G. *High Temperature Superconductivity, An Introduction.* New York: Acdemic Press Inc., 1992.

Craik, D. J. *Structure and Properties of Magnetic Materials.* London: Pion Ltd., 1971.

Hinken, J. H. *Superconductor Electronics—Fundamentals and Microwave Applications.* New York: Springer-Verlag, 1989.

Ishikawa, Y. and N. Miura. *Physics and Engineering Applications of Magnetisms.* New York: Springer-Verlag, 1991.

Kittel, C. *Introduction to Solid State Physics,* 3d ed. New York: John Wiley & Sons, Inc. 1966.

Kresin, V. Z., and S. A. Wolf. *Fundamentals of Superconductivity.* New York: Plenum Press, 1990.

Omar, M. A. *Elementary Solid State Physics.* Reading, Mass.: Addison-Wesley Publishing Company, 1975.

**EXERCISES**   ## 5.2   Magnetic Properties of Solids

1. Name the different types of magnetic dipoles present in solids. Which of these dipoles are responsible for magnetization found in permanent magnets?
2. Compute the magnetic dipole moment due to an electron circulating around a proton/nucleus at the speed of light, i.e., $3 \times 10^8$ m/s. The Bohr radius of the orbit is 0.053 nm.
3. Determine $\Delta \mu_m / H$ in Cu; $\chi_e$ of Cu $= 1 \times 10^{-5}$.
4. Determine the magnetic energy per unit volume stored in a piece of solid if the relative permeability is $\mu_r = 100$ at a field intensity of $1 \times 10^5$ A/m.

### 5.3   Sources of Magnetization

5. Estimate the contribution to diamagnetic permeability in Na. Assume that $N = 1 \times 10^{29}/m^3$, $n_a = 1$, and the electron radius is $1 \times 10^{-10}$ m.
6. Estimate the contribution to paramagnetic permeability in a solid if $N = 1 \times 10^{26}/m^3$, $n'_a = 10$, $\mu' = \mu_B$, and $T = 300$ K.
7. List the different angular momentum magnetic moments for orbiting electrons in an atom.
8. How many electrons are there in an atom if $\tilde{n} = 4$?
9. Using Table 5.3, plot the magnetic moment of the transition metal ions versus their atomic numbers. What can you deduce from the observed relationship?
10. During Pauli magnetization, if only those electrons within an energy shell (level) of $\pm 5kT$ of the Fermi level can take part, compute the ratio of these electrons to the total number of electrons present in the solid. The Fermi level is 1.0 eV, and $T = 300$ K.
11. Can you give reasons why Co has a higher Curie temperature than the other elements (see Table 5.5)?
12. Explain the effects of nonmagnetic ions in an alloy of magnetic and nonmagnetic elements.

13. If within a domain wall each electron spin is rotated by 9° from its neighbor, how many electron spins are involved in the formation of the domain wall?

14. Using Eq. (22), determine the domain wall thickness and the minimum energy required to form the domain wall if $A_m = 1 \times 10^{-10}$ J/m and $K_m = 60$ J/m$^2$.

15. What are the implications of a finite switching time for the domain wall in a ferromagnetic solid? Suggest how the switching time of a transformer core affects the operation of the transformer.

16. What are the advantages of using a ferrite rather than a transition metal as magnetization material?

## 5.4   Magnetic Anisotrophy and Invar Alloys

17. What is meant by magnetostriction? Replace Eq. (26) by a similar expression in terms of Young's modulus and the applied stress.

18. Using Table 5.6, compute the applied stress in $K_\sigma = K_1$. Assume $\epsilon_{st} = 1 \times 10^{-7}$.

19. List the most important criteria used to select magnetic materials for a transformer core.

20. Define magnetovolume effect and explain why it occurs only in some solids and not in others. In a given Invar alloy, $A_v = 2 \times 10^{22}$ m$^2$/A$^2$. What is the range of magnetization if $\Delta V/V$ is not to exceed $\pm 10\%$?

## 5.5   Amorphous Magnetic Materials

21. Give examples of transition metals and rare-earth metals that are used in the formation of amorphous magnetic solids.

## 5.6   Hard and Soft Magnets

22. What are the specific properties of a soft magnet and a hard magnet?

23. Compute the power efficiency of a motor if the input current is 10 mA, the magnetic flux density is 1 T, the shaft velocity is 1 m/s, the cross-sectional area of the wire forming the motor is $1 \times 10^{-9}$/m$^2$, and the wire conductivity is $1 \times 10^7$ S/m.

24. Compute the eddy current induced in the magnetic core of a transformer if the current loop is 0.05 m $\times$ 0.05 m, the wire resistance is 1 k$\Omega$, and $dB/dt = 100$ T/s.

## 5.7   Magnetic Devices and Applications

25. Briefly describe the operation of magnetic recording.

26. Determine the induced voltage in a recording head if the number of windings is 100, the speed of the passing tape is 10 m/s, and $d\phi/dx = 1 \times 10^{-6}$ T·m.

27. In a magnetic core memory with a core radius of 1 μm, compute the current density needed to reverse the magnetization of the core if the initial magnetization, $M$, is $1 \times 10^6$ A/m. Assume that the magnetic susceptibility of the core is 200.

28. Suggest how the write process is carried out in a magneto-optical recording system.

29. Name two applications of nuclear magnetic resonance, or NMR.

## 5.8   Properties of Superconductors

30. Determine the ratio of the currents through two superconductor coils connected in parallel if coil A has 10 windings and coil B has 20 windings.

31. Explain the Meissner effect in superconductivity.

32. Using the value of $\lambda_L$ of In from Table 5.10, compute the number of superconducting carriers present in In if $m_s$ and $q_s$ have values that are the same as the free electrons.

33. Using Eq. (38), compute the fractional volume of a superconducting Al wire that experiences flux penetration (with $B > 0.1B_0$) if $T = 0.8\ T_{cr}$. The radius of the wire is 1 μm.

34. Assuming that there are $1 \times 10^{29}$ electrons/m$^3$ in a superconductor, estimate the distance between the Cooper pairs if, between each pair, there are $1 \times 10^4$ electrons.

35. Using Eq. (39), compute the fractional decrease in the binding energy of Cooper pairs in In if $T = T_{cr}/3$.

36. Estimate the real part of the surface resistance, $R_s'$, of Al at 1 MHz if $T = T_{cr}/2$. Compare it with the value of $R_s$ when Al is in the normal (nonsuperconducting) state. $\sigma(Al) = 3.65 \times 10^7$ S/m.

37. Using the result of Exercise 32, determine the critical current density in In. The Fermi level of In is 8.6 eV and $\Delta^* = 2$ meV.

38. Distinguish between Type I and Type II superconductors.

## 5.9   Applications of Superconductors

39. Briefly describe the current-voltage characteristics of an SIS junction.

40. Show that an SIS detector with a sensitivity of $2 \times 10^4$/V translates into a detection-frequency limit of 12 GHz.

41. Suggest how the SIS detector may be used as a microwave mixer.

42. Show that the time rate of change of the phase difference of electrons on both sides of the insulator in a Josephson junction is given by $4\pi q V_0/h$, where $V_0$ is the applied dc voltage, $q$ is the electron charge, and $h$ is Planck's constant (make use of the relationship $h\omega = 2\pi E$, where $E$ is the energy of the Cooper pair). Estimate the angular frequency, $\omega$, when the smallest $V_0 = 10$ mV (i.e., set $m' = 1$ in Eq. (56)).

43. Describe how the Josephson junction is used as a voltage standard.

44. Describe the operation of a SQUID (superconducting quantum interference device).

45. From Eq. (59), show that the maximum value of $I_g'$ is given by Eq. (60).

46. Determine the magnetic-field intensity needed to suspend a superconducting slab to a height of 10 cm if the magnet weighs 10 g and has a volume of $1 \times 10^{-6}$ m$^3$. The thickness of the slab is 1 cm.

## 5.10   Superconducting Materials

47. Describe the problems related to the properties of thin-film superconductors.

48. Briefly describe the methods used to prepare HT superconductors.

# CHAPTER 6

# Micromachines, Sensors, and Packaging Materials

## 6.1   INTRODUCTION

The ability of a technologist to provide a solution to a problem depends in part on the kind of tools available. For many years, hardware tools were developed mainly to deliver power, and major design efforts were concentrated on applying mechanical principles and improving machine efficiency. A typical example is the design of tackles, pulleys, and gearboxes, all of which are essential to aid in transferring heavy weights and moving mechanical parts. The invention of the electrical motor is quite likely one of the most significant developments in electromechanics. With the recent arrival of the information age, the concept of tools also changed considerably. Our ability to handle large forces is no longer a prerequisite to engineering design when compared with the need for control-lability and integration. Power consumption has become a critical issue in large-scale systems, and hardware is often made with smaller mechanical parts, thus imposing a far greater demand on precision machining and molding. The concept of micromachines emerged, following the same line of development as miniaturization in very-large-scale integrated circuits (VLSICs). Controllability, in general, requires our ability to determine accurately the exact conditions of a machine or a system. This is, of course, possible only if we have the appropriate sensors to measure the physical parameters and the ability to process information quickly and to translate the processed data to the actuators. The solutions to controllability and integration can be found in control and systems theories, and the signal-conditioning hardware can be built using microcircuits. Sensing, however, depends on the availability of the required sensor. Many microsensors and micromachines are built using Si technology; in addition to miniaturization, these devices are made at a relatively low cost. In many instances, micromachines and microsensors have already demonstrated clear advantages over the conventional machines and devices that they replaced.

In this chapter, we study micromachines and sensors separately. The former includes most miniaturized mechanical or electromechanical devices that already have counterparts in conventional machines. They are designed under the same principles, with the exception that their constructions are based on materials and processes being used in the fabrication of microcircuits. Sensors are structures and devices that make use of materials in microelectronics to facilitate signal detection. The materials used are mostly semiconductors and insulators. In a sensor design, the operation is often developed from first principles, and because of the diversity of the different micromachines and sensors, it is more difficult to provide a systematic classification and treatment. We simply consider these devices based on their structures and their applications.

## 6.2 MICROMACHINING

**Micromachining** is a multidisciplinary subject that requires both a good knowledge of engineering design and an understanding of chemistry and materials science. Although Si fabrication technology is well established, the adaptation of the technology to the making of mechanical structures and electromechanical devices is new and requires many developmental skills. The most extensively studied device in micromachining is probably the integrated pressure sensor, and fairly sophisticated microelectronic circuits have been added to make it work effectively. Other important devices that have been fabricated include acceleration sensors, disposable blood-pressure sensors, microgrippers, minute motors and parts, and various forms of actuators.

### Processing Technology

**Integrated circuits (IC) technology** provides an ideal environment for the construction of micromachines, sensors, and actuators. Si, in particular, offers the advantages of low cost, amenity to miniaturization, relative ease of forming different shapes and patterns, and suitability for batch-mode fabrication. In the following sections, we study how micromachining using Si is possible.

*Lithography and Patterning*  **Lithography** is one of the most important steps in microstructure fabrication. In many ways, the process is like printmaking, which is the origin of its name. To form a **microstructure** in a Si wafer, several lithographic steps are required. In the majority of these cases, the pattern is formed either on an oxide layer (usually $SiO_2$) or on a metal layer. Just as in photographic printing, artwork has to be prepared that contains the physical dimensions of the pattern to be copied. The pattern can be initially drawn on a piece of reticle and transferred to a more permanent photographic glass plate (the negative). Alternatively, the pattern may be directly written onto the wafer—or, more exactly, onto a coating (resist) on the wafer—using an E-beam source (no negative is required in this case). Very fine resolution lines with a width down to a fraction of a micrometer can be drawn over a large plate area (up to tens of square centimeters), and even finer lines are possible using an X-ray source. The finished photographic (glass) plate is called a *mask*. Most processes require a mask set of several (glass) plates, which can be aligned using a marker (printed on each of the masks). The transfer of the pattern from the (glass) plates onto the Si wafer also requires a mask aligner, and the patterns appearing in the masks are usually projected optically. To retain the optical image, a light-sensitive substance called a **resist** is initially coated on the wafer. A resist is an organic chemical whose properties can change when exposed to light, usually an ultraviolet (UV) light. The wavelength of the light dictates the resolution of the pattern. Two types of resists are available. A **positive resist** is one that, when exposed to light, becomes hardened and soluble in a developer solvent; but it is otherwise resistant to etching by chemicals. The selective removal of the resist using lithography allows windows to be formed and exposes the underlying oxide or metal layer. When no longer protected by the resist, the oxide or the metal layer can easily be removed using a solution of either HF (for $SiO_2$) or HCl (for a metal). The etching process is shown in Fig. 6.1. As observed, once the Si is exposed, further etching of the Si is also possible. A **negative resist** works in the opposite way: The exposed resist becomes insoluble to the developer solvent, but the unexposed resist can be developed and washed away (for window formation).

In addition to serving as a protective surface layer during selective etching, a resist is sometimes used to form a metallization pattern on a Si wafer. In the process, a positive resist is first coated on the Si wafer and the required pattern is lithographically transferred onto the resist. Except for the part of the resist with the pattern, the rest is exposed and developed. A thin metal layer is then deposited on the entire wafer, and the developed resist is removed by chemical etching. Physically, this produces a liftoff effect on the overhanging metal, and only the metal covering the pattern remains. The process is called **liftoff** and is frequently used in the formation of metal interconnections in microcircuits. Figure 6.2 shows the key steps in the liftoff process.

**Fig. 6.1**
Oxide etching using positive resist

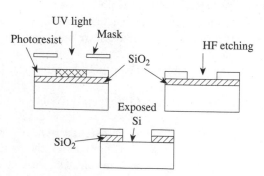

**Fig. 6.2**
Metallization of Si using
negative resist and liftoff

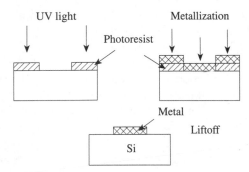

Example 6.1 Compare the ideal feature size that may be achieved if the light sources used in lithography are (a) a UV light, (b) a 50-eV E-beam, and (c) an X-ray source.

*Solution:*
 (a) UV light has a wavelength of around 400 nm.
 (b) An E-beam with an energy of 50 eV will have a (de Broglie) wavelength of $h/$momentum $= h/\sqrt{2Em_0} = 6.6 \times 10^{-34}$ J·s/
$\sqrt{2 \times 50 \text{ eV} \times 1.6 \times 10^{-19} \text{ V/eV} \times 0.91 \times 10^{-30} \text{ kg}} = 3.4 \times 10^{-8}$ m $= 34$ nm.
 (c) An X ray has a wavelength of the order of 1 nm. §

Example 6.2 In proximity contact printing lithography, the quality of the image depends on a quality factor, $Q_F = w/\sqrt{2\lambda_L d}$, where $w$ is the line-to-line spacing in meters, $\lambda_L$ is the wavelength of light in meters, and $d$ is the spacing separating the mask and the wafer in meters. Compare the quality factor, $Q_F$ (to be minimized for best resolution), in the following two cases: (a) $w = 1$ µm, $\lambda_L = 430$ nm, $d = 20$ µm, and (b) $w = 1$ µm, $\lambda_L = 430$ nm, $d = 10$ µm.

*Solution:*
 (a) $Q_F = 1$ µm$/\sqrt{2 \times 0.43 \text{ µm} \times 20 \text{ µm}} = 0.241$
 (b) $Q_F = 1$ µm$/\sqrt{2 \times 0.43 \text{ µm} \times 10 \text{ µm}} = 0.341$ §

***The Chemistry of Etching*** Chemical etching is one of the most direct ways of removing unnecessary Si, and by selective etching, it is possible to produce microstructures with the desired shapes and patterns. Chemical etching can be isotropic or anisotropic, depending on the crystal orientation. Most etchants are temperature and dopant dependent; i.e., their etch will change with temperature and with the dopant density. Table 6.1 shows the properties of the more common etchants used for removing Si.

**Table 6.1** Properties of the common etchants for the removal of Si

| Etchant (aqueous) | Anistropic | Masking Materials |
|---|---|---|
| EDP (Ethylene, diamine, pyrocatechol) | Yes | $SiO_2$, Cr, Au |
| KOH | Yes | $SiO_2$ |
| HNA (HF, $HNO_3$, $CH_3OOH$) | Yes | $Si_3N_4$, Au |

Of the three etchants listed in Table 6.1, EDP is most commonly used, since it produces the best effect—including a smooth surface after etching. KOH, on the other hand, etches quickly but gives a less smooth surface. In addition, KOH also removes $SiO_2$, which may be the masking material. In general, etching of Si involves an oxidation-reduction process, starting with an injection of holes into Si from the etchant solution or an external source, such as a bias voltage. The holes attract the $OH^-$ ion present in the etchant, and oxidation takes place over the Si surface. Si, being one of the reactants, dissolves in the solution. The reaction proceeds steadily as long as the etchants are replenished. As in any other chemical process, Si etching is temperature dependent, and the etch rate can vary over many orders of magnitude by changing the temperature.

Figure 6.3 shows the types of grooves that may be formed on a Si wafer by etching along the different crystallographic directions with and without external agitation. In Si, the etch rates in the $<1\ 0\ 0>$ and $<1\ 1\ 0>$ directions are usually much faster than in the $<1\ 1\ 1>$ direction due to the differences in the atomic densities. Etching along the $<1\ 0\ 0>$ direction produces an inverse pyramidal groove, which terminates with the exposure of the $(1\ 1\ 1)$ planes. A similar etching along the $<1\ 1\ 0>$ direction produces almost vertical walls, as shown in Fig. 6.4. The technique is frequently used to form very deep vertical grooves or holes in a Si wafer. Undercutting during etching is less common with a concave corner, but it frequently occurs with convex corners, and a smoothing out of the convex curvature can be expected.

Figure 6.5 shows the etch rates in Si using different etchants. As observed, both

**Fig. 6.3**
Holes and grooves formed in a Si wafer

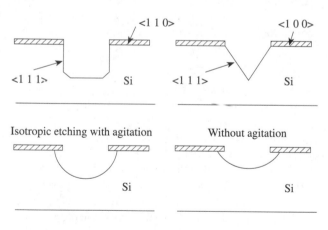

**Fig. 6.4**
Grooves formed in Si by etching along the $<1\ 1\ 0>$ direction

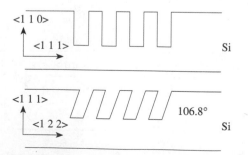

Fig. 6.5

Relative etch rates for
different etchants

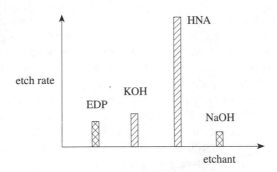

KOH and EDP have fairly high etch rates, but the highest is HNA (HF/NHO$_3$). For KOH below a certain solution concentration, the Si surface after etching can become quite rough, and a KOH concentration in excess of 30% is normally needed to give a smooth surface. HNA acts as an etchant of Si only when it is in an oxidizing state. The addition of dopants, such as B, to Si will normally slow down the etching process when the dopant density exceeds $1 \times 10^{25}/m^3$. Figure 6.6 shows the etch rate in Si using an EDP solution as a function of the B density. The transition occurs at roughly $2 \times 10^{25}/m^3$ when the edge rate begins a sharp drop. This process is often used as a means of terminating the etching (**etch stop**). Figure 6.7 shows the etch rate versus temperature for Si using an EDP solution.

Fig. 6.6

Etch rates versus boron
density

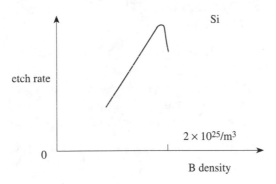

Fig. 6.7

Etch rates at different
temperatures

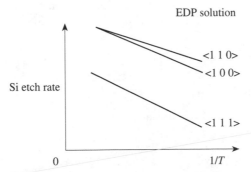

(*Source* for Figs. 6.6 and 6.7: H. Seidel, "The Mechanism of Anisotropic Silicon Etching and Its Relevance for Micromachining," in R. S. Muller, R. T. Howe, S. D. Senturia, R. L. Smith, and R. M. White, eds., *Microsensors*. © 1991 IEEE Press. Reprinted with permission of IEEE.)

Electrochemical etching is an effective means of accelerating the etch rate in Si. In electrochemical etching, the etchant solution is HF. During etching, the Si substrate is positively biased, and the holes are accumulated at the Si surface (where a nearby negative electrode is placed). The high concentration of holes accelerates the oxidation process, and the HF then removes the $SiO_2$. Electrochemical etching also provides a convenient way to selectively remove highly doped surface layers that are unprotected. In the process, the current in the electrodes will initially be quite large, since the more resistive substrate is shunted by the highly doped surface layer. Etching will terminate once the surface layer is removed and the current starts to flow exclusively through the substrate.

A popular means of establishing a built-in etch-stop marker, particularly in the formation of a thin diaphragm, is to introduce a sacrificial p-type layer on an n-type substrate. The setup is shown in Fig. 6.8. A **p-n junction** is first formed on the n-type substrate from which the p-type layer is to be removed. A reverse bias across the junction (with the negative voltage terminal connected to the p-type surface layer) allows holes to be accumulated near the surface of the p-type layer, and this accelerates the etching. Etching stops once the n-type substrate is exposed, since anodic oxidation will proceed over the exposed n-type Si. To ensure a proper bias voltage, a three-electrode configuration is sometimes used. The additional electrode is placed close to the n-type substrate to ensure that the anodic potential is reached. Right at the point when etch stop occurs, the electrode current will increase sharply. This current surge conveniently serves as an etch-stop marker. Very thin Si diaphragms can be formed in this way with a surface roughness on the order of 0.1 μm.

**Example 6.3**     Suggest how $HNO_3$ as an etchant goes into solution to form ions useful to the etching process.

*Solution:*     $HNO_3$ and $H_2O$ react in the following manner: $HNO_3 + H_2O \Rightarrow HNO_2 + 2OH^- + 2h^+$ (holes). Holes and $OH^-$ ions will take part in the oxidation and reduction process, resulting in the removal of the Si.     §

**Example 6.4**     If a 120-ml etchant solution contains 30 ml of 6 $N$ $HNO_3$, compute the $OH^-$ ion concentration using Example 6.3.

*Solution:*     There are $2 \times 6 N$ (= 12 $N$) of $OH^-$ions in the 30-ml solution. After having been diluted to 120 ml, the concentration becomes 3 $N$, and there are $3 \times 6.02 \times 10^{23}$ ions/mol = $18.1 \times 10^{23}$ ions/mol.     §

**Fig. 6.8**
Setup for electrochemical etching

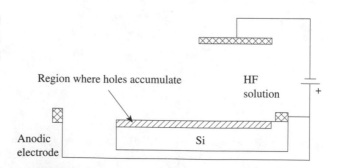

**Example 6.5**   A *p*-type sacrificial surface layer on an *n*-type substrate has an exponentially decaying dopant (acceptor) profile, and the surface acceptor density is $3 \times 10^{20}/m^3$. If the decay length, $x_0$, is 3 $\mu$m, what would the required substrate dopant (donor) density be if 10 $\mu$m of Si were to be removed? Assume all dopants are ionized.

*Solution:*   For etch stop to occur after the removal of 10 $\mu$m of surface Si, the junction interface must be located 10 $\mu$m from the Si surface. At this point, $N_D = N_A = N_{A0}\exp(-x/x_0) = 3 \times 10^{20}/m^3 \times \exp(-10 \ \mu m/3 \ \mu m) = 1.07 \times 10^{19}/m^3$. §

*Dry Etching*   **Dry etching** requires the use of etchants in the form of reactive ions in a plasma created by a voltage discharge. A **plasma** is a collection of positive and negative ions and free electrons, as shown in Fig. 6.9, and it can be created in a confined chamber at a reduced pressure using either dc or ac voltage discharge. The ions present in the plasma are particularly reactive and will readily take part in the oxidation and reduction process used to remove Si or $SiO_2$. $CF_4$–$O_2$ is frequently chosen as an etchant, and the typical etch rate is of the order of hundreds of nanometers per minute. The volatile byproducts in the reaction are pumped out in a reduced atmosphere. In **reactive ion etching** (RIE), an RF voltage source is used to create the plasma, and the Si wafer sits on top of the positive RF electrode. RIE is primarily anisotropic, and etching is along the direction of the movement of the ions. Anisotropy can be reduced in a wafer by doping the latter heavily with donor atoms. Because of the anisotropic properties, RIE is very popular and can be used to form very deep submicrometer-size grooves with almost vertical walls in a Si substrate.

**Example 6.6**   Suggest the type of chemical reactions that are present in RIE of Si using $CF_4$–$O_2$.

*Solution:*   The following reactions take place during RIE: (1) $CF_4 \Rightarrow C + F_4$; (2) $C + O_2 = CO_2$; (3) $Si + F_4 = SiF_4$. These reactions result in the removal of Si in the form of $SiF_4$. §

**Example 6.7**   A metal line has a cross section that is an isosceles trapezoid with a base width of 0.5 $\mu$m, a flat surface of 0.4 $\mu$m, and a height of 0.2 $\mu$m. It is covered by a uniform oxide 0.5 $\mu$m thick. If the etching of the oxide is isotropic and the etch rate is 0.1 $\mu$m/s normal to the flat surface, how long will it take to expose the metal sidewall?

**Fig. 6.9**
A cross section of a plasma tube

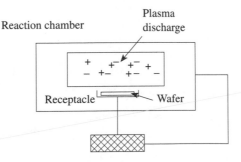

***Solution:***   The slope at the edge of the metal line is (0.5 μm − 0.4 μm)/
(2 × 0.2 μm) = 0.5. This corresponds to an angle θ = 76°. If etching is normal
to the flat surface, the time to expose the metal sidewall is (0.5 μm − 0.4
μm)/(cos(76) × 0.1 μm/s) = 4.1 s.   **§**

Note that etching 0.5 μm of the oxide in the horizontal plane will take 5 s (a
somewhat longer time).

## Formation of Microstructures

***Review of Mechanical Properties***   The force acting on a homogeneous microstructure is normally expressed in terms of the stress $\sigma_s$ in pascals (Pa), which is given by
the applied force in newtons divided by the area of cross section, $A_{cs}$ (m²). A force of
1 N applied over an area of 1 m² will give a stress of 1 Pa. Since all solids are to a
greater or lesser extent elastic, the microstructure will suffer either compression or
elongation (depending on the direction of the applied force). The strain, $\epsilon_{st}$, is given
by the change in length divided by the initial length, and the ratio $\sigma_s / \epsilon_{st}$ is called
the Young's modulus, $E_Y$. The Young's modulus is a measure of the elasticity of the
solid—i.e., how likely it is that the solid will produce a compression or an elongation.
Figure 6.10 gives a typical stress-strain diagram of a metal. From the figure, $E_Y$ is
given by the slope in the linear region (up to the point called the proportional limit).
The rest of the curve represents the inelastic region, and the metal will eventually
fracture beyond a point called the *ultimate stress*. Most of the time, microstructures are
allowed to operate only within the elastic region. In addition to the axial compression or
elongation along the direction of the applied force, lateral compression or elongation
produces what is known as *lateral strain*. The Poisson ratio, $\ddot{y}$, is given by the ratio of
the lateral strain to the axial strain. Thus, the change in the volume of a microstructure,
$\Delta V$ (m³), is given by

$$\Delta V = V_0[(1 + \epsilon_{st})(1 - \ddot{y}\epsilon_{st})^2 - 1] \tag{1}$$

where     $V_0$ is the original volume in m³.

To a first order of approximation,

$$\Delta V/V_0 = \epsilon_{st}(1 - 2\ddot{y}) \tag{2}$$

where     $\Delta V/V_0$ is sometimes called the *dilatation*.

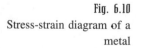
**Fig. 6.10**
Stress-strain diagram of a
metal

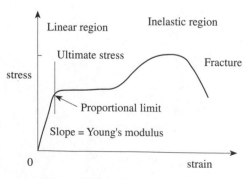

Analogous to axial stress and axial strain, shear stress and shear strain can be similarly defined. The shear stress is applied parallel to the surface of the microstructure, and this will result in a change in the shape of the microstructure. The shear strain, which is often measured by the angle in radians, relates to the distortion. This is illustrated in Fig. 6.11.

***Holes and Grooves***   The simplest way to form holes and grooves in Si is by chemical etching. The masking material is usually $SiO_2$. Pyramidal grooves can be etched in Si along the $<1\,0\,0>$ direction using a standard etchant, as shown in Fig. 6.12. If the vertex of the groove penetrates through the entire wafer, a small square hole will be formed on the underside. The size of the hole may be controlled by adjusting the composition of the etchant solution and the temperature. Finer control is achieved if a thinned wafer is used. Such a technique has been applied to form ink-jet nozzles with a hole size as small as $20\ \mu m^2$.

Holes with vertical walls are formed if a $<1\,1\,0>$-oriented Si wafer is used. Grooves are formed if the holes do not penetrate through the wafer. The depth of a groove can be defined by a highly doped $p^+$ buried layer in the wafer forming an etch-stop marker, as shown in Fig. 6.13. Grooves formed on Si wafers have been used to make precision moldings, miniature circuit boards, and even optical benches.

**Fig. 6.11**
Shear stress and strain

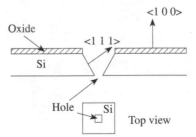

**Fig. 6.12**
Side view and top view of an etched hole in Si

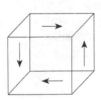
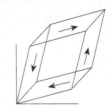

**Fig. 6.13**
Etch stop in Si using a B-doped layer

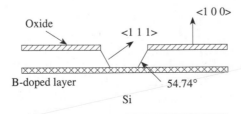

**Example 6.8**   If the etch rate of Si in EDP is 0.75 μm/min at room temperature, how long does it take to form a hole 5 μm deep?

*Solution:*   The time it takes to form a hole 5 μm deep is 5 μm/0.75 μm/min = 6.7 min.   §

*Cantilever Beams*   **Cantilever beams** are usually built on top of a cavity, as shown in Fig. 6.14. In Si, a cavity is formed by etching in the <1 0 0> direction through a sacrificial surface Si layer. To define the cavity, the initial Si surface is first doped with a high concentration of dopants, such as B, which forms the etch stop. A sacrificial Si layer is then grown epitaxially on top to a thickness equal to the depth of the cavity, as shown in Fig. 6.15. To form the cantilever beam, the surface of the sacrificial layer is oxidized, and a protective metal layer corresponding to the dimensions of the beam is deposited. The whole wafer is then etched in hot EDP until the cavity is formed and the undercutting has cleared the cavity beneath the cantilever beam.

The depth of the cavity corresponds to the thickness of the sacrificial layer, and fairly vertical walls can be formed if the side walls are oriented in the <1 1 0> direction. Typical dimensions of a cantilever beam are 100 μm long, 15 μm wide, and 0.5 μm thick. The top metal layer can be a Cr–Au alloy deposited to a thickness of about 15 nm. Electrostatic deflection of the beam is possible when a bias voltage is applied to the surface metal and the $p^+$ layer. For a small applied voltage, the beam deflection varies approximately as the square of the applied voltage. In most cases, a beam deflection in excess of a few micrometers can produce instability. In addition to $SiO_2$, other types of insulating films, including polysilicon, can be used

**Fig. 6.14**
Cantilever beam and cavity

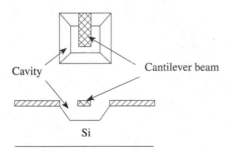

Cavity                    Cantilever beam

Si

**Fig. 6.15**
The substrate for cavity formation

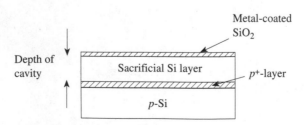

Depth of cavity

Metal-coated $SiO_2$

Sacrificial Si layer

$p^+$-layer

$p$-Si

to form the beam. Mechanical **resonance** of the beam occurs at a frequency $\nu_R$ (Hz), which is given by

$$\nu_R = 0.162(t_h/l^2)\sqrt{(E_Y/\rho_0)} \tag{3}$$

where $t_h$ is the beam thickness in m,

$l$ is the length of the beam in m,

$E_Y$ is the Young's modulus of the beam N/m², and

$\rho_0$ the density of the beam in kg/m³.

**Example 6.9** Assume that the beam deflection, $\Delta y$, of a $Si_3N_4$ cantilever beam varies as the square of the applied voltage, $V_a$, according to the equation, $\Delta y = \epsilon_0 V_a^2 l^4/(E_Y d^2 t_h^3)$, where $\epsilon_0$ is the free space permittivity, the length of the beam is $l = 80$ μm, Young's modulus is $E_Y = 3.85 \times 10^{11}$ N/m², the electrode spacing is $d = 10$ μm, and the beam thickness is $t_h = 100$ nm. Compute the voltage required to give a beam deflection of 5 μm.

***Solution:*** The required voltage is

$$V_a = \sqrt{\Delta y E_Y d^2 t_h^3/(\epsilon_0 l^4)}$$

$$= \sqrt{5 \times 10^{-6}\text{ m} \times 3.85 \times 10^{11}\text{ N/m}^2 \times (1 \times 10^{-5}\text{ m})^2 \times (1 \times 10^{-7}\text{ m})^3/(8.85 \times 10^{-12}\text{ F/m} \times (8 \times 10^{-5}\text{ m})^4)}$$

$$= 24\text{ V.} \quad §$$

**Example 6.10** Compute the resonant frequency of a $Si_3N_4$ cantilever beam 80 μm long and 100 nm thick if the density of the beam is $3.1 \times 10^3$ kg/m³.

***Solution:*** Based on Eq. (3),

$$\nu_R = 0.162(t_h/l^2)\sqrt{E_Y/\rho_0}$$

$$= 0.162 \times (1 \times 10^7\text{ m}/(8 \times 10^{-5}\text{ m})^2) \times \sqrt{3.85 \times 10^{11}\text{ N/m}^2/3.1 \times 10^3\text{ kg/m}^3}$$

$$= 2.8 \times 10^4/\text{s} = 28\text{ kHz.} \quad §$$

***Suspended Membranes and Diaphragms*** Suspended membranes, such as polyimide membranes, can be formed on a Si substrate by anisotropic etching. The **membrane** to be formed is first deposited on the substrate, usually a thinned Si diaphragm. The **diaphragm** is then removed from the back by etching, except for a supporting structure, as shown in Fig. 6.16. Once the diaphragm is removed, the membrane is

**Fig. 6.16**
A membrane with a rim support

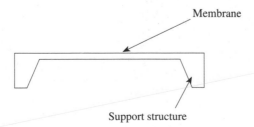

Membrane

Support structure

freed. Suspended membranes with a thickness between 1 and 15 μm can be formed for an area of 25 mm². These membranes are frequently used in the fabrication of pressure sensors and capacitive sensors. In general, a Si membrane has a number of advantages over bulk Si. These include a smaller parasitic effect, a better thermal insulation, and possible illumination through the back surface when optical sensors are formed on top of the membrane.

**Example 6.11**      How long does it take for a 15-μm-thick Si membrane to be formed on a wafer 300 μm thick if the etch rate is 0.75 μm/min?

*Solution:*   The time required is $t = (300 \ \mu\text{m} - 15 \ \mu\text{m})/0.75 \ \mu\text{m/min} = 380 \ \text{min} = 6.3 \ \text{h}.$   **§**

# 6.3   POLYSILICON FILMS USED IN MICROSTRUCTURES

Polysilicon films deposited using low-pressure chemical vapor deposition (LPCVD) are used to form resistors in the making of pressure sensors. These films are usually placed on oxidized Si wafers and are doped with *p*-type impurities. The resistivity can vary over a wide range of values and is always higher than that of single-crystal Si. These films are also temperature sensitive. In a pressure sensor, polysilicon is deposited as resistors, which are arranged in the form of a bridge circuit. Upon an applied stress, the resistance will change; the fractional change, $\Delta R/R$, is measured in terms of a **gage factor** $g$ (dimensionless), which is given by

$$g = (\Delta R/R)/\epsilon_{\text{st}} \tag{4}$$

where    $\epsilon_{\text{st}}$ is the strain.

A polysilicon pressure sensor is more sensitive than a metal sensor, since the maximum value of $g$ in polysilicon is around 30, almost 15 times that of a metal. The gage factor is also temperature dependent, and the temperature coefficient in polysilicon is about $-1 \times 10^{-3}/\text{K}$.

Figure 6.17 shows the structure of a polysilicon pressure gage formed on a Si membrane. Two sets of resistors (with different geometries) are deposited on the membrane in the form of a **Wheatstone bridge circuit,** and an unbalanced voltage

**Fig. 6.17**
Cross section of pressure
sensor formed on a Si
membrane

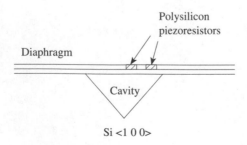

Polysilicon
piezoresistors

Diaphragm

Cavity

Si <1 0 0>

**Fig. 6.18**
Voltage-pressure
characteristics of a
pressure sensor with its
equivalent circuit

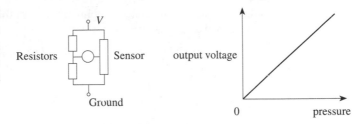

will be observed under an applied stress. Figure 6.18 shows the equivalent circuit together with the output (voltage versus pressure) characteristics of the sensor. A typical membrane has an area of 1 mm × 1 mm and a thickness of 30 μm. The same pressure sensor can also be used for temperature sensing, provided that the temperature range is not too large, since polysilicon becomes unstable at a high temperature.

**Example 6.12**   Determine the lateral $\Delta R/R$ for a rectangular block of polysilicon if the thickness of the resistor, $t$, is reduced by 10% under a given stress. Assume that both the length and the width of the block were initially equal to $100t$.

*Solution:*   Assuming that the block is incompressible, the length/width ratio of the block after compression equals $\sqrt{100t \cdot 100t \cdot t/0.9t} = 105.4t$. The value of the lateral $\Delta R/R$ is $[105.4t/(105.4t \cdot t) - 1/((1 - 0.1)t)]/(105.4t/(105.4t \cdot t)) = 0.11$. §

## 6.4   APPLICATIONS OF MICROSTRUCTURES

### Ink-Jet Nozzles

Ink-jet nozzles are made by anisotropically etching a Si wafer in the <1 0 0> direction until the wafer is etched through and an array of holes is formed. The holes are squares and have typical dimensions of 20 μm × 20 μm. It is possible to estimate the hole dimension, $l$ (m), using the following geometrical relationship:

$$l = L - 2t_h/\tan(\Theta) \tag{5}$$

where   $L$ is the mask dimension in m,

$t_h$ is the thickness of the wafer in m, and

$\Theta$ is the angle between the <1 0 0> and <1 1 1> planes in degrees.

To operate as an ink-jet nozzle, the ink jet from a reservoir under high pressure is injected through the holes onto a piece of paper in front of the nozzle. Under a small pressure disturbance, the ink drops in the jet stream will break up into small droplets if the emission rate exceeds $10^5$ drops per second. The droplets are charged through an electrode near the nozzle and are electrostatically deflected to strike the paper at the desired locations. An ink-jet printer is an improvement over the standard dot-matrix printer, which has a much poorer resolution.

**Example 6.13**     If a square hole at the bottom of a Si diaphragm 50 μm thick has a side of 20 μm, what is the size of the hole at the top surface?

*Solution:*     Based on Eq. (5), the size of the hole on the top of the wafer $L$ is $l +$ $2t_h/\tan(\Theta) = 20$ μm $+ 2 \times 50$ μm $\times \tan(54.74) = 161.4$ μm.     §

## Multichip Module

With micromachining, two or more Si wafers can be stacked together to form a **multichip module.** The chip-to-chip connection is achieved by using Hg to fill a set of alignment holes formed in the wafers. The holes are sealed at the top and at the bottom with small Si plugs, as shown in Fig. 6.19. Because of the same thermal expansion coefficient in the wafers, the thermal stress within the module will be very small. Direct bonding of two Si wafers together is also possible in a process called *silicon fusion bonding* (SFB). Sometimes, fusion is achieved through an intermediary thin film such as $SiO_2$. Two Si wafers, when bonded together, can have much the same mechanical properties and electrical properties as a single wafer, and the technique can be used either to reinforce the mechanical strength of a membrane or to match structures and devices formed by incompatible processes. SFB is difficult to do well and requires a temperature of approximately 1100 °C.

**Example 6.14**     Estimate the resistance of a Hg connection path between two wafers in a module if each Si wafer is 300 μm thick. The hole diameter is 10 μm. The resistivity of Hg is 96.8 μΩ·cm.

*Solution:*     Since $R = \rho L/A_{cs}$, for Hg, $R = 96.8 \times 10^{-8}$ Ω·m $\times 2 \times 3 \times$ $10^{-4}$ m/$(3.14 \times (1 \times 10^{-5}$ m/2$)^2) = 740$ Ω.     §

## Gas Chromatography Chip

**Gas chromatography** is an important chemical testing technique and is used to identify different types of gases present in a gas mixture. The principle is to separate gases of different molecular weights by their different flow rates in a microstructure conduit. To reduce the volume of the required gas mixture and integrate the necessary microelectronics, a gas chromatography chip implemented using micromachining

**Fig. 6.19**
A two-wafer Si module

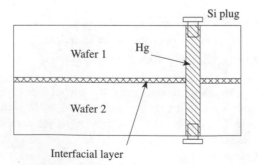

Interfacial layer

(*Source:* R. S. Muller, R. T. Howe, S. D. Senturia, R. L. Smith, and R. M. White, eds., *Microsensors* (Piscataway, N.J.: IEEE Press, 1991). © 1991 IEEE. Reprinted with permission.)

is very attractive. A schematic of the microstructure designed by J. Jerman and S. Terry is shown in Fig. 6.20. The conduit is filled with a chromatography liner and is 1.5 m long, with a cross section of 40 $\mu$m $\times$ 150 $\mu$m. The gas mixture is injected at one end of the conduit (the input), and a gas detector consisting of a resistance gas-flow sensor is located at the output end. The differential arrival times of the gases are detected by the changes in the resistance measured in the gas-flow sensor. Control valves are needed at the gas input and during purging. In the reported device, up to 10 different gases were identified with an accuracy of 10 ppm. Such a microstructure gas chromatography chip, if successfully mass-produced at low cost, can potentially serve numerous applications in medicine, environmental control, and food processing.

**Example 6.15** Estimate the length of a conduit required to distinguish between $CH_4$ and $CF_4$ if the sensor at the output of the conduit can resolve a time difference of only 1 ms. As an approximation, assume the gas-flow rate is inversely proportional to the square root of the molecular mass of the gas and the flow rate for $CH_4$ is $1 \times 10^{-7}$ $m^3$/s. The diameter of the conduit is 1 $mm^2$.

*Solution:* Since the flow rates are inversely proportional to the square root of the molecular mass, the flow rate of $CF_4$ is $1 \times 10^{-7}$ $m^3$/s $\times \sqrt{10/42} =$ $0.49 \times 10^{-7}$ $m^3$/s. The required length of the conduit is $L =$ $(1 \times 10^{-7}$ $m^3$/s $- 0.49 \times 10^{-7}$ $m^3$/s$) \times 1 \times 10^{-3}$ s/$(1 \times 10^{-6}$ $m^2) =$ $0.51 \times 10^{-4}$ m, or 51 $\mu$m. §

## Miniature Cooler

The idea of cooling a Si chip using **Joule-Thomson cooling** was demonstrated by W. A. Little. A heat-exchange system built with a microstructure can provide cooling when compressed air is allowed to expand within the system. The entire system, including an expansion chamber, can be implemented using a microstructure, as shown in Fig. 6.21. For a 25-cm-long conduit with a cross-sectional diameter of 100 $\mu$m, cooling rate down to 77 K can be as short as a few seconds with a compressed air pressure of 1000 psi (1 psi = $6.90 \times 10^3$ Pa). The experimental conduit used is an elongated groove etched in a Si wafer and sealed by an anodically bonded glass plate (see Section 6.11). This type of low-cost on-chip cooling system

**Fig. 6.20**
A schematic for a chromatography chip

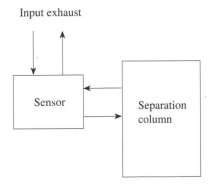

Input exhaust

Sensor

Separation column

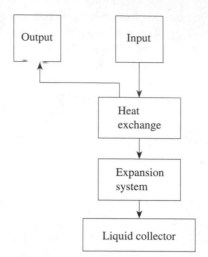

Fig. 6.21
A microstructure cooling
system

can be applied to IR detectors, power transistors, high-density digital ICs, and even high-temperature superconducting devices.

**Example 6.16**     Two side-by-side rectangular tubes in a Si wafer have a length of 25 cm and a cross section of 100 μm × 100 μm. Estimate the heat transfer between the tubes if one of them is cooled to 77 K. The wall thickness between tubes is 25 μm. The ambient temperature is 300 K. Thermal conductivity of Si is 150 W/m·K.

*Solution:*     The heat flow between the two tubes is $Q' = -k'A_{cs}\Delta T/\Delta x$, where $k'$ is the thermal conductivity. Therefore, $Q' = -150$ W/m/K × 0.25 m × 1 × $10^{-4}$ m × (300 K − 77 K)/(50 × $10^{-6}$ m) = 167 W.   **§**

## X-Ray Masks

A Si diaphragm can be used to form an **X-ray mask,** because Si is relatively transparent to X rays. The diaphragm is made by forming a heavily doped p-type surface layer onto an n-type Si substrate. The p-type layer essentially defines the mask thickness (1 to 5 μm). A thin sheet of metal, such as Au, is deposited on the p-side, and the pattern on the X-ray mask is transferred to the Au layer by lithography. The Si wafer is then etched from the bottom side until a thin diaphragm corresponding to the thickness of the p-type layer is left behind. Etch stop is essentially automatic once the p-type layer is reached. As X rays do not penetrate Au but pass readily through Si, an X-ray mask is formed. For a p-type layer doped with B, since the B atoms enter the lattice substitutionally, the tension developed in the Si lattice due to the smaller B atoms keep the diaphragm relatively flat. This is an important factor to consider in mask-making. X-ray masks are used in submicrometer lithography for the fabrication of ULSIC (ultra-large-scale integrated circuits).

**Example 6.17**     Compare the attenuation of X rays through 5 μm of Si and a similar layer of Au. Assume that the absorption coefficient for Au is 3/μm and for Si is 8 × $10^6$/ μm.

***Solution:***   The fraction of X rays penetrating 5 μm thick of Au is $\exp(-\alpha^* x) =$ $\exp(-3/\mu m \times 5\ \mu m) = 3.05 \times 10^{-7}$. Si should have close to 100% transmission, since the X-ray wavelength is too short for any significant absorption to occur.   **§**

## 6.5   HEAT AND TEMPERATURE SENSORS

A hot object will emit light naturally when it is at a temperature above that of its surrounding. This phenomenon is called **blackbody radiation.** During blackbody radiation, the absorption rate is equal to the emission rate (see Sections 4.4 and 4.5) and the emission intensity, $I(v)$ ($J/m^2$), is given by

$$I(v) = A_{sp}/[B_{21}[(B_{12}/B_{21})\exp(hv/(kT)) - 1]] \tag{6}$$

where   $h$ is Planck's constant,

   $v$ is the light (photon) frequency in Hz,

   $k$ is the Boltzmann constant, and

   $T$ is the absolute temperature in K.

In arriving at Eq. (6), we used Eqs. (16) and (21) of Chapter 4 and assumed the Boltzmann relationship: $N_2/N_1 = \exp(-hv/(kT))$. If $B_{12} = B_{21}$—i.e., the probabilities of the upward transition and the downward transition are the same—Eq. (6) becomes Planck's law, provided $A_{sp}/B_{21} = 8\pi\eta_s 3hv^3/c^3$, where $c$ is the velocity of light and $\eta_s$ is the index of refraction. Figure 6.22 shows a plot of the intensity of the blackbody radiation spectrum at 300 K. Note that the radiation spectrum decreases to zero at both ends of the spectrum and the peak intensity at 300 K is in the far infrared.

**Fig. 6.22**
Blackbody radiation
spectrum

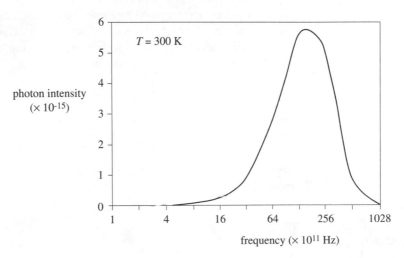

## Heat Sensors

There are many different types of solid-state heat sensors, the majority of which are made of thin films. These sensors will respond to blackbody radiation coming from a hot object. Heat sensors are divided into **quantum detectors,** which respond only to a specific range of photon energies, and **energy detectors,** which respond to a continuum of photon energies. An example of a quantum detector is a photocell, and an example of an energy detector is a thermocouple. Quantum detectors are discussed in Section 6.8 and also in Chapter 4. The sensitivity of the energy detectors is normally quite low. Heat sensors are characterized by their responsivity, which includes parameters such as the output voltage per incident power, the response time, and the energy threshold. The **response time** of a heat detector depends on the thermal mass of the detector itself, whereas the **energy threshold** is quite specific to the type of detector. The following are some commonly used energy detectors.

*Bolometers*  A **bolometer** is an energy detector, and its operation depends on the change in the resistance of the device during heating. Both metal and semiconductor bolometers are used, although semiconductor bolometers, with the larger change in the carrier densities, are more sensitive. Thin-film bolometers are usually made with metals, and their response time is on the order of 5 ms. The temperature dependence of the resistance, $R(T)$, of a metal ($\Omega$) is approximately given by

$$R(T) = R_0(1 + a_1T + a_2T^2 + a_3T^3 + \cdots) \tag{7}$$

where   $R_0$ is the resistance at $T = 0$ °C in $\Omega$,

$a_1, a_2$, etc., are the temperature coefficients, and

$T$ is temperature in °C.

Pt, Ni, and Cu are the more common metals used for making bolometers. Most bolometers are used within a temperature range of 0 °C to 100 °C, which requires only two of the terms in Eq. (7). More sophisticated bolometers are made with Pt wires, which can measure a temperature range from −183 °C to 630 °C. Thin-film Pt resistors have an accuracy and repeatability of less than 0.01 °C. Semiconductor bolometers, on the other hand, have a negative temperature coefficient, since the (intrinsic) carrier density increases with increasing temperature. Their temperature dependence is nonlinear and is approximately given by

$$R(T) = R_0\exp[\beta^*(1/T - 1/T_0)] \tag{8}$$

where   $R_0$ is the resistance of the bolometer at $T = T_0$ in $\Omega$, and

$\beta^*$ is a constant dependent on the energy gap of the semiconductor in K.

Semiconductor bolometers are called **thermistors;** typical constructions are shown in Fig. 6.23. Most thermistors operate within a temperature range from

**Fig. 6.23**
Packaged thermisters

50 to 100 °C. There are numerous other applications for a bolometer in addition to temperature measurement. These include pneumography, velocity measurement, thermodilution flow measurement, liquid-level sensor, altimeter, etc. In pneumography, the thermistor is placed in the outer nasal passage to detect the temperature difference between the inspired cool air and the expired warm air. This, in turn, is used to obtain the breathing rate. Velocity measurement depends on convective cooling of a heated thermistor suspended in a fluid. The reduction in temperature is a measure of the air flow. Thermodilution flow measurement depends on the change in the temperature of blood in the ventricle when a cold saline solution is injected into the atrium. The temperature reduction recorded by an invasive thermistor is a means of measuring the blood flow. As a liquid-level sensor, the thermistor is initially submersed in a fluid, and its temperature is lowered due to the greater thermal conductivity of the fluid. When the fluid level drops below the thermistor, the temperature goes up, which reduces the resistance of the thermistor. A thermistor is used as an altimeter by placing the thermistor at the surface of a liquid in an open container. As the liquid boils, the boiling point depends on the atmospheric pressure (which varies with the altitude), and the output of the thermistor is a measure of the altitude. It is possible to measure altitudes from sea level to 37,500 m.

**Example 6.18**   Assume $a_1$ for Pt is given by 0.00392/°C. To a first order of approximation, estimate the temperature when the resistance of a Pt bolometer is increased by 50%.

*Solution:*   Based on Eq. (7), $R(T) \approx R_0(1 + a_1T)$. Since $R(T)/R_0 = 1.5$, $T = 1.5/a_1 = 1.5/0.00392/°C = 383$ °C.   §

**Example 6.19**   If the resistance of a semiconductor bolometer has been reduced to 10% of its initial value at 300 K, suggest at what temperature the measurement is taken. Assume $\beta^* = 1.27 \times 10^4$ K.

*Solution:*   Based on Eq. (8), $R(T)/R_0 = \exp[\beta^*(1/T_1 - 1/T_0)] = 0.1$. Since $T_0 = 300$ K, $T_1 = 1/[1/300\text{ K} + 1/\beta^*\ln(0.1)] = 1/(1/300\text{ K} + \ln(0.1)/(1.27 \times 10^{-4}\text{ K})) = 355$ K.   §

*Thermocouples*   The principle behind the operation of a **thermocouple** is the thermoelectric effect; i.e., a voltage is produced when the two junctions of a thermocouple are at different temperatures (see Section 2.6). The thermoelectric voltage, $\Delta V$ (V), is given by

$$\Delta V = K_{12}(T_1 - T_2) \tag{9}$$

where   $K_{12}$ is the Seeback coefficient in V/°C, and

   $T_1$ and $T_2$ are the temperatures of the junctions in °C.

Often $T_2$ is the temperature of an ice bath and $T_1$ becomes proportional to $\Delta V$. In practice, the temperature-voltage (T-V) relationship of a thermocouple is nearly linear, and the exact values are usually tabulated in a handbook. The requirement that $T_2$ be set equal to 0 °C is not always necessary, and some thermocouples can have a simulated cold junction that serves as the reference. Such a thermocouple, however, will have a lower accuracy ($\pm 1$ °C). The time constant of most thermocouples is usually quite long (of the order of milliseconds). Integrated thermocouples are formed using thin films of Bi–Sb or Au–polysilicon deposited on a Si diaphragm.

In these thermocouples, the cold junction has to be shielded from direct radiation, and a resistive temperature sensor provides temperature correction to the cold junction. A **thermopile** is a heat sensor made with a number of thermocouples connected in series. It gives a larger voltage output.

**Example 6.20**   A Pt–Pt thermocouple has an output voltage of 18 mV for a temperature difference of 1500 °C. Transform these values to microvolts per degree Celsius.

*Solution:*   The temperature-to-voltage conversion factor of a Pt–Pt thermocouple is 18 mV/1500 °C = 12 $\mu$V/°C.   §

*Junction Thermometers*   A *p-n* junction forms a good temperature sensor since for a given current passing through it, the voltage across the device is approximately proportional to the absolute temperature. The temperature, $T_j$ (K), of a *p-n* junction is given by

$$T_j = A + BV_a \tag{10}$$

where   $A$ and $B$ are constants in units of K and K/V, respectively, and

$V_a$ is the voltage drop across the *p-n* junction in V at a given current flow.

By characterizing the $T$-$V_a$ (temperature-voltage) characteristics of a *p-n* junction, the junction can be used to measure temperature. The useful temperature range for a **junction thermometer** is between 40 and 400 K. Bipolar transistors can also be used to measure temperature in a similar manner by making use of the emitter-base junction.

**Example 6.21**   Assuming the ideal diode equation is given by $I = I_0(\exp(qV_a/(kT)) - 1)$, show how it correlates with Eq. (10).

*Solution:*   Based on the given equation, $qV_a/(kT) = \ln(I/I_0 + 1)$, or $T = qV_a/[k \ln(I/I_0 + 1)] = A + BV_a$. When correlated to Eq. (10), $A = 0$.   §

*Ferroelectrics*   A ferroelectric solid can be used as a heat sensor. As discussed in Chapter 2, a ferroelectric solid will retain a remnant polarization even after the applied voltage is removed. A common ferroelectric material is $BaTiO_3$. To generate the remnant polarization, the ferroelectric solid has to be polarized (or poled) at a high temperature. Remnant polarization is sensitive to temperature change, and a differential change in the remnant polarization is a measure of a temperature change. A **pyroelectric heat sensor** consists of two poled ferroelectric slabs, one with a

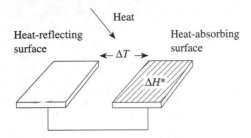

**Fig. 6.24**

A pyroelectric heat sensor

Heat-reflecting surface

Heat

$\Delta T$

Heat-absorbing surface

$\Delta H^*$

heat-reflecting surface and another with a heat-absorbing surface, as shown in Fig. 6.24.

The difference in heat absorption will produce different temperatures and polarization voltages in the two slabs. The differential configuration shown in the figure is particularly effective in removing any residual voltage not related to the temperature change in the ferroelectric slabs. Quantitatively, the temperature gradient, $\Delta T$ (K), between the ferroelectric slabs is given by

$$\Delta T = \Delta H^*/(A_s h_1 \rho_0 s_p) \tag{11}$$

where    $\Delta H^*$ is the change in heat energy in J,

$A_s$ is area of the slab in $m^2$,

$h_1$ is thickness of the slab in m,

$\rho_0$ is the density of the ferroelectric slab in $kg/m^3$, and

$s_p$ is the specific heat of the ferroelectric slab in J/kg·K.

**Example 6.22**   A pyroelectric sensor measures a temperature difference of 10 K. Estimate the amount of heat falling on the sensor per unit area if the slabs are 2 mm thick. Assume that the density of the slabs is $2.3 \times 10^3$ $kg/m^3$ and the specific heat is $0.7 \times 10^3$ J/kg·K.

***Solution:***   Based on Eq. (11), $\Delta H^* = \Delta T A_s h_1 \rho_0 s_p = 10\ \text{K} \times 2 \times 10^{-3}\ \text{m} \times 2.3 \times 10^3\ kg/m^3 \times 0.7 \times 10^3\ \text{J/kg·K} = 3.2 \times 10^4$ J, or 32 kJ.   §

## 6.6   DISPLACEMENT AND FLOW SENSORS

### Resistive Strain Gage

A **resistive strain gage** is one of the simplest displacement sensors. It relies on the resistance change of the sensor as a result of stress or displacement and can be used for measuring force, pressure, and acceleration. There are basically two types of resistive strain gages: metal-semiconductor gage and elastic gage.

Metal-semiconductor gages are more suitable for measurement of small displacements because of their plasticity, whereas elastic gages can withstand significant elongation. Thus, **elastic gages** are suitable for both static as well as dynamic measurements. In the presence of an applied force, the fractional change in the resistance, $\Delta R/R$, of a rectangular structure is given by

$$\Delta R/R = \Delta\rho_0/\rho_0 + \Delta L/L - \Delta A_{cs}/A_{cs} \qquad \text{(12)}$$

where   $\rho_o$ is the initial resistivity in $\Omega$,

   $L$ is the initial length in m,

   $A_{cs}$ is the initial cross-sectional area in $m^2$, and

   $\Delta$ stands for the incremental changes in the parameters.

If the structure is cylindrical, then Eq. (12) becomes

$$\Delta R/R = \Delta\rho_0/\rho_0 + \Delta L/L(1 + 2\ddot{y}) \qquad \text{(13)}$$

where   $\ddot{y}$ is the Poisson ratio $(= -(\Delta D/D)/(\Delta L/L))$, and

   $D$ is the diameter of the cylinder in m.

The performance of the different gages is measured in terms of a parameter known as the *gage factor, g,* given by

$$g = (\Delta R/R)/(\Delta L/L) = 1 + 2\ddot{y} + (\Delta\rho_0/\rho_0)/(\Delta L/L) \qquad \text{(14)}$$

Table 6.2 shows the gage factors for some typical metals and semiconductors. As observed in the table, the gage factors for semiconductors are significantly higher than those for metals because of the piezoresistive effect. For more sensitive measurements, the strain gage should be arranged in a bridge circuit (see Fig. 6.18). The bridge configuration will automatically allow for temperature compensation.

**Table 6.2**   Gage factors for some typical metals and semiconductors

| Solid | Gage Factor |
| --- | --- |
| Constantin $(Ni_{45}-Cu_{55})$ | 2.1 |
| Ni | $-12$ to $-20$ |
| Si (*p*-type) | 100 to 170 |
| Si (*n*-type) | $-100$ to $-140$ |
| Ge (*p*-type) | 102 |
| Ge (*n*-type) | $-150$ |

(*Source:* R. S. C. Cobbold, *Transducers for Biomedical Measurements: Principles and Applications,* © 1974 John Wiley & Sons, Inc. Reprinted by permission of John Wiley & Sons, Inc.)

Figure 6.25 shows the Si gage factor as a function of doping and the crystal orientation. Semiconductor gages have a large gage factor, but they are also more sensitive to a temperature change. They are frequently formed on Si diaphragms, and the piezoelectric resistors can be formed either by diffusion or vapor-phase deposition (VPD).

Elastic gages are frequently used in biomedical applications. A typical device consists of a rubber tube filled with Hg or a conducting electrolyte. The ends of the tube are sealed and terminated with conducting electrodes. As the tube stretches, the resistance also changes. It can be shown that the fractional change in resistance, $\Delta R/R$, is given by

$$\Delta R/R_0 = 2L/L_0 + (\Delta L/L_0)^2 \tag{15}$$

where     $R_0$ is the initial resistance of the elastic gage in $\Omega$,

$L_0$ is the initial length of the rubber tube in m, and

$\Delta$ stands for the incremental changes in the parameters.

**Example 6.23**  Compare the fractional change in resistivity if the dimensions of a rectangular structure and a cylindrical structure are each reduced by 10%.

*Solution:*  Based on Eq. (12), for a rectangular structure $\Delta R/R = \Delta\rho_0/\rho_0 + \Delta L/L - \Delta A_{cs}/A_{cs} = 0 - 0.1 + 0.1^2 = -0.09$. Based on Eq. (13), for a cylindrical structure $\Delta R/R = \Delta\rho_0/\rho_0 + \Delta L/L(1 + 2\ddot{y}) = 0 - 0.1 \times (1 - 2 \times 0.1/0.1) = 0.1$.  §

**Example 6.24**  Compute the gage factors for the structures in Example 6.23.

*Solution:*  Based on Eq. (14), $g = (\Delta R/R)/(\Delta L/L)$. For a rectangular structure, $g = -0.09/(-0.1) = 0.9$. For a cylindrical structure, $g = -0.1/0.1 = -1$.  §

## Linear Variable Differential Transformer (LVDT)

A **linear variable differential transformer** (LVDT) is an electromechanical device that produces an electric output proportional to the displacement of a movable core.

**Fig. 6.25**
Gage factor versus doping in Si for different crystal orientations

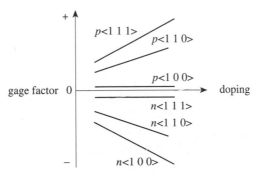

(*Source:* R. S. C. Cobbold, *Transducers for Biomedical Measurements: Principles and Application,* © 1974 John Wiley & Sons, Inc. Reprinted by permission of John Wiley & Sons, Inc.

The core is inserted in a cylinder consisting of a magnetically coupled primary coil and two secondary coils symmetrically placed on either side of the primary coil, as shown in Fig. 6.26. The primary coil is excited by a sinusoidal signal between 100 and 1000 Hz. In the null position, the movable core is located in the center of the cylinder, and the induced voltages in the secondary coils exactly cancel each other. Any displacement of the movable core causes an output voltage offset in the form of an amplitude-modulated (AM) signal. The output can then be rectified and is directly proportional to the displacement. The LVDT has a number of advantages, including a long mechanical life due to minimum friction between the movable core and the cylinder, high resolution, linearity, and good input-output isolation.

## Piezoelectric Transducer

**Piezoelectricity** is electrical polarization due to the displacement of positive and negative charges or ions in a solid as a result of an applied force. Piezoelectricity exists only in anisotropic crystals with a nonsymmetric crystal structure. Materials such as quartz, ZnO, ammonium dihydrogen phosphate, $BaTiO_3$, lead zirconate titanate (PZT), and polyvinylidene fluoride ($PVF_2$) are known to exhibit piezoelectricity. The output of a **piezoelectric sensor** is usually a voltage. Piezoelectricity exists only at a temperature below a critical temperature called the *Curie temperature.* Some piezoelectric solids exist in the form of ceramics ($BaTiO_3$), and others exist as polymers ($PVF_2$). Unlike the crystalline piezoelectric solids, the noncrystalline solids have to be heated to a high temperature and then cooled in the presence of a large electric field (10 MV/m or higher) to generate the piezoelectric effect. This process is known as *poling,* and the temperature required for PZT is 365 °C. Piezoelectric sensors normally operate in a frequency range from 10 Hz to 50 kHz and can withstand a pressure between 5 and 10,000 psi. Most piezoelectric sensors are used in the dynamic mode—i.e., in the sensing of vibrations or impulses. A typical device construction is shown in Fig. 6.27.

A number of parameters are useful in the characterization of piezoelectricity. Among them are the piezoelectric charge constant, $a_{piez}$, and the piezoelectric force constant (electric field per unit stress), $g_{piez}$. Piezoelectricity depends on the asymmetry in the crystal structures and the direction in which force is applied. Piezoelectric parameters are, therefore, given for the different crystal directions. Typical values are listed in Table 6.3.

**Fig. 6.26**
A linear variable
differential transformer

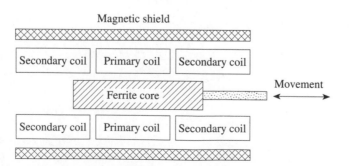

Fig. 6.27

A piezoelectric transducer

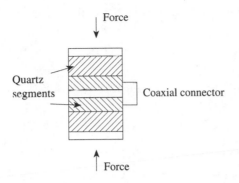

Force

Quartz
segments

Coaxial connector

Force

Table 6.3   Parameters measured in the common piezoelectric solids

| Solid | $a_{piez33}$* <br> (C/N $\times$ $10^{-12}$) | $g_{piez31}$* <br> (V·m/N $\times$ $10^{-3}$) |
|---|---|---|
| BaTiO$_3$ | 149 | 14 |
| PZT | 285/374 | 26.1/24.8 |
| Quartz | 2.3 | 58 |
| PVF$_2$ | 39 | 190 |
| Li$_2$SO$_4$ | 16 | 175 |
| PbSO$_4$ | 85 | 42.5 |

*The numerical subscripts indicate the modes of deformation.
(*Source:* R. E. Boltz and G. I. Tuve, *CRC Handbook for Applied Engineering Science,* 2d ed. Boca Raton: CRC Press, 1973.)

Figure 6.28 shows an equivalent-circuit model of a piezoelectric transducer. Piezoelectricity is modeled as capacitors and leakage resistors in series with a charge generator. Assuming that the induced charge, $Q$ (C), is proportional to the applied force, $F$ (N), then $Q$ is given by

$$Q = a_{piez}F \tag{16}$$

where   $a_{piez}$ is the piezoelectric charge constant in C/N (see Table 6.3).

Furthermore, if the applied force is also proportional to the displacement, $x$ (m), then

$$Q = k_p x \tag{17}$$

where   $k_p$ is a constant in C/m.

Fig. 6.28

An equivalent circuit for a
piezoelectric transducer

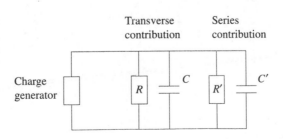

Transverse
contribution

Series
contribution

Charge
generator

$R$   $C$   $R'$   $C'$

The current through the capacitor, $I_{cap}$, is then given by

$$I_{cap} = \Delta Q/\Delta t = k_p \, \Delta x/\Delta t \tag{18}$$

From Fig. 6.28, it can be shown that

$$C \, dV_0/dt = I_{cap} - V_0/R \tag{19}$$

where    $V_0$ is the measured voltage across the capacitor in V,

   $I_{cap}$ is the current through the capacitor in A,

   $C$ is the capacitance in F, and

   $R$ is the leakage resistance in $\Omega$.

Combining Eqs. (18) and (19), it can be shown that in the frequency domain, the capacitor voltage, $V_0(j\omega)$ (V), is given by

$$V_0(j\omega)/X(j\omega) = K_s \, j\omega\tau/(1 + j\omega\tau) \tag{20}$$

where    $j \, (= \sqrt{-1})$ is the imaginary number,

   $\omega$ is the angular frequency in rad/s,

   $K_s = k_p/C$, and

   $\tau = RC$.

The transfer characteristics of a piezoelectric transducer (Eq. (20)) plotted as a function of frequency are shown in Fig. 6.29. Note that the flat region in the figure provides a frequency-independent response. Transducers have an internal leakage current, and the signal is best maintained if the time constant $(= RC)$ of the piezoelectric solid is large. This is usually achieved by increasing $R$ rather than $C$, since any increase in $C$ will reduce the midfrequency gain (when $j\omega\tau = 1$). Any increase in $R$, however, increases the output impedance of the transducer and makes the signal transfer to the output more difficult. At high frequency, a piezoelectric transducer will also exhibit a resonance effect and can be used as a resonator in an analog filter.

Fig. 6.29

Transfer characteristics versus frequency of a piezoelectric transducer

$V_0(j\omega) / X(j\omega)$

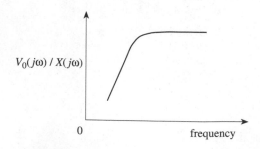

**Example 6.25** Based on Eq. (19), comment on the steady-state solution of the output from a piezoelectric transducer and the limit when $R$ is large.

***Solution:*** Equation (19) suggests $C\, dV_0/dt = I_{cap} - V_0/R$. In the steady state, $V_0 = I_{cap}R$, and the piezoelectric sensor behaves like a dc current source. When $R$ is large, there is no leakage current. $V_0 = 1/C \int [I_{cap}]\, dt$, and the sensor is capacitive. **§**

## Displacement and Velocity Sensors and Accelerometers

Displacement sensors are varied, ranging from a potentiometer-type displacement sensor to the LVTD-type sensor mentioned earlier. The cross section of a **capacitance displacement sensor** is shown in Fig. 6.30. The Si membrane is formed by micromachining, and the differential capacitance is measured when there is a difference in the pressure across the membrane. Velocity sensors, on the other hand, measure the rate of change of displacement rather than the displacement itself, and most of them rely on a changing magnetic flux to produce a voltage output. A typical example is a **tachometer,** which measures angular velocity, as shown in Fig. 6.31. The dc tachometer uses a permanent magnet that physically rotates with the moving object to be measured and generates an induced dc voltage across a sensing coil. In a similar ac setup, the permanent magnet is replaced by a primary coil excited by an ac voltage. The coupling between the primary coil and the secondary coil varies with the speed of the rotor.

**Fig. 6.30**
A membrane-type capacitance pressure sensor

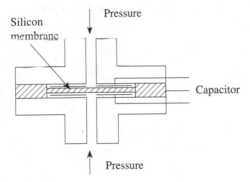

**Fig. 6.31**
A tachometer

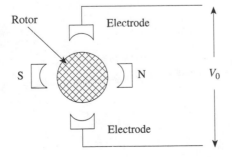

An **accelerometer** measures the acceleration or deceleration of a moving object. The basic construction of a mechanical accelerometer is shown in Fig. 6.32. The acceleration is directly transformed to the displacement in the movable mass. Piezoelectric crystals are also used in accelerometers. In this case, the mass, the damper, and the spring are replaced by a piezoelectric crystal. Compression in the crystal produces a voltage change proportional to the acceleration. A stack of several piezoelectric crystal plates can be used to increase the output voltage. Cantilever beams have also been used in accelerometers. The force applied to the tip of the cantilever beam generates a defection, which can be measured either with a piezoelectric sensor attached to the beam or a capacitance sensor. For typical dimensions, the capacitance of the beam is between 10 and 100 fF.

**Example 6.26**    Compute the capacitance of a $SiO_2$ cantilever beam if the dimensions are 105 μm × 25 μm × 0.5 μm. The permittivity of $SiO_2$ is $34.51 \times 10^{-12}$ F/m.

***Solution:***    The capacitance of the $SiO_2$ beam is $C_0 = \epsilon_0 lw/d = 34.51 \times 10^{-12}$ F/m $\times 105 \times 10^{-6}$ m $\times 25 \times 10^{-6}$ m$/0.5 \times 10^{-6}$ m $= 17.94 \times 10^{-14}$ F $= 179.4$ fF.    **§**

## Thermal Convection Flowmeter

**Thermal flowmeters** are frequently used to measure the velocity of a fluid. In a thermal convection sensor, this is done by monitoring the amount of heat supplied to a thermistor that is cooled by fluid, as shown in Fig. 6.33.

The thermistor is normally heated to a temperature higher than the ambient temperature to produce a greater temperature gradient. The rate of heat transfer by the fluid is given by King's law:

$$P' = A_s(T_{\text{th}} - T_f)(C_1 + C_2\sqrt{v}) \tag{21}$$

**Fig. 6.32**
A mechanical
accelerometer

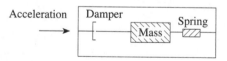

**Fig. 6.33**
A setup for measuring
fluid flow

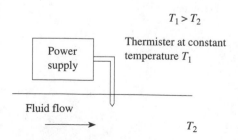

where $P'$ is the rate of heat transfer in W,

$A_s$ is the surface area of the thermistor in m$^2$,

$T_{th}$ is the temperature of the thermistor in K,

$T_f$ is the temperature of the fluid in K,

$v$ is the velocity of the fluid in m/s, and

$C_1$ and $C_2$ are empirical constants.

At equilibrium, $P' = I^2R$. $P$ is the power supplied to the flowmeter, $I$ is the current (A), and $R$ is the resistance of the thermistor ($\Omega$). Substituting $P'$ into Eq. (21) gives

$$v = K_0(I^2 - K_1)^2 \tag{22}$$

where $K_0$ and $K_1$ are constants provided $T_{th} - T_f$ is a constant.

To maintain the constant temperature difference, a control circuit, as shown in Fig. 6.34, is required. The circuit will increase the current flow into the thermistor once the temperature drops (see Example 6.27). By monitoring the current passing through the thermistor and calibrating the system, the flow rate of the fluid can be computed. Typical constructions of the thermistor probes are shown in Fig. 6.35.

**Fig. 6.34**
A constant temperature
(resistance) circuit

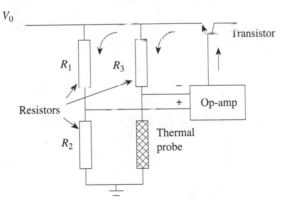

**Fig. 6.35**
Thermal convection
probes

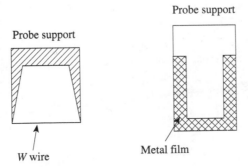

The thermistor probes can be in the form of a wire mounted on a support structure or a metal film deposited on a Pyrex substrate.

**Example 6.27**    In a thermal convection flowmeter, suggest how the resistance of the heating element may be kept constant by the circuit shown in Fig. 6.34.

*Solution:*    In Fig. 6.34, the op-amp acts as an isolation amplifier. If the probe resistance increases due to cooling, the differential input to the op-amp also increases. This results in a larger output voltage and a larger current through the probe. The heating effect increases its temperature. There is, however, a limit on the ability to self-correct in a feedback system.    §

## Constant-Heat-Infusion Flowmeter

The mass flow of a fluid can be determined by injecting a known amount of heat into the fluid and measuring the temperature change at a given location downstream. This is done by monitoring the temperature at two points in the flow using thermistors, as illustrated in the setup shown in Fig. 6.36.

The average mass flow, $F_m$ (kg/s), is given approximately by

$$F_m = Q_h / [s_f(T_{ds} - T_{us})] \qquad (23)$$

where    $Q_h$ is the rate of heat added in J/s,

$s_f$ is the specific heat of the fluid in J/kg·K,

$T_{us}$ is the upstream temperature in K, and

$T_{ds}$ is the downstream temperature in K.

Equation (23) assumes that heat loss in the system in negligible and the value of $F_m$ is not a function of the separation between the two probes.

**Example 6.28**    Compute the mass flow in a **constant-heat-infusion flowmeter** if a heater of 10 W is used, the specific heat of the fluid is 4.2 kJ/kg·K, and the temperature difference between upstream and downstream is 5 °C.

*Solution:*    From Eq. (23), $F_m = Q_h / [s_f(T_{ds} - T_{us})] =$ 10 W/(4.2 × 10³ J/kg·K × 5 K) = 0.44 gm/s.    §

**Fig. 6.36**
A constant-heat-infusion
flowmeter

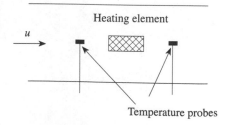

## Rotating Flowmeters

A **full-bore turbine flowmeter,** as shown in Fig. 6.37, measures the average volume flow of a fluid. The fluid flow drives the turbine in the conduit, and the rotor speed is monitored by a magnetic sensor. The volume flow rate, $V_F$ ($m^3$/s), is given approximately by

$$V_F = v_r D^3 f(v_r, D, \eta_v) \tag{24}$$

where   $v_r$ is the rotor speed in m/s,

$D$ is diameter of the conduit in m, and

$f(v_r, D, \eta_v)$ is a function of $v_r$, $D$, and the viscosity, $\eta_v$, of the fluid.

Fluid-flow measurements using a rotor depend to a large extent on the design of the rotor and its speed linearity.

## Differential Pressure Flowmeter

The pressure difference between two points along the flow path of a fluid is related to the flow rate. In the case of the **Pitot tube,** as shown in Fig. 6.38, the static pressure of the fluid, $p_s$ (Pa), is measured by a hole placed perpendicular to the fluid flow, whereas the dynamic pressure, $p_d$ (Pa), is measured with a hole facing the flow.

**Fig. 6.37**
A rotating flowmeter

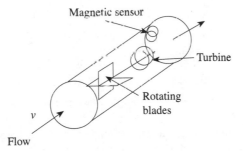

**Fig. 6.38**
A Pitot tube

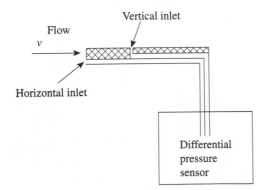

Assuming an incompressible fluid, the pressure difference, $\Delta p$ ($= p_d \quad p_s$), according to Bernoulli's equation is given by

$$\Delta p = \Gamma_f v^2/2 \tag{25}$$

where   $\Gamma_f$ is the fluid mass density in kg/m³, and

   $v$ is the fluid speed in m/s.

The Pitot tube is frequently used to measure air speed in wind tunnels. It is, nevertheless, more suitable for measuring high-speed fluid flow.

**Example 6.29**   Compute the speed of water if the pressure difference in a Pitot tube is 0.03 Pa. Water density is 0.023 kg/m³.

*Solution:*   From Eq. (25), $v = \sqrt{2\Delta p/\Gamma_f} = \sqrt{2 \times 0.03 \text{ Pa}/0.023 \text{ kg/m}^3}$ m/s
$= 1.6$ m/s.   §

## Electromagnetic Flowmeter

An **electromagnetic flowmeter** is used to measure the flow rate of a conductive fluid. Its operation depends on the fact that a conductive fluid moving perpendicular to an applied magnetic field will induce a transverse voltage on the walls of the conduit. The value of the voltage is linearly proportional to the fluid velocity. This is quite similar to Hall effect (see Section 2.6), which is used to measure the carrier density in a solid. Figure 6.39 shows a schematic of the setup. The velocity of the fluid, $v$ (m/s), is given by

$$v = \mathbf{B}L/V_{\text{tr}} \tag{26}$$

where   $\mathbf{B}$ is the magnetic induction in T,

   $L$ is the distance in the flow under the applied magnetic field in m, and

   $V_{\text{tr}}$ is the transverse voltage measured in a pair of electrodes in V.

Both dc and ac measurements can be made. Dc measurement is less effective because it is affected by the polarization of the electrodes. Ac measurement, on the other

**Fig. 6.39**
An electromagnetic
flowmeter

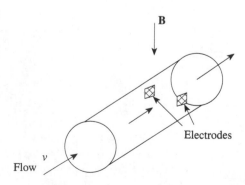

hand, can generate spurious effects due to the induced magnetic induction on the measuring wires.

**Example 6.30**   Determine the flow rate in an electromagnetic flowmeter if the magnetic flux density is 1 T, $d$ is 1 m, and $V = 500$ mV.

*Solution:*   From Eq. (26), $v = \mathbf{B}d/V_{tr} = 1\text{ T} \times 1\text{ m}/0.5\text{ V} = 2$ m/s.   §

## Ultrasonic Flow Sensors

Ultrasonic flow sensors are frequently used to measure fluid flow in biological and industrial systems. **Ultrasonic sensors** measure the sound waves emerging from an acoustic source by generating an electrical signal proportional to the energy in the sound waves. In fluid-flow measurements, the signal frequency is above the audible range. There are two ways that acoustic signals are used for measuring fluid velocities. The first method relies on the fact that acoustic signals, when propagating through a moving medium, acquire effective velocity equal to the algebraic sum of the sound velocity and that of the medium. A device based on this principle is called a **transit-time flowmeter.** The second method makes use of the Doppler shift principle; a **Doppler shift flowmeter** measures the velocity of the scattered or reflected sound waves from particles suspended in the fluid. The shift in the frequency of the sound waves is related to the velocity of the fluid. A number of piezoelectric transducers can be used, especially those with a high **Q-factor.** For this reason, PZT is most commonly used. In an ultrasonic flow measurement, the piezoelectric material is normally poled and molded into the form of a disc or a plate with a thickness equal to one-fourth the wavelength of the sound wave. Metal electrodes are then deposited on the opposite faces to give the effect of a resonator. As a transmitter, the piezoelectric transducer is driven by an oscillator circuit. A receiving sensor, on the other hand, is connected to an RF amplifier. Piezoelectric transducers using this type of construction suffer from diffraction of the acoustic beam. In the near-field region, the beam is approximately uniform up to a distance $d_{nf}$ from the source. The distance (m) is given by

$$d_{nf} = D^2/4\lambda_a \qquad\qquad (27)$$

where   $D$ is the transducer diameter in m, and

$\lambda_a$ is the wavelength of the sound waves in m.

In the far-field region, the beam diverges at an angle $\phi_{ff}$ and is given approximately by

$$\phi_{ff} = \sin^{-1}(1.2\lambda_a/D) \qquad\qquad (28)$$

Most ultrasonic transducers operate within the near-field region in order to maintain a high signal resolution. By operating at a higher frequency, the near-field region can be extended, provided that the absorption of the ultrasound in the fluid is small. In the case of blood-flow measurements, this requirement presents a problem since the absorption coefficient of ultrasound in blood increases with frequency. Therefore, other types of methods, such as the Doppler shift method, will have to be used. In the Doppler shift method, the backscattered power actually increases with the signal frequency up to the fourth power.

Figure 6.40 shows the measurement setup for the transit-time flowmeter. The piezoelectric transducers can be located either internally within the pipe or outside it. In the case of externally placed transducers, the ultrasound will have a velocity, $v^*$ (m/s), given by

$$v^* = v_a + v \tag{29}$$

where    $v_a$ is the velocity of the ultrasound in space in m/s, and

$v$ is the velocity of the fluid in the path of the ultrasound in m/s.

Thus, for downstream measurement, the transit time, $\tau_{ds}$ (s), is given by

$$\tau_{ds} = d^*/(v_a + v \cos(\Theta)) \tag{30}$$

where    $\Theta$ is the angle between the direction of the ultrasound and the direction of the fluid flow in degrees, and

$d^*$ is the separation between transducers in m.

A correction factor, $A_{cf}$, in the velocity of the fluid flow is used since the fluid flow is not necessarily uniform across the pipe. The relationship between $v$ and an average flow rate, $\hat{u}$ (m/s), is given by

$$v = A_{cf}\hat{u} \tag{31}$$

where    $A_{cf} = 1.33$ for laminar flow and $A_{cf} = 1.07$ for turbulent flow.

Transit-time flowmeters can also be used in an upstream configuration. The measured transit time will, of course, be longer, since the returned sound waves are against the flow. In industry, transit-time flowmeters are used to measure the flow rates of water, milk, treated sewage, etc. The fact that ultrasound sensors are noninvasive makes the method very desirable for medical applications. In the case when transducers are placed within a pipe, the transit time for downstream measurement is equal to $d^*/(v_a + \hat{u})$.

**Fig. 6.40**
A transit-time flowmeter

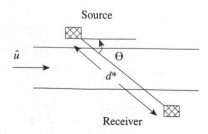

**Example 6.31** Estimate the distance of the near-field region if the transducer has a diameter of 1 cm and the sound wave is 50 MHz. Assume that the sound velocity is 1500 m/s.

*Solution:* From Eq. (27), $d_{nf} = D^2/4\lambda_a = (0.01 \text{ m})^2 \times 5 \times 10^7/$ $(4 \times 1.5 \times 10^3 \text{ m}) = 0.83 \times 10^{-6}$ m, or 0.83 μm. §

The Doppler shift method uses either a continuous-wave ultrasound source or short ultrasound pulses. In both cases, for a receding receiver, the Doppler frequency, $\nu_D$ (Hz), is given by

$$\nu_D = 2\nu_0 \nu \cos(\Theta)/v^* \tag{32}$$

where $\nu_0$ is the frequency of the ultrasound in Hz,

$v$ is the velocity of the scattering centers in the fluid in m/s,

$v^*$ is the velocity of sound in the fluid in m/s, and

$\Theta$ is the angle between the sound path and the direction of fluid flow in degrees.

Figure 6.41 shows the typical setup used for the Doppler shift measurement. In the measurement of biological signals, significant cleaning of the signals is often required because the in-vivo system is full of spurious noise sources. Figure 6.42 shows some typical blood flow measurements (before and after signal filtering). The continuous-wave Doppler shift method is not capable of determining the flow direction. The pulsed Doppler shift flowmeter uses pulsed signals instead of continuous-

**Fig. 6.41**
A Doppler flowmeter

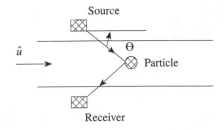

Source

$\hat{u}$

$\Theta$

Particle

Receiver

**Fig. 6.42**
Typical Doppler shift measurements from blood vessels

Signal superimposed on artifacts due to blood vessel wall movement

Signal after low-pass filtering showing periodic echo

**Fig. 6.43**
A pulsed Doppler
flowmeter

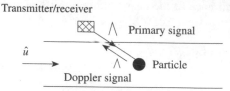

**Fig. 6.43**
A pulsed Doppler
flowmeter

wave signals. Figure 6.43 shows a typical setup when the source is also used as the receiver. The principle of operation is similar to the continuous-wave operation, except that the response consists of a collection of reflections both from the wall of the pipe and from the individual particles in the fluid. Ideally, the pulses should be very short to cover the near-field region; the accepted pulse width is about 1 μs at a few megahertz. When the source is also used as the receiver, very good isolation is required to make sure that the much stronger input signal does not contaminate the weaker signal in the response. The main advantage of a pulsed Doppler flowmeter is that it provides a fluid-flow profile rather than an average flow rate. The size of the transducer can be very small in order to fit it into a catheter tip. The small size of a microsensor can offer a significant advantage in the case of biomedical applications.

**Example 6.32**    In a Doppler shift flow measurement, if the Doppler frequency shift is 1 kHz, estimate the fluid velocity if the acoustic source has a frequency of 5 MHz and the transducers are at an angle of 45° to the length of the pipe. The average velocity of sound in the fluid is 1500 m/s.

*Solution:*    Based on Eq. (32), $v = v^* v_D /(2v_0\cos(\Theta)) = 1500$ m/s $\times 1 \times 10^3/\text{s}/(2 \times 5 \times 10^6/\text{s}/\sqrt{2}) = 0.21$ m/s.    §

## 6.7   CHEMICAL SENSORS

### Gaseous Sensors

The purpose of **gaseous sensors** is to detect gases in whatever amount is present. Most gaseous sensors are chemical sensors, and they are used in industry, in health care, and for environmental control. The majority of gaseous sensors are placed within a reaction chamber to allow for real-time control of chemical reactions. Such sensors have been found to be extremely effective in controlling the operation of engines, the monitoring of chemical reactions, and even the timely delivery of intravenous drugs. In real life, there are many different situations when gases have to be identified and measured. The simplest type of solid-state gaseous sensors relies on the changes in the physical properties of the sensors to detect the gases present. Two of the more popular sensors are chemovoltaic sensors and chemoresistive sensors.

*Chemovoltaic Sensors*    In **chemovoltaic sensors,** the chemical changes are reflected in the form of a voltage difference. An example is a sensor used to measure

$O_2$ partial pressure in the exhaust gas of a gas engine. It is well known that the performance of a gas engine depends on the combustion efficiency, which is related to the output $O_2$ partial pressure. Thus, by monitoring the output $O_2$ partial pressure and providing control to the fuel input, both fuel economy and performance can be realized. The $O_2$ partial pressure at the output of the engine is usually measured relative to the atmospheric partial pressure of $O_2$ at the input. One way to measure this partial-pressure difference is to transform the partial pressure to ion concentrations and then measure the resulting potential difference. A particular requirement is that the sensor must be able to withstand the high temperature of the exhaust-gas system. The solution to this problem is found in the *solid-state electrolytic cell*. A schematic of the cell and its operation are shown in Fig. 6.44. The pair of Pt electrodes serves as the ionizing agent, converting the O atoms to their ions. The electrolyte is simply a mixture of $ZrO_2$ and $Y_2O_3$. $ZrO_2$ forms the conducting medium, whereas $Y_2O_3$ is added to enhance the conductivity by increasing the O vacancies and their stability. The cell basically measures the ion concentration gradient through the chemical potential established across the electrolyte. The equation governing the chemical process is given by

$$P_{low}/P_{high} = k_1\exp(-4\Delta V/V_t) \tag{33}$$

where    $P_{low}$ is the $O_2$ partial pressure at the output in atmospheres (1 atm = $1.01 \times 10^5$ Pa),

$P_{high}$ is the atmospheric $O_2$ partial pressure in atm,

$k_1$ is a constant,

$\Delta V$ is the voltage due to the ion-concentration gradient in V, and

$V_t = kT/q$.

Equation (33) can be rewritten as

$$\Delta V = k_2\ln(P_{low}) + k_3 \tag{34}$$

where    $k_2$ and $k_3$ are constants in V.

Therefore, by measuring the voltage across the Pt electrodes, the $O_2$ partial pressure at the output can be estimated. One major concern with this type of gaseous sensor is the response time. The response time depends primarily on the ionization

**Fig. 6.44**
A chemovoltaic sensor

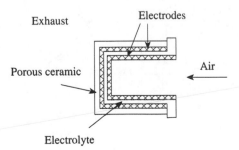

Exhaust

Electrodes

Porous ceramic

Air

Electrolyte

of the O atoms and the establishment of an ion-concentration gradient across the electrolyte. As far as the materials are concerned, a set of finely textured Pt electrodes and a conductive electrolyte with plenty of O vacancies are required.

**Example 6.33**   From Eqs. (33) and (34), relate $k_3$ to $k_1$.

*Solution:*   Based on Eq. (33), $P_{low}/P_{high} = k_1 \exp(-4\,\Delta V/V_t)$ and $\Delta V = -(V_t/4)\ln(P_{low}) + (V_t/4)\ln(k_1 P_{high})$. Thus, $k_2 = -V_t/4$ and $k_3 = (V_t/4)\ln(k_1 P_{high})$.  §

*Chemoresistive Sensors*   **Chemoresistive sensors** respond directly to changes in resistance. Similar to a chemovoltaic sensor, a chemoresistive sensor also requires Pt electrodes to generate the $O^{2-}$ ions and a medium permeable to $O^{2-}$ ions. The main difference, however, is that the resistance of the sensor is measured instead of the output voltage. The proper choice of a suitable medium is equally important. Let us consider $TiO_2$ as the medium. At a high temperature, the following equilibrium reaction takes place:

$$O^{2-} + Ti^{4+} \Leftrightarrow Ti^{2+} + \frac{1}{2}O_2 \text{ (gaseous)} \qquad (35)$$

As observed, the $O_2$ partial pressure directly affects the $Ti^{2+}$ ion concentration, and a current flows as the electrons hop from the $Ti^{2+}$ sites to the $Ti^{4+}$ sites. By monitoring the level of the current, the $O_2$ partial pressure can be measured. There is, however, one problem with the chemoresistive sensor. Although the conduction process is well defined, the density of the $Ti^{2+}$ ions is very sensitive to temperature. To single out the effect due to oxygen, a differential pair of sensors is used. These sensors may be arranged in the form of a bridge, as shown in Fig. 6.45.

**Example 6.34**   In an engine exhaust-gas system, if $P_{low}$ varies between $10^{-2}$ and $10^{-20}$ atm, estimate the voltage change in a chemovoltaic sensor operating at 700 K.

*Solution:*   Based on Eq. (34), $\Delta V = k_2 \ln(P_{low}) + k_3$ and $k_2 = -V_t/4 = -15$ mV at 700 K. $\Delta V_2 - \Delta V_1 = -15 \times 10^{-3} \text{ V}[\ln(10^{-20}) - \ln(10^{-2})] = 0.745$ V.  §

**Fig. 6.45**
A bridge circuit for a chemoresistive sensor

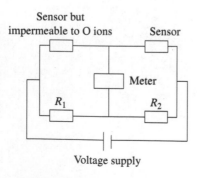

Sensor but impermeable to O ions

Sensor

Meter

$R_1$

$R_2$

Voltage supply

A device that offers significant potential for use as an integrated chemical sensor is the ion-sensitive field-effect transistor (ISFET). The key element of the ISFET is a surface insulator layer (either $SiO_2$, $Si_3N_4$, or other metal oxides) deposited on a semiconductor substrate. The overall structure is quite similar to the MOSFET (described in Chapter 3), except that the surface metal layer is missing. When the device is exposed to a chemical solution containing ions, the properties of the insulator-semiconductor interface change due to the charge-imaging effect. Most frequently, this change is either in the resistance of the charge-induced channel or in the capacitance of the depletion layer. When the device is properly calibrated, it can be used to measure pH (ion concentration) values varying between 2 and 10 for ions including $K^+$, $Ca^{3+}$, $Cl^-$, and $I^-$.

**Example 6.35**   In an ISFET, if the substrate dopant density is $1 \times 10^{22}/m^3$ and a surface charge density of $1 \times 10^{17}/m^2$ is found in the insulator, what will be the induced channel resistance? The area of the ISFET is 200 $\mu$m $\times$ 200 $\mu$m and the channel length is 10 $\mu$m. Assume the semiconductor mobility to be 0.15 $m^2/V{\cdot}s$.

*Solution:*   The thickness of the induced channel is
$1 \times 10^{17}/m^2/(1 \times 10^{22}/m^3) = 1 \times 10^{-5}$ m.
The channel resistance is $R_c = \rho L/A_s = 10 \times 10^{-6}$ m/
$(1 \times 10^{22}/m^3 \times 1.6 \times 10^{-19}\,C \times 0.15\ m^2/V{\cdot}s \times 10 \times 10^{-6}\,m \times 200 \times 10^{-6}\,m) =$
21 $\Omega$.   §

# 6.8   OPTICAL SENSORS

**Optical sensors** are sensors that respond to electromagnetic radiation, including high-energy radiation such as X rays and gamma rays, UV and visible light, infrared (IR) light, and heat. To quantify light or radiation, we define the steradian (sr), or a unit of the solid angle. The steradian, $\Omega_s$, for a point source located at a distance $r$ (m) from where the radiation is measured is defined as

$$\Omega_s = A_s/r^2 \qquad (36)$$

where   $A_s$ is the area in $m^2$ on a projected spherical surface where the radiation is measured.

Thus, the total radiation flux from a point source amounts to $4\pi$ sr. The radiant refers to the radiation energy from a source and is given in joules. Radiation (or radiant) flux, on the other hand, is the radiation flow per unit area per unit time. It is measured in watts. Finally, the radiant intensity (W/sr) refers to the radiant flux per unit solid angle.

The application of the optical sensors requires some understanding of the radiation source. The most direct radiation source is the sun, and it has the output spectrum shown in Fig. 6.46. An artificial light source similar to the sun is the heated tungsten filament. It radiates at a temperature between 2200 and 3000 K. Figure 6.47 shows its output spectrum. In addition to these light sources, **light-emitting diodes** (LEDs) and lasers are two important sources of radiation. Their operations are described in Chapter 4. Finally, arc discharge and fluorescent lamps are also used as light sources. Arc discharge can provide a high radiant output from a small area and in the frequency range from 200 nm to 360 nm. Examples of the common arc-discharge vapors include Hg, Na, and Xe.

Optical sensors are often designed to have a specific response for a given range of radiation. The selection of the spectral content can be achieved using a light filter and a monochromator. **Light filters** are divided into high-pass, low-pass, bandpass, band-reject, and neural density. **Neural density filters** are partially coated with silver to attenuate equally at all wavelengths (primarily in the visible and the IR range). Figure 6.48 shows some typical filter characteristics. In addition to normal

**Fig. 6.46**
The solar spectrum

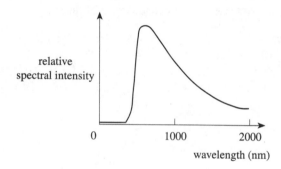

**Fig. 6.47**
Spectral response of a heated tungsten filament

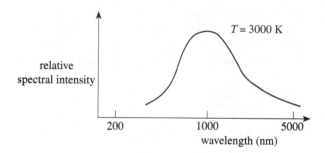

**Fig. 6.48**
Transmission characteristics of light filters

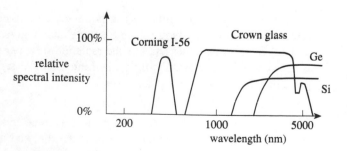

spectral filters, there are the **interference filters,** which are transparent to specific wavelengths of light due to constructive interference. These filters, however, are also transparent to the harmonics of the wavelengths. An interference filter has a very narrow bandwidth, as low as 4 nm, and a transmission up to 90%. A **monochromator,** on the other hand, is not a light filter, and it works on a different principle. A monochromator primarily makes use of a set of diffraction gratings to disperse light into its spectral components. The dispersed light is spread out, and only a small portion is allowed to exit the monochromator through a slit. The spectral resolution of a monochromator varies inversely with the width of the slit, and a resolution as fine as 0.5 nm can be achieved. The light sources mentioned so far are dc sources. In some instances, ac sources are required to minimize noise and drift; ac light can be generated using a chopper, which operates in a manner similar to the rotating blades of an electric fan. Choppers can operate at a frequency up to several kHz.

The figure of merit of an optical sensor is measured by its **sensitivity** (similar to responsivity). Sensitivity is measured by the dependence of the output signal on the radiant power. It is particularly important in measurement systems when the signal level is low. For the characterization of detectors, two other parameters are also of interest. The **detectivity,** $D*$ (see Section 4.6), essentially measures the time response and the noise level in the system. A high $D*$ implies a low-noise, efficient detector system. **Quantum efficiency** (see Section 4.6) is defined as the effective number of conduction electrons created per incident photon. It can be expressed in terms of the transit time of the electrons (in the detector) divided by their lifetime.

**Example 6.36**   A photodetector of dimensions 1 cm $\times$ 1 cm $\times$ 0.2 cm generates a current of 10 $\mu$A for a photon flux of $1 \times 10^{16}/\text{m}^2 \cdot \text{s}$. Calculate the quantum efficiency.

*Solution:*   The amount of charge that leaves the detector is $10 \times 10^{-6}/(1.6 \times 10^{-19} \times 0.01 \times 0.002)/\text{m}^2 \cdot \text{s} = 3.12 \times 10^{17}/\text{m}^2 \cdot \text{s}$. The quantum efficiency is $3.12 \times 10^{17}/(1 \times 10^{16}) = 31.2$.   **§**

## Photon Detectors

**Photoconductivity** is most widely used for photon detection. Both intrinsic and extrinsic semiconductors are used as photon detectors. In an intrinsic semiconductor, the radiation energy has to exceed the value of the energy gap to generate a response, whereas in an extrinsic semiconductor, a smaller energy is required. In either case, carriers are created, and the conductivity goes up with the excitation. The photocurrent is linear with the light intensity at low intensity but becomes nonlinear if the intensity is high. Many photodetectors made with extrinsic semiconductors have a long response time due to the presence of traps. For high sensitivity, detectors are cooled to a lower temperature to reduce the internal noise. Some of the well-known detectors include CdS, CdSe, PbS, and PbSe. Both CdS and CdSe have a sharp response in the visible and in the IR range, whereas the PbS and PbSe detectors are used mainly in the near IR. The response time of detectors can vary from a few microseconds to several seconds.

Photovoltaic devices and photodiodes (similar to the *p-n* junction) are also used for light detection. Photodiodes, including Schottky photodiodes, are widely used for radiation detection because of the strictly linear relationship between the output current and the photon flux. **p-i-n diodes** (*p-n* junctions with an intrinsic layer in the junction to increase light absorption) operating in reverse bias can provide a significant gain in the photocurrent as well as a very fast response time due to a lack of the minority carrier storage effect. A gain as high as 100 and a response time less than 0.5 ns can be achieved. An even higher gain can be achieved if avalanche breakdown and carrier multiplication occur in the high-field region (intrinsic region); these devices are called *avalanche photodiodes* (APD). Schottky photodiodes (metal-semiconductor junctions), on the other hand, are more suitable for high-speed, low-light-intensity operation. They have a small dark current and can operate up to a frequency of several gigahertz. Phototransistors, like photodiodes, are also used for light detection. They can often achieve a gain of 1000 or higher.

The most sensitive light detector is the **photomultiplier tube** (PMT). In principle, PMTs can detect the presence of a single photon. The PMT consists of a photocathode, which serves as an emitter. When the photon energy exceeds the workfunction of the photocathode, electrons are emitted, and they are accelerated in a very high potential gradient (up to 1500 V). Several dynodes at successively higher potentials are placed in the path of the electrons in between the photocathode and the anode, as shown in Fig. 6.49. As the electrons strike the dynodes, secondary emission takes place. This multiplication effect can increase the photocurrent by as much as a factor of $10^8$. The response of the PMT is relatively linear.

Bolometers and thermopiles are also designed for light-intensity measurements, even though, in principle, they are temperature or heat detectors. In these devices, the areas exposed to the radiation are usually darkened to maximize the absorption process. The time response of these devices is of the order of tens of milliseconds. Pyroelectric sensors (see Section 6.5) can also be used to measure light intensity.

Examples of the applications of light detectors include the *smoke density monitor* and the *nephelometer.* A **smoke-density monitor** is placed directly across from a light source and keeps track of the amount of light absorbed. A **nephelometer,** on the other hand, is mounted perpendicular to the light beam. The scattered light in the presence of dense smoke activates the detector. Nephelometers are frequently used as household smoke detectors.

**Fig. 6.49**
A photomultiplier tube

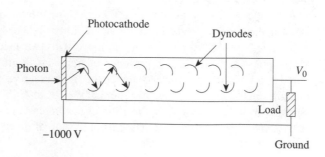

**Example 6.37** A PMT has 12 dynodes. If the number of electrons emitted from each dynode is increased by a factor of 4 and the loss is negligible, what is the multiplication factor, *M*, of the PMT?

*Solution:* The multiplication factor $M$ is $4^{12} = 1.67 \times 10^7$. §

## 6.9 DESIGN OF A PRESSURE TRANSDUCER

Piezoresistive transducers are used in numerous applications where pressure sensing is needed. Most piezoresistive sensors are made with Si because of the well-established properties of Si and the compatibility with IC technology. The making of piezoresistive sensors requires the use of micromachining, lithography, etching, ion implantation, and contact formation. The main structure in a pressure sensor is a diaphragm formed by micromachining. The requirements for the diaphragm include (1) a precise thickness, (2) uniformity, and (3) a smooth surface. The control of the thickness of the diaphragm is particularly important, since its sensitivity varies as the square of the diaphragm thickness. The diaphragm is produced by lithography and etching from the back side of the wafer. The number of masks in the mask set is between 8 and 12 (a modest number by IC standards). Thickness control is achieved using a $p^+$ etch-stop layer, since most etchants will slow down significantly with a highly doped *p*-type layer.

Electrical contacts between the piezoelectric resistors and the outside world is through metallization on the diaphragm and the support structure. Bonding pads are deposited on the support structure, to which thin Al or Au wires can be attached. Al is used because of the lower cost, but Au is a more passive element. Since metallization is the final step in the fabrication process, after metallization, the die is often covered with a protective silicone-based gel that does not affect the performance of the sensor.

Piezoresistive sensors are generally fabricated by sealing two wafers together. The top wafer has the bridge resistors formed on it and a backside etch that creates the diaphragm. It is then sealed to a second wafer, creating a reference cavity in between. An absolute sensor requires a sealed cavity, whereas in a differential sensor, the cavity is vented through the second wafer to where pressure is to be measured. The bonding of the two wafers can be done either by glass sealing or electrostatic sealing in vacuum (see Section 6.10). *Glass sealing* involves the use of low-temperature glass frits heated to form a smooth layer around the wafers. The wafers are then put into position and reheated to reflow the glass for bonding. *Electrostatic sealing* requires the use of Pyrex as an intermediary, and molecular bonding is achieved by passing a large current through the two wafers. Some sensors have to be attached to a mount, and the materials requirements on the mount can be quite severe, particularly the thermal compatibility of the mount with the wafer. Often, the mount will experience essentially the same pressure as the sensors (say up to 100 psi, or $6.89 \times 10^5$ Pa). The most popular way to attach a sensor is to use a compliant mount that does not transmit stress between the sensor and the mount. Most sensor mounts are made of a rubber-based material called RTV, and Au is used as the bonding material. During the formation, when the Au–Si eutectic temperature is reached, Au will bond very well to Si, and the sensor can be rigidly attached to the mount.

The packaging of sensors involves two additional types of mounts. The sensor can be either mounted on a printed-circuit board or to a pipe structure. The printed-circuit board is small and has fittings for pressure connections and electrical connections. Most printed-circuit boards have fixed dimensions and are used mainly for low-pressure measurements ($< 100$ psi). Pipe mounting relies on a threaded pipe connection to the system and can measure very high pressure (as high as 10,000 psi, or $6.90 \times 10^7$ Pa). These sensors are more rugged and are often placed outdoors.

The formation of piezoresistors requires $p$-type implants into $n$-type Si substrates. The $p$-$n$ junction so formed has to be isolated using a reverse bias between the $p$-type surface layer and the substrate (the negative terminal of the power supply is connected to the $p$-type surface layer). Leakage current through the $p$-$n$ junction causes errors in the measurement, and it increases with increasing temperature. The temperature-dependent error can, in fact, be separated into (1) an offset effect, which is independent of the applied pressure, and (2) an effect due to span, which is directly related to the applied pressure. Span $S$ (sensitivity to the applied pressure, $V \cdot N/m^2$) is related to the gage factor, $g_{\text{piez}}$ (see Table 6.3), of the piezoresistors through the following relationship:

$$S = g_{\text{piez}} V_{\text{br}} p_a \tag{37}$$

where    $V_{\text{br}}$ is the bridge voltage in V, and

            $p_a$ is the applied pressure in Pa.

The temperature effect on $S$ can be compensated for by changing the parameters on the right-hand side of Eq. (37) to maintain $S$ independent of temperature.

Figure 6.50 shows the temperature dependence of $g_{\text{piez}}$. If the temperature sensitivity of the resistors is larger than that of the gage factor, then a constant current through the resistors will increase the temperature coefficient (TC) of the resistors (which increases with increasing temperature) until it exactly balances the TC of the gage factor. Other techniques are also available to achieve temperature compensation. These usually involve using an active circuit, such as the one shown in Fig. 6.51. As temperature goes up in the circuit, the voltage across the Zener diode goes down and a higher current passes through the bridge (note that the voltage drop across the op-amp is essentially zero). The end result is that the TC of the resistors goes up to match that of the gage factor. In the case when the TC of the resistors is less than that of gage factor, then a thermistor with a negative TC can be put in series

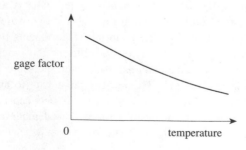

**Fig. 6.50**
Temperature dependence of the gage factor

**Fig. 6.51**

An active temperature-
compensation circuit

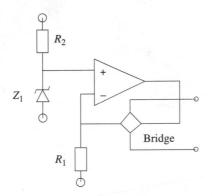

with the sensor and the power rail so that the overall TC is the same as that of the gage factor. The thermistor may be a *p-n* junction diode.

Offset temperature compensation is due to a number of electromechanical properties of the sensor. It could be due to (1) resistor mismatch, (2) inferior resistor tracking, (3) self-heating and temperature change related to pressure transients, (4) temperature gradient in the sensor, (5) differential impedance loading at the output, (6) built-in strain, or (7) mismatch in the thermal expansion coefficients. Additional compensation and calibrations are often necessary.

Simulation is often used in sensor design. With the current numerical computation techniques and computing power, it is not difficult to simulate the entire device and perform a spatial and time domain analysis. The main advantage of simulation is that once the model of the sensor is developed and verified, it is quite easy to alter the design parameters and to carry out local and global optimization.

For the diaphragm with dimensions $L$ and $W$ (m), as shown in Fig. 6.52, the deflection of the diaphragm, $W(x', y')$, is given by

$$W(x', y') = W(0, 0)[((y')^2 - 1)^2((x')^2 - 1)^2 \cdot \qquad \text{(38)}$$
$$[1 + C_1(y')^2 + C_2(x')^2 + C_3(y')^2(x')^2]$$

where  $x' = 2x/L$ and $y' = y/W$, as illustrated in Fig. 6.52,

$W(0, 0) = C'W^4 \, \Delta p/D,$

$\Delta p$ is the pressure difference across the plate in N/m$^2$,

$D = E_Y \, t_h^3/12(1 - \ddot{y}^2)$

**Fig. 6.52**

Dimensions of a
diaphragm for simulation

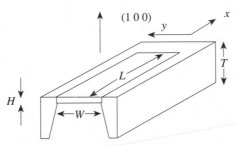

(Reprinted from D. T. Tandeske, *Pressure Sensors: Selection and Application,* 1991, p. 106, by courtesy of Marcel Dekker Inc.)

**Fig. 6.53**

Fitting coefficients versus $L/W$

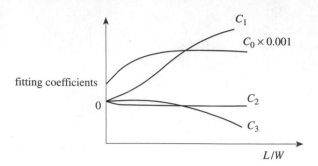

**Fig. 6.54**

Stress pattern on a diaphragm

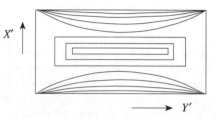

(Figs. 6.53 and 6.54 are reprinted from D. T. Tandeske, *Pressure Sensors: Selection and Application*, 1991, pp. 108 and 110, by courtesy of Marcel Dekker Inc.)

where $E_Y$ is Young's modulus, $t_h$ is the diaphragm thickness in m, $\ddot{y}$ is the Poisson ratio, and $C'$, $C_1$, $C_2$, and $C_3$ are the fitting coefficients.

Equation (38) models the deflection of the diaphragm if the coefficients $C'$, $C_1$, $C_2$, and $C_3$ are known. Figure 6.53 shows their dependence on the ratio $L/W$. A complete simulation of the sensor can produce an overall stress pattern similar to what is shown in Fig. 6.54. One important aspect in sensor design involves the output sensitivity. A study of the output sensitivity using sensors with different diaphragm thicknesses has shown that the agreement between experiment and simulation can be greater than 90%. The success of these simulation results has made simulation tools very useful in the study of the performance of piezoelectric sensors.

As in any other product, specifications on sensor performance are important to the users, and they offer a summary of the features of the sensors. Because of the large number of parameters that may be specified, we shall only mention two of them, namely, the pressure range and the TC, which we believe are more important.

1. *Pressure range:* Specifications on the pressure range are very important and should include the offset value, the span, stability and repeatability, the hysteresis, and the linearity.
2. *Temperature coefficient* (TC): Specifications on the temperature coefficient should include offset TC, span TC, and overall TC.

## 6.10  GROUNDING, SHIELDING, AND INTERFERENCE

The operation of microsensors and micromachines often involves very small electrical or magnetic signals that are in close proximity to each other and are subject to contamination by feed-through and unwanted interference. When an electric signal is generated inside a microsensor or a micromachine, an internal voltage reference point in the system is required to minimize spurious effects. Such a reference point is called the **ground point** and should be as close to the signal source as possible. Of course, this is not always feasible if a number of signal sources are present. A ground point also serves as an electromagnetic voltage or current sink, in which unconnected and yet electrically active regions in the microsensor or micromachine will be connected and held at a fixed potential. This ensures that those regions will not function as a radiator or receiver of electromagnetic radiation and is particularly important in the suppression of RF noise. An unwanted voltage can also arise from a junction of two dissimilar metals acting as a thermocouple. The thermoelectric voltage, however, will be eliminated if the metals are grounded. In addition to grounding the floating parts in a sensor system, it is also desirable to use coaxial cables for the input and the output. A coaxial cable with its shielding maintained at the input voltage will prevent leakage current along the signal path and remove contamination by common-mode signals—i.e., signals that have the same sign as the input.

## Electric Shielding

When a sensor or a circuit is placed inside a metal casing, there is capacitive coupling between the casing and the different parts of the sensor or its support circuits. If there is an ac signal present within the system, the casing will radiate electromagnetic noise unless it is grounded. Grounding the casing also minimizes the capacitance between the sensor and the casing. This is known as **electric shielding.** Figure 6.55 shows the effect of grounding in an electrically active system.

In addition to minimizing the capacitive effect, electrical isolation between a sensor and its circuits (such as a pre-amp) can be crucial when small signals are involved. Figure 6.56 shows how a shunt current across the input of an operational amplifier (op-amp) can create an unwanted dc offset voltage. The unwanted signal is eliminated by having good isolation across the positive and the negative terminals of the op-amp. This happens when the amplifier has a high input impedance. Similar isolation can be achieved using isolation transformers or an optical coupler. In an isolation transformer, the input signal is modulated by a carrier signal at a given frequency and is then fed to the primary coil. The secondary coil picks up the modulated signal and demodulates it. Gain is achieved by using coils with different

**Fig. 6.55**

A circuit showing the effect of two different ground points

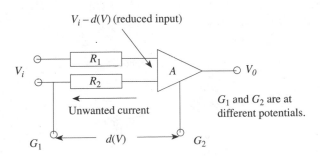

**Fig. 6.56**
A circuit showing how a shunt current can produce a dc offset

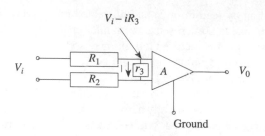

numbers of windings. An isolation transformer has the advantage that the input circuit (or the sensor) may be floating, which is often the case for a medical or biological sensor. An optical coupler consists of a light-emitting diode (LED) and a phototransistor. Because the electrical signal is transformed into an optical signal and then back into an electrical signal again, excellent isolation and no feed-through will be observed. Figure 6.57 shows the schematic of a typical optical coupler.

**Example 6.38**    Estimate the capacitance between a grounded circuit board and the base of a floating chassis if the circuit board has a dimension of 20 cm × 20 cm and is 2 cm above the chassis.

*Solution:*    We assume the circuit board is closest to the base of the chassis, and we simply consider the capacitance to be due to a parallel-plate capacitor. The capacitance is $C = \epsilon_0 A_{cs}/d = 8.85 \times 10^{-12}$ F/m × $(0.2 \text{ m})^2/0.02$ m = $1.77 \times 10^{-11}$ F, or 17.7 pF.   §

## Magnetic Shielding

Micromachines and microsensors with small moving parts are particularly sensitive to magnetic interference. For instance, a changing magnetic field will generate an induced voltage across a conductor, including the case when a moving armature forming an enclosed conducting loop sweeps through a dc magnetic field (Faraday's law). To suppress the induced voltage and the resulting eddy current, **magnetic shielding** is applied and the area enclosed in any current-carrying loop in the system must be minimized (see Chapter 5). In wiring, for instance, the use of twisted wire pairs and coaxial cables is a very effective way to minimize magnetically induced current.

**Example 6.39**    In the presence of a dc magnetic induction of 0.1 T, a metal bar resting on a U-shaped metal frame with the arms 10 cm apart moves at a constant speed of 10 cm/s. Estimate the induced voltage, $V_{\text{ind}}$.

**Fig. 6.57**
An optical coupler

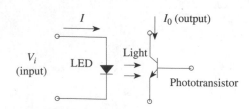

*Solution:* We assume the current loop to be rectangular. The induced voltage is $V_{ind} = \mathbf{B}d(A_{cs})/dt = 0.1 \text{ T} \times 0.1 \text{ m} \times 0.1 \text{ m/s} = 1 \text{ mV}.$ **§**

## 6.11 PACKAGING MATERIALS

**Packaging** is the last step in the fabrication of a microsensor or a micromachine. Nevertheless, packaging is extremely important, since it is the key linkage between a working device and the outside world. Normally, packaging in microsensors and micromachines includes metallization and encapsulation. Packaging is required to provide electrical interfacing to the other components in the system. It also provides a platform for support, a pathway for heat transfer, and protection against the environment so that the microsensor or micromachine can be handled safely. As far as signals are concerned, packaging should be neutral and passive and should provide no attenuation or time delay to the output. For increasingly more complex microsensor systems, packaging is a demanding problem, and much effort is spent to improve the related technologies. The simplest packaging structure involves a container with an isolated platform on which the microsensor or micromachine is to be mounted. The platform should have the necessary bonding pads for interconnection to the microsensor or the micromachine.

Figure 6.58 shows a typical **can package** with a device wire-bonded to it. Before the can is sealed, it is usually filled with an inert gas to minimize surface contamination and corrosion. A second form of package is called a **lead-frame package** and is shown in Fig. 6.59. The device is bonded to the center of the lead

**Fig. 6.58**
A TO5 can header with can cover removed

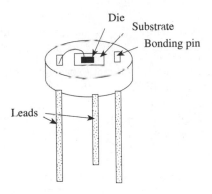

Die
Substrate
Bonding pin
Leads

**Fig. 6.59**
A lead-frame package

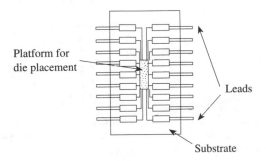

Platform for die placement
Leads
Substrate

frame, and the whole package is potted into a polymer or glass capsule. Both these packaging methods are simple and geared for automated assembly. In complex microsensors and micromachines when heat dissipation is important, the device is usually bonded to a much larger ceramic substrate, and the interconnections are achieved by wire-bonding, similar to the case of a can package. Ceramic substrates allow for more complex wiring and interconnections, which can include the placement of several devices or chips on a single ceramic substrate.

## Substrate Materials

The key requirement of a substrate is that it should be thermally and electrically conducting. In addition, the substrate is to be bonded to a support structure, which may be a material such as glass or ceramic. To provide good adhesion, the substrate and the support structure must be thermally compatible. Table 6.4 shows the thermal properties of some typical substrate materials.

Metallic alloys of Fe, Ni, and Co can usually be designed to match the substrate thermal expansion coefficient. As far as substrate materials are concerned, $Al_2O_3$ (alumina) offers the lowest cost, whereas BeO (beryllia) offers the highest thermal conductivity. The effect of impurities in the substrate is usually unimportant, except for the presence of radioactive elements. A small trace of a radioactive element such as U (uranium) can produce a particle flux of the order of 0.01 particles/cm$^2$·h, and a single particle is sufficient to generate as many as $10^6$ electron-hole pairs. This is enough to change the binary state of a memory cell.

**Example 6.40**     If the thermal properties of metallic alloys are linearly dependent on the properties of the constituent elements, suggest how you will mix Si and Au to match the thermal expansion coefficient of $Al_2O_3$ during packaging. The thermal expansion coefficients of Si and Au are $2.6 \times 10^{-6}$ /K and $14.3 \times 10^{-6}$ /K, respectively.

*Solution:*   From Table 6.4, the thermal expansion coefficient of $Al_2O_3 = 6.5 \times 10^{-6}$/K. Let the fraction of Si be $x$ and the fraction of Au be $1 - x$. The requirement is that $6.5 = 2.6x + 14.3x(1 - x)$. This equation gives $x = 0.46$. To achieve approximately the same thermal expansion coefficient, the alloy should have 46% Si and 54% Au.   §

**Table 6.4**     Thermal properties of some typical substrate materials

| Material | Thermal Conductivity (W/m·K) | Thermal Expansion Coefficient ($\times 10^{-6}$ /K) |
| --- | --- | --- |
| Glass | 0.001 | 2.0 |
| Ceramic | 0.003 | 4.2 |
| $Al_2O_3$ | 0.2 | 6.5 |
| BeO | 2.0 | 6.5 |
| Epoxy | 0.8 | 53 |
| $Si_3N_4$ | 0.25 | 3.1 |

(*Source:* C. R. M. Grovenor, *Microelectronic Materials,* Bristol and Philadelphia: Adam Hilgar, 1989. Reprinted by permission of IOP Publishing Ltd.)

**Example 6.41**    A U atom produces a particle flux, $F$, of 0.01 particles/h, and each particle produces $1 \times 10^6$ electrons in Si. If there is 1/1000 ppm of U in Si, what will be the value of the dark current per unit surface area (due to the radiative generation of carriers)? Assume the thickness of the Si wafer, $t_h$, is 300 μm and the mass density of U is $2.33 \times 10^3$ kg/m$^3$.

*Solution:*    The number of U atoms present per unit area is $n_U = 1 \times 10^{-9} N_{AG} \rho_0 t_h$ /atomic weight $= 1 \times 10^{-9} \times 6.02 \times 10^{23} \times 2.33 \times 10^3$ kg/m$^3 \times 3 \times 10^{-4}$ m/28.06 g $= 1.5 \times 10^{13}/\text{m}^2$. The dark current density is $I_d / A_{cs} = nqF = 1.5 \times 10^{13}/\text{m}^2 \times 1 \times 10^6 \times 1.6 \times 10^{-19}$ C $\times$ 0.01/3600/s $= 6.6 \times 10^{-6}$ A/m$^2$, or 6.6 μA/m$^2$.    §

## Materials Bonding

The types of bonding required in packaging are quite diverse, and they can involve metal-to-metal bonding and metal-to-substrate bonding. In general, there are two major kinds of bonding in the packaging operation, van der Waals bonding and chemical bonding. Van der Waals bonds are usually quite weak and require very clean surfaces. Polymer-metal adhesion is due to this kind of bonding. Chemical bonds are much stronger, and they exist in metal-to-metal bonding and between glasses, metal oxides, and ceramics. These bonds usually involve the molten phase of the two bonding materials. A **solder** is sometimes used to fill the interface spacing in the bond to ensure close contact. The bonding between a metal and an oxide glass is sometimes more difficult to achieve. One possibility is to have an interface layer of a metallic oxide. As in the case of Ni on $Al_2O_3$, $NiAl_2O_4$ is formed, and this material bonds very well to either Ni or alumina. Noble metals such as Au and Pt also readily bond to ceramic materials through similar intermediaries. Polymers and reactive metals such as Ni and Cr will bond reasonably well through interface oxides, whereas Cu and a polymer form much weaker bonds. Most polymers and ceramics are bonded by van der Waals bonds. The selection of metal-to-metal contacts has to be very carefully designed because of the corrosion effect present in dissimilar metals, especially in the presence of moisture. Most solder materials are sufficiently massive that it takes corrosion a long time to totally destroy the contact.

## Chip Bonding to Substrates

Chips are bonded to the substrate using either metal solders or polymer adhesives. The main concern is thermal compatibility. Solders usually require a heating cycle to wet the back surface of the chip as well as the top surface of the substrate. Usually both of these surfaces are coated with metal layers to ensure good adhesion. Au-plated metal layers are used to avoid surface oxidation. The cheapest bonding method is to use a glass or a polymer adhesive. Silver deposits are added to reduce the resistivity. Table 6.5 shows the thermal properties of conducting materials commonly used in bonding.

Wire bonding involves the use of fine metal wires between metallization formed on the bonding pads and the package pins or the leadframes. Normally some form of heat, pressure, and ultrasound is used to cause plastic flow at the metal-metal

Table 6.5    Electrical properties of materials used in bonding

| Solid | Melting Temperature (°C) | Thermal Expansion Coefficient ($\times 10^{-6}$ /K) |
|---|---|---|
| Si | 1410 | 3 |
| Ni | 1453 | 13 |
| Cr | 1857 | 6 |
| Cu | 1083 | 16.6 |
| Au | 1064 | 14.2 |
| Pb Solder | 327 | 29 |

(*Source:* R. C. Weast, *CRC Handbook for Physics and Chemistry,* 66th ed., Boca Raton: CRC Press Inc., 1985/86.)

interface. Ultrasound has the effect of breaking down any interface oxide layer. To provide good compatibility with standard metallization materials, both Au and Al are extensively used in wire bonding. **Tape bonding** is an alternative method. In this case, a polymer sheet with a contact array is directly bonded on the chip surface. Pieces of metal bumps are placed between the metallization and the contact array, and the whole assembly is soldered together. **Flip-chip mounting** uses a similar principle, except that a solder is used instead of metal bumps.

**Example 6.42**    Compute the difference in the resistances between Au and Al wire bonding if the average separation between the bonding pad and the package pin is 5 mm and the wire diameter is $2.5 \times 10^{-2}$ cm. Assume the resistivities of Au and Al are $2.0 \times 10^{-8}$ $\Omega \cdot$m and $2.5 \times 10^{-8}$ $\Omega$, respectively.

***Solution:***    Since $R = \rho L/A_{cs}$, $\Delta R = L/A_{cs}\Delta\rho = 5 \times 10^{-3}$ m/ $(3.14 \times (2.5 \times 10^{-4}$ m/2)$^2$) $\times (2.5 \times 10^{-8}$ $\Omega \cdot$m $- 2 \times 10^{-8}$ $\Omega \cdot$m$) =$ $1.96 \times 10^{-2}$ $\Omega$.  **§**

## Field-Assisted Thermal Bonding

Very often microstructures formed on Si wafers will need some form of encapsulation or hermetic seals. This result can be conveniently achieved by a process called **field-assisted thermal bonding.** In this case, a glass slide is placed over a polished Si wafer previously thermally oxidized, and the assembly is heated to 400 °C. A high voltage of about 1200 V is applied between the Si and the glass slide, as shown in Fig. 6.60. Since the negative bias voltage is applied to the glass slide, positive

**Fig. 6.60**
Field-assisted thermal bonding

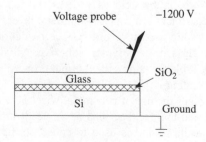

ions are depleted near the glass/Si interface, resulting in a very high electric field across the interface. This produces a strong and uniform hermetically sealed bonding. The seal is not affected by the enclosed circuitries and the metallization under the seal. Field-assisted thermal bonding has been found to provide good protection for enclosed microstructures.

**Example 6.43**    Compute the electric field in field-assisted thermal bonding if the spacing between the glass slide and the substrate is 100 µm.

*Solution:*    The electric field is $\acute{E} = V/d = 1200$ V/$(1 \times 10^{-4}$ m) = 12 MV/m.  §

## Solder and Potting Materials

Soldering of a microelectronic package can be automated using what is called solder dip, or solder reflow. In the **solder dip technique,** the whole assembly is immersed in a moltened solder, and the soldering material sticks only to the metallic joints between the pins and the substrate contacts. The ceramic substrate material will not be coated. **Solder reflow** requires the deposition of a solder material in between the contact areas, and a heating cycle melts the solder. The choice of a suitable **potting material** is also of interest. The potting material has to be insulating and bond well to both metals and ceramics. The potting process involves the flow of a highly fluid thermoplastic or thermosetting material over the chip, and the substrate (together with the potting material) forms a hardened block after cooling. The potting temperature is usually above 150 °C. The material should have a low viscosity and good thermal compatibility with the substrate in order to form a leakproof interface. Polymers such as epoxy resins, however, are not effective when bonded to ceramics.

## Hybrid Circuits

**Hybrid circuits** allow the addition of passive components on a given substrate. This often gives shorter interconnections and a better packing density. In a hybrid circuit, the passive components are deposited in the form of thick films using a silkscreen printing process. The inks used are a mixture of glass particles, metals, and metal oxide particles suspended in an organic compound. The ink is filtered through a photolithographic mask, and the pattern in the mask is printed on the substrate. The ink is then dried to remove the organic components and fired at 700 to 900 °C. A high density of metallic components in the ink forms the conducting paths in the thick film, and the oxide provides a good adhesion to the ceramic substrate. Resistors are formed by an ink with a lesser metallic content. Sometimes, resistors have to be trimmed to provide the required values. Thin-film hybrid circuits are developed by depositing different layers of conductors and insulators on the substrate. The components are defined using lithography. Conducting tracks are formed by the deposition of reactive metals such as Cr, Ti, and Ni on the ceramic substrate prior to the subsequent deposition of Cu, Al, or Au as the top layer. Resistors can be formed by metals such as Ni−Cr, Ta, or $SnO_2$. Dielectric and capacitor layers are formed by $SiO_2$, $TaO_2$, and $Al_2O_3$. The control of the performance of a thin-film hybrid circuit is actually quite difficult, and they are frequently replaced by thick-film hybrid circuits.

## Printed Circuit Boards (PCBs)

**Printed circuit boards,** or PCBs, must be insulating, able to resist the soldering temperature, water resistant, strong, and low in cost. PCBs are made by passing woven glass cloth through an epoxy resin solution and then removing the solvent. The board is sometimes made with several layers laminated together. To provide the metal interconnections, a copper layer coated to a thickness of 30 μm or so is deposited. Resist chemistry is used for patterning the interconnections. The process is closely analogous to that used in standard photography. **Through-hole technology** (THT) and **surface-mount technology** (SMT) are more frequently used for component assembly. SMT is an advancement over the more conventional THT because no holes are drilled in SMT and a much higher packing density can be achieved. The components are simply stuck on the substrate and soldered through.

## GLOSSARY

| | |
|---|---|
| **micromachining** | the making of small mechanical components and machines using IC fabrication technology. |
| **integrated circuit (IC) technology** | the processes used to produce microcircuits on semiconductor wafers. |
| **lithography** | the step in an IC process required to transfer a pattern from a mask onto a resist. |
| **microstructure** | a miniature mechanical part built in a semiconductor wafer. |
| **resist** | a light-sensitive polymer used in the transfer of a pattern from a mask to a wafer. |
| **positive resist** | a coating that becomes soluble in a developer solvent when exposed to light and hardened but is otherwise resistant to etching. |
| **negative resist** | a coating that becomes insoluble in a developer solvent when exposed to light and hardened but is otherwise susceptible to etching. |
| **liftoff** | the patterning of a surface metal film by removing the underlying resist. |
| **etch stop** | a process in which a heavily doped buried layer is formed inside a wafer and used to terminate etching. |
| **p-n junction** | a device formed by p-type semiconductor and an n-type semiconductor. |
| **dry etching** | etching using reactive ions in a plasma created by a voltage discharge. |
| **plasma** | a collection of confined positive and negative ions and free electrons. |
| **reactive ion etching** | etching using an RF voltage source to create a plasma over the sample, which also sits on top of the positive RF electrode. |
| **cantilever beam** | a semiconductor beam or an oxide beam suspended in a cavity. |
| **resonance** | the case when the frequency of the driving force is the same as the natural vibrational frequency of the structure. |
| **membrane or diaphragm** | an extremely thin semiconductor or oxide layer. |
| **gage factor** | a measure of the ratio of the fractional change in the resistance to the strain. |

| | |
|---|---|
| **Wheatstone bridge circuit** | a set of balanced resistors connected in a bridge configuration used to determine an unknown resistance. |
| **ink-jet nozzles** | an array of holes used in a printer head to produce alphanumeric characters. |
| **multichip module** | a form of packaging when two or more wafers are stacked together to reduce the interconnections. |
| **gas chromatography** | a chemical technique used to identify different types of gases present in a mixture. |
| **Joule-Thomson cooling** | a technique for cooling down a system through gas expansion. |
| **X-ray mask** | a mask made on semiconductor (usually Si) diaphragms used for submicrometer lithography in IC fabrication. |
| **blackbody radiation** | energy emission from an object at a finite temperature assuming the object has 100% emissivity. |
| **quantum detector** | a detector that responds to a specific range of photon energies. |
| **energy detector** | a detector that responds to a continuum of photon energies. |
| **response time** | the time delay before a detector produces a full output. |
| **energy threshold** | the minimum input energy needed for a detector to respond to an input. |
| **bolometer** | an energy detector whose operation depends on a change in the resistance during heating. |
| **thermistor** | a temperature sensor whose output depends on a change in the resistance of the device. |
| **thermocouple** | a temperature-measuring device whose operation depends on the thermoelectric effect. |
| **thermopile** | a device made with a number of thermocouples connected in series. |
| **junction thermometer** | a $p$-$n$ junction used for temperature measurements. |
| **pyroelectric heat sensor** | a device that measures the change in the remnant polarization due to a temperature difference. |
| **resistive strain gage** | a displacement sensor whose operation depends on a change in the resistance of the strain gage. |
| **elastic gage** | a gage that measures the resistance change in a rubber tube filled with Hg or a conducting electrolyte as the tube changes shape under an applied pressure. |
| **linear variable differential transformer** | an electromechanical device that produces an electric output proportional to the displacement of a movable core. |
| **piezoelectricity** | electrical polarization due to the displacement of positive and negative charges or ions in a solid as a result of an applied force. |
| **piezoelectric sensor** | a pressure sensor whose operation depends on the piezoelectric effect. |
| **capacitance displacement sensor** | a sensor that measures the differential capacitance across a membrane. |
| **tachometer** | a sensor that measures the angular velocity of a rotating object. |
| **accelerometer** | a sensor that measures the acceleration or deceleration of a moving object. |
| **thermal flowmeter** | a sensor used to measure the velocity of a fluid. |
| **constant-heat-infusion flowmeter** | a sensor used to measure the mass flow of a fluid by injecting a known amount of heat into the fluid and measuring the temperature change. |
| **full-bore turbine flowmeter** | a device that measures the average volume flow of a fluid. |
| **Pitot tube** | a device used to measure the fluid speed in wind tunnels. |
| **electromagnetic flowmeter** | a device used to measure the flow rate of a conductive fluid. |
| **ultrasonic sensor** | a device that measures the sound waves emerging from an acoustic source by generating an electrical signal proportional to the energy in the sound waves. |

**transit-time flowmeter**     a device that measures the fluid flow by measuring the transit time for sound waves to be backscattered to the source.

**Doppler shift flowmeter**     a device that measures the shift in frequency of the sound waves when the latter are scattered or reflected from particles suspended in a fluid.

**$Q$-factor**     a measure of the gain in a system.

**gaseous sensor**     a detector to sense the presence of gases.

**chemovoltaic sensor**     a device that measures the chemical change in the form of a voltage difference.

**chemoresistive sensor**     a device that measures the chemical change in terms of a change in the resistance.

**optical sensor**     a sensor that measures electromagnetic radiation.

**light-emitting diode**     a *p-n* junction designed to emit electromagnetic radiation.

**light filter**     a partially transparent glass filter that removes a portion of the electromagnetic radiation.

**neutral density filter**     a light filter that is partially silvered to attenuate equally at all wavelengths.

**interference filter**     a light filter that is transparent to a narrow band of light near a specific wavelength due to constructive interference.

**monochromator**     a device that can produce tunable light of a very narrow bandwidth.

**sensitivity**     a measure of the dependence of the output signal on the radiant power.

**detectivity**     a measure of the time response and the noise level in the signal.

**quantum efficiency**     a measure of the effective number of conduction electrons produced per incident photon.

**photoconductivity**     the change in conductivity due to light absorption.

**$p$-$i$-$n$ diode**     a *p-n* junction with an extended intrinsic region in the middle of the device.

**photomultiplier tube**     a light detector with a high gain due to secondary emission at some internal dynodes (secondary electrodes).

**smoke-density monitor**     a smoke detector that makes use of light attenuation in the presence of smoke.

**nephelometer**     a smoke detector that measures the scattered light in the presence of dense smoke.

**ground point**     the voltage reference in a system held at the lowest potential to minimize spurious effects.

**electric shielding**     the means used to minimize electrical interference in a system.

**magnetic shielding**     the means used to minimize magnetic interference in a system.

**packaging**     the last step in the fabrication or manufacturing process.

**can package**     a form of packaging in which the device or circuit is enclosed in a metal can.

**lead-frame package**     a form of packaging in which the device or circuit is bonded to a lead frame and the whole assembly is potted into a polymer or glass capsule.

**solder**     a metal or metallic alloy used to electrically connect metal wires and bonding pads together.

**tape bonding**     bonding that uses a polymer sheet with a set of contact arrays that are directly bonded on the chip surface.

**flip-chip mounting**     a chip-mounting technique that uses a solder instead of metal bumps, as in tape bonding.

**field-assisted thermal bonding**     a form of glass seal bonding that applies a very high electric field across the glass layer during the formation of the seal.

**solder dip technique**     a technique of soldering by dipping the whole assembly in molten solder.

**solder reflow**     a technique in which solder is deposited in between the contact areas, and a heating cycle is used to melt the solder.

**potting material**     a thermoplastic or thermosetting material used for the encapsulation of a chip.

| | |
|---|---|
| **hybrid circuit** | a circuit that includes passive components on a given substrate. |
| **printed circuit board** | a plastic board coated with a copper layer, on which metallic interconnections can be formed and devices mounted. |
| **through-hole technology** | a technology for mounting components and circuits on a circuit board. |
| **surface-mount technology** | a technology for mounting devices and circuit components by direct attachment to the substrate. |

**REFERENCES**    Bryzek, J., K. Petersen, and W. McCulley, "Micromachines on the March," *IEEE Spectrum* (May 1994):20–31

Muller, R. S., R. T. Howe, S. D. Senturia, R. L. Smith, and R. M. White. *Microsensors.* Piscataway, N.J.: IEEE Press, 1991.

Petersen, K. E. "Silicon as a Mechanical Material," *Proc. IEEE* 70, no. 5 (1982):420–57.

Tandeske, D. T. *Pressure Sensors Selection and Applications.* New York: Marcel Dekker, Inc., 1991.

Tompkins, W. J. and J. G. Webster. *Interfacing Sensors to The IBM PC.* Upper Saddle River, N.J.: Prentice-Hall, Inc., 1988.

**EXERCISES    6.2  Micromachining**

1. In lithography, if the minimum feature to appear in the final pattern on a wafer is 1 μm, what is the minimum dimension that should appear in the artwork assuming a 100X reduction in dimension size?

2. If a particular process requires four masks and the maximum overall tolerance in alignment is 0.2 μm, what will be the maximum alignment tolerance between any two masks?

3. Distinguish between the properties of positive and negative resists. If we want to cut open a 100-μm × 100-μm window through a $SiO_2$ layer deposited on top of Si, how would you achieve this using positive resists?

4. The etch rates of Si using EDP at 100 °C are 400 μm/h in the <1 1 0> direction, 350 μm/h in the <1 0 0> direction, and 0.7 μm/h in the <1 1 1> direction. Is there any correlation between these etch rates and the crystal planes?

5. It has been observed that the etch rate of Si has a temperature dependence of the form $\exp(-E_a/kT)$, where $E_a$ is the activation energy and $T$ is the absolute temperature (see Fig. 6.7). Determine $E_a$ in the <1 1 1> direction if the etch rate is 0.05 μm/h at 50 °C and 0.7 μm/h at 100 °C.

6. The etch rate of Si in the presence of B with a concentration $C_B$ is given by $E_r = E_{r0}/[1 + C_B/C_0)^4]$, where $E_{r0}$ is the etch rate at low B concentration and $C_0 \sim 3 \times 10^{25}$ /m³. Determine the B concentration when $E_r/E_{r0} = 0.001$.

7. A Si wafer has a thickness of 100 μm, and a square hole 10 μm × 10 μm is to be formed in the base of the wafer. If etching is in the <1 0 0> direction, determine the size of the window (in the oxide layer). The angle of the side wall is 54.74°.

8. A cantilever beam made with $SiO_2$ is designed to have a resonant frequency of 10 kHz. Determine the aspect ratio, $t_h/l^2$, as given in Eq. (3). Young's modulus for $SiO_2$ is $0.73 \times 10^{11}$ N/m$^2$, and the mass density $\rho_0$ for $SiO_2$ is 2.2 kg/m$^3$.

## 6.3   Polysilicon Films Used in Microstructures

9. Determine the gage factor, $g$, of a polysilicon sample if the fractional change in sample resistance is $1 \times 10^{-2}$ at a strain of $5 \times 10^{-4}$.

10. Suggest how the gage factor, $g$, of a doped polysilicon sample varies with temperature and the dopant density.

## 6.4   Applications of Microstructures

11. In the formation of ink-jet nozzles in <1 0 0> silicon, estimate the mask dimensions in order to achieve a hole size of 10 μm × 10 μm if the thickness of the wafer is 250 μm and the angle between the <1 0 0> plane and the <1 1 1> plane is 54.74°.

12. What are the dimensions of the droplets emerging from a nozzle if a single ink drop emerging from the hole (10 μm × 10 μm) is divided into $10^5$ drops? Assume spherical dimensions for the droplets.

13. In an ink jet, if the spacing between the holes is 100 μm, how far should the paper be placed away from the holes if the droplets from the neighboring holes were to overlap? Assume that the maximum angle at which the droplets emerge from the nozzle is 54.74° with the horizontal plane.

14. Estimate the amount of heat dissipated per Hg connection in a multichip module if the hole diameter is 10 μm and the thickness of the silicon wafers is 600 μm. The resistivity of Hg is 96.8 μΩ·cm, and the voltage drop across the connection is 10 V.

15. Estimate the amount of chip area required to place a conduit with a length of 1.5 m and a cross-sectional area of 40 μm × 150 μm on a 4-in. silicon wafer. Suggest what type of layout you would use.

16. What is the range of wavelengths used in X-ray lithography?

17. An X-ray mask is made with W deposited on a silicon diaphragm. For a diaphragm diameter of 1 cm, suggest what may be the maximum deflection in the diaphragm if there is a temperature change of 50 °C. The thermal expansion coefficient of W and Si are $4.5 \times 10^{-6}$ /°C and $2.6 \times 10^{-6}$ /°C, respectively.

## 6.5   Heat and Temperature Sensors

18. Give examples for a quantum detector and a heat sensor.

19. A bolometer is to operate between $-183$ °C and 630 °C and requires an accuracy of better than $\pm 5\%$. What should be the maximum value of $a_2$ if only two of the temperature-dependent terms appearing in Eq. (7) are to be used? Assume $a_1 = 1 \times 10^{-7}$.

20. Assuming that the temperature dependence of the carrier density in a semiconductor has the form $n = n_0 \exp(-E_g/(2kT))$, arrive at an expression for $R(T)$ similar in form to Eq. (8). If $T = 300$ K, estimate the value of $R_0$ appearing in Eq. (8) for Si. Assume that the sample resistivity is 100 Ω·m, $L = 1$ mm, and $A_{cs} = 1 \times 10^{-6}$ m$^2$.

21. To achieve a higher sensitivity for a thermistor, explain whether you would use a semiconductor with a larger or a smaller energy gap.

22. If you are required to measure temperature up to 1000 °C and need the best sensitivity, what type of thermocouple would you select? Make a plot of the T-V (temperature-voltage) relationship between −100 °C and 1000 °C for the thermocouple you have chosen.

23. In Example 6.21, if $I_0$ is proportional to $T^n$, where $n$ is a constant, how would you modify Eq. (10) to account for this temperature dependence?

24. Name a material other than $BaTiO_3$ that is ferroelectric.

25. A pyroelectric sensor is used to measure transient heating from a light source. If the temperature difference measured between two slabs (one is exposed to light and the other is sheltered) is 10 °C, compute the heat energy deposited on the hotter slab. Assume that the slab has a dimension of 1 cm × 1 cm × 2 mm and its weight density is $1 \times 10^3$ kg/m$^3$. The specific heat of the slab is $0.5 \times 10^3$ J/Kg·K.

## 6.6　Displacement and Flow Senors

26. If blue light with a wavelength of 0.35 μm is used in Exercise 25 and the duration of heating is 1 s, assuming a 100% light-to-energy conversion efficiency, determine the photon flux.

27. Determine the Poisson ratio of a metallic cylinder if $D$ is 1 cm and $L$ is 5 cm. Assume an incompressible solid and $\Delta L/L = +1\%$.

28. Explain the dopant dependence and the dependence on the crystal orientations in the gage factor for Si (see Fig. 6.25).

29. If the secondary coils in a linear variable differential transformer (LVDT) has an inductance that is linearly proportional to the extent the ferrite core moves into/out of these coils, estimate the fractional change in output voltage if the core is shifted by 10% to one side away from its equilibrium position.

30. Estimate the amount of charge present in a piece of PZT if a force of 1 N is applied in the (33) mode.

31. Show that Eq. (20) will result in a curve of the form given in Fig. 6.29.

32. Using the gage factor of p-type Ge listed in Table 6.2, estimate the fractional change in resistance in a bridge-type sensor if the diaphragm has a diameter of 2 mm and a peak displacement of 0.1 mm.

33. A capacitive-type displacement sensor is made up of three parallel plates 1 mm away from each other. If the center plate is displaced by 0.02 mm toward one of the endplates, compute the fractional change in differential capacitance of the sensor.

34. A mechanical accelerometer translates the applied acceleration to a force, as shown in Fig. 6.32. If the mass is 0.1 kg and the applied force is 0.1 N, what will be the displacement of the spring if it has a force constant of $1 \times 10^{-3}$ N/m?

35. Derive expressions for the constants $K_0$ and $K_1$ appearing in Eq. (22).

36. An experiment is carried out to calibrate a thermal convection flowmeter. Using Eq. (21), compute $C_1$ and $C_2$ if the temperature difference between the thermistor and the fluid is 10 °C when the fluid-flow rate is 1 m/s and 20 °C when the fluid-flow rate is 0.3 m/s. The power delivered to the thermistor is 0.4 mW and the total surface area of the thermistor is 0.0002 m$^2$.

37. In the constant-temperature circuit shown in Fig. 6.34, $R_1$, $R_2$, $R_3$, and the resistance of the thermal probe are 1 kΩ. The initial value of $V_0$ is 5 V. If the resistance of the probe is increased to 1.05 kΩ due to cooling, compute the current increase in the probe if the amplifier has a gain of 50.

**38.** Compute the mass flow rate in a constant-heat-infusion flowmeter, as shown in Fig. 6.36, if 0.1 mW is delivered to the system, resulting in a temperature increase of 10 °C. The specific heat of the fluid is 5 kJ/kg·K.

**39.** From Eq. (24), suggest, to a first order of approximation, the dependence of the function $f(v_r, D, \eta_v)$ on $v_r$, $D$, and $\eta_v$.

**40.** Using Eq. (25), estimate the wind speed if there is a pressure difference of 1 Pa. The mass density of air is 1.2 kg/m$^3$.

**41.** If the fluid velocity in an electromagnetic flowmeter is 0.01 m/s, the length of the conduit is 1 m, and the applied magnetic field is 0.1 Wb, compute the transverse voltage.

**42.** Relate Eq. (26) to the equation used to describe the Hall effect in Chapter 2.

**43.** Based on Eqs. (27) and (28), sketch the near-field and far-field acoustic beam for an ultrasonic flow sensor.

**44.** Compute the transit time in a ultrasound flowmeter system if the average fluid flow rate is 10 m/s for laminar flow. The separation between the transducers is 1 m and the velocity of sound in air is 330 m/s. Assume the angle between the direction of the ultrasound and the direction of fluid flow is 30°.

**45.** What will be the velocity of the scattering centers if the Doppler frequency is twice the frequency of the ultrasound, assuming the angle between the sound path and the direction of the fluid flow is 30°? The average velocity of sound in the fluid is 1 m/s.

## 6.7    Chemical Sensors

**46.** In a chemovoltaic sensor, determine the constant $k_1$ if $P_{low}/P_{high}$ in Eq. (33) is 0.1 and $\Delta V = 50$ mV at $T = 300$ K.

**47.** Compute the change in resistivity in a chemoresistive sensor if conductivity is primarily due to Ti ions and their concentration is $1 \times 10^{18}$ ions/m$^3$. The effective mobility of the Ti ions is assumed to be 0.1 cm/s.

## 6.8    Optical Sensors

**48.** Compute the number of steradians from a point light source if the surface area of an object placed 1 m away from the point source has an area of 5 cm$^2$ and its normal is tilted at an angle of 10° from the direction of the light beam.

**49.** Describe the different types of light filters used to change the characteristics of light beams.

**50.** Explain the characteristics of a typical detector that has a high detectivity.

**51.** The increase in the photoconductivity of a detector is less than what we expect from a measurement of the incident light flux. Give a possible reason why this may be so in terms of the electrical properties of the detector.

**52.** In a photomultiplier tube, what is the overall gain if the multiplication effect for each electrode is 100 and there are 10 electrodes (dynodes)?

## 6.9    Design of a Pressure Transducer

**53.** Using the following equation relating the applied pressure to $d$, the deflection in the diaphragm, what should be the area of the diaphragm to ensure that the maximum deflection, $d_{max}$, is less than 1 mm? The diaphragm thickness is $t_h = 10$ μm, the radius of the diaphragm is 1 mm, and the peak applied pressure, $p_{max}$, is 1 Pa ($= 0.145 \times 10^{-3}$

psi). $p_a a^2/(dt) = 5 \times 10^7 + 1.8 \times 10^{10} \times (d/a)^2$, where $p_a$ is the applied pressure in Pa, $d$ is the deflection of the diaphragm in μm, and $a$ is the radius of the diaphragm in μm.

54. In the formation of the bridge resistors for a pressure sensor, if we assume that the implant energy is directly proportional to the implant range and if at an implant energy of 100 KeV, the implant depth is 10 μm, determine the implant energy required to achieve a resistor value of 1 kΩ if the length of the resistor is 1000 μm and the width is 20 μm. The resistivity of the implant resistors is $1 \times 10^{-4}$ Ω·m.

55. For the resistors specified in Exercise 54, if the average leakage current density into the substrate is $1 \times 10^{-6}$ A/m$^2$ at an applied voltage of 2 V, what will be the measured resistance of the bridge resistors in the presence of leakage?

56. Physically, explain the parameter span given by Eq. (37).

57. How does the active temperature compensation circuit shown in Fig. 6.51 operate?

## 6.10   Grounding, Shielding, and Interference

58. If two locations within a system have ground points that are at slightly different potentials, suggest how can you unify the two ground points to a single ground.

59. Distinguish between an ac ground point and a dc ground point.

60. Estimate the capacitance per unit length between two coaxial metallic cylinders with radii equal to $a$ meters and $b$ meters, respectively.

61. Name the advantages of using an optical coupler.

62. In the space (= 0.01 m) between two circular magnetic poles with a radius of 0.1 m, the magnetic field is found to build up linearly with time. Compute the maximum allowable rate of change of the magnetic field if the breakdown field is $2 \times 10^6$ V/m.

## 6.11   Packaging Materials

63. Choose the material listed in Table 6.5 that is thermally most compatible with Si.

64. Assume a particle flux of 0.01 particles/cm$^2$·h in a solid and that each particle creates $1 \times 10^6$ electron-hole pairs. Estimate the current density if this occurs within a reverse-bias $p$-$n$ junction.

65. Describe how wire bonding is formed.

66. What is the role of the applied electric field in field-assisted thermal bonding?

# Index